Speckle Phenomena in Optics

Speckle Phenomena in Optics

Theory and Applications

Joseph W. Goodman

ROBERTS & COMPANY
Englewood, Colorado

Roberts and Company Publishers
4950 South Yosemite Street, F2 #197
Greenwood Village, CO 80111
USA

Internet: www.roberts-publishers.com
Telephone: (303) 221-3325
Facsimile: (303) 221-3326

Ordering information:

Telephone: (800) 351-1161
Facsimile: (516) 422-4097
Internet: www.roberts-publishers.com

Distributed in Europe by Scion Publishing Ltd.

Internet: www.scionpublishing.com
Telephone: 44 (0) 1295 722873
Facsimile: 44 (0) 1295 722875

Publisher: Ben Roberts
Designer: Mark Ong at Side by Side Studios
Production Management: Side by Side Studios
Compositor: Asco Typesetters
Copyeditor: Lee A. Young

© 2007 by Ben Roberts & Company

Reproduction or translation of any part of this work beyond that permitted by Section 107 or 108 of the 1976 United States Copyright Act without permission of the copyright owner is unlawful. Requests for permission or further information should be addressed to the Permissions Department at Roberts and Company Publishers.

Library of Congress Cataloging-in-Publication Data

Goodman, Joseph W.
 Speckle phenomena in optics : theory and applications / Joseph W. Goodman.
   p. cm.
 Includes bibliographical references and index.
 ISBN 0-9747077-9-1
 1. Speckle. I. Title.
 QC427.8.S64G66 2006
 535′.32—dc22   2006028325

10 9 8 7 6 5 4 3 2

To Michele, with thanks for all the joy she brings us.

The Author

Joseph W. Goodman came to Stanford in 1958 as a graduate student, and remained there his entire professional career. He was the primary dissertation advisor for 49 graduate students, many of whom are now prominent in the field of optics. He held the William Ayer Chair in Electrical Engineering at Stanford, and also served in several administrative posts, including Chair of the Department of Electrical Engineering, and Senior Associate Dean of Engineering for Faculty Affairs. He is now the William Ayer Professor Emeritus.

His work has been recognized by a variety of awards and honors, including the F.E. Terman Award of the American Society for Engineering Education; the Dennis Gabor Award of the International Society for Optical Engineering (SPIE); the Max Born Award, the Esther Beller Hoffman Award and the Ives Medal from the Optical Society of America; and the Education Medal of the Institute of Electrical and Electronics Engineers. He is a member of the National Academy of Engineering and has served as president of the Optical Society of America and the International Commission for Optics.

Preface

Writing this book was a labor of love! In 1963, after finishing my Ph.D. research in the field of radar countermeasures, the first subject in optics that I studied in detail was the subject of speckle. So I began my career in optics with speckle, and it has been gratifying to return to this subject again after more than 40 years.

This book is directed towards a sophisticated audience with a good grasp of Fourier analysis, and a previous exposure to the broad concepts of statistics and random processes. It is suitable for a graduate textbook or a professional reference book. After the introductory chapter, the next three chapters deal with the theory of speckle, while the last last five chapters deal with what I consider to be application areas.

The field of speckle is a broad one, as evidenced by the breadth of the subjects covered here. Inevitably I have failed to reference all those who deserve credit for contributing to this field, and to those I have omitted, I offer my apologies.

This book has been several years in the making, partially because in mid course I was diverted to writing a third edition of *Introduction to Fourier Optics*. I have learned a lot that I didn't know about speckle in the process, and I have many people to thank for this education. To begin, I owe a tremendous debt to Pierre Chavel, of the Institut d'Optique, for reading the entire book and making suggestions and corrections that improved the work immensely. I thank Ben Roberts, of Roberts & Company Publishers, for having faith that a book on this subject would be appropriate for his small publishing company. I also thank Sam Ma, who also did yeoman's duty seeking out misprints and errors. I thank Lee Young, the copy editor who kept my writing clear, consistent, and in good English.

On a finer-grain level, I thank the following for their help with the material in various chapters:

Chapter 4 Kevin Webb, Isaak Freund and Mark Dennis;

Chapter 6 Kevin Webb, James Bliss, Daniel Malacara, Jahja Trisnadi, Michael Morris, Raymond Kostuk, Marc Levinson, and Christer Rydberg;

Chapter 7 Moshe Nazarathy and Amos Agmon;

Chapter 8 Mitsuo Takeda and James Wyant;

Chapter 9 Michael Roggemann and James Fienup; and

Appendix A Rodney Edwards.

All of these individuals helped improve the book. Of course, responsibility for errors rests with me.

Finally, I want to thank my wife, Hon Mai, who did not complain about the many hours I spent writing this book, and in fact encouraged me to push forward on many occasions.

Contents

Preface ix

1 **Origins and Manifestations of Speckle** 1
    1.1   General Background 1
    1.2   Intuitive Explanation of the Cause of Speckle 2
    1.3   Some Mathematical Preliminaries 5

2 **Random Phasor Sums** 7
    2.1   First and Second Moments of the Real and Imaginary Parts of the Resultant Phasor 8
    2.2   Random Walk with a Large Number of Independent Steps 10
    2.3   Random Phasor Sum Plus a Known Phasor 13
    2.4   Sums of Random Phasor Sums 17
    2.5   Finite Number of Equal Length Components 17
    2.6   Nonuniform Distribution of Phases 19

3 **First-Order Statistical Properties** 25
    3.1   Definition of Intensity 25
    3.2   First-Order Statistics of the Intensity and Phase 27
            3.2.1   Large Number of Random Phasors 27
            3.2.2   Constant Phasor plus a Random Phasor Sum 30
            3.2.3   Finite Number of Equal-Length Phasors 34
    3.3   Sums of Speckle Patterns 37
            3.3.1   Sums on an Amplitude Basis 38
            3.3.2   Sum of Two Independent Speckle Intensities 38

		3.3.3 Sum of N Independent Speckle Intensities	42
		3.3.4 Sums of Correlated Speckle Intensities	44
	3.4	Partially Polarized Speckle	47
	3.5	Partially Developed Speckle	50
	3.6	Speckled Speckle, or Compound Speckle Statistics	53
		3.6.1 Speckle Driven by a Negative Exponential Intensity Distribution	54
		3.6.2 Speckle Driven by a Gamma Intensity Distribution	55
		3.6.3 Sums of Independent Speckle Patterns Driven by a Gamma Intensity Distribution	57

4 Higher-Order Statistical Properties of Speckle 59

	4.1	Multivariate Gaussian Statistics	59
	4.2	Application to Speckle Fields	60
	4.3	Multidimensional Statistics of Speckle	62
		4.3.1 Joint Density Function of the Amplitudes	64
		4.3.2 Joint Density Function of the Phases	65
		4.3.3 Joint Density Function of the Intensities	68
	4.4	Autocorrelation Function and Power Spectrum of Speckle	73
		4.4.1 Free-Space Propagation Geometry	73
		4.4.2 Imaging Geometry	80
		4.4.3 Speckle Size in Depth	82
	4.5	Dependence of Speckle on Scatterer Microstructure	84
		4.5.1 Surface vs. Volume Scattering	85
		4.5.2 Effect of a Finite Correlation Area of the Scattered Wave	85
		4.5.3 A Regime where Speckle Size Is Independent of Scattering Spot Size	90
		4.5.4 Relation between the Correlation Areas of the Scattered Wave and the Surface Height Fluctuations—Surface Scattering	92
		4.5.5 Dependence of Speckle Contrast on Surface Roughness—Surface Scattering	98
		4.5.6 Properties of Speckle Resulting from Volume Scattering	102
	4.6	Statistics of Integrated and Blurred Speckle	105
		4.6.1 Mean and Variance of Integrated Speckle	106
		4.6.2 Approximate Result for the Probability Density Function of Integrated Intensity	111
		4.6.3 "Exact" Result for the Probability Density Function of Integrated Intensity	113
		4.6.4 Integration of Partially Polarized Speckle Patterns	118
	4.7	Statistics of Derivatives of Speckle Intensity and Phase	120
		4.7.1 Background	121
		4.7.2 Parameters for Various Scattering Spot Shapes	123

	4.7.3	Derivatives of Speckle Phase: Ray Directions in a Speckle Pattern	124
	4.7.4	Derivatives of Speckle Intensity	127
	4.7.5	Level Crossings of Speckle Patterns	130
4.8		Zeros of Speckle Patterns: Optical Vortices	133
	4.8.1	Conditions Required for a Zero of Intensity to Occur	133
	4.8.2	Properties of Speckle Phase in the Vicinity of a Zero of Intensity	133
	4.8.3	The Density of Vortices in Fully Developed Speckle	136
	4.8.4	The Density of Vortices for Fully Developed Speckle Plus a Coherent Background	138

5 Optical Methods for Suppressing Speckle 141

- 5.1 Polarization Diversity 142
- 5.2 Temporal Averaging with a Moving Diffuser 143
  - 5.2.1 Background 143
  - 5.2.2 Smooth Object 149
  - 5.2.3 Rough Object 151
- 5.3 Wavelength and Angle Diversity 153
  - 5.3.1 Free-Space Propagation, Reflection Geometry 154
  - 5.3.2 Free-Space Propagation, Transmission Geometry 164
  - 5.3.3 Imaging Geometry 167
- 5.4 Temporal and Spatial Coherence Reduction 170
  - 5.4.1 Coherence Concepts in Optics 170
  - 5.4.2 Moving Diffusers and Coherence Reduction 172
  - 5.4.3 Speckle Suppression by Reduction of Temporal Coherence . 175
  - 5.4.4 Speckle Suppression by Reduction of Spatial Coherence .. 178
- 5.5 Use of Temporal Coherence to Destroy Spatial Coherence 185
- 5.6 Compounding Speckle Suppression Techniques 186

6 Speckle in Certain Imaging Applications 187

- 6.1 Speckle in the Eye 187
- 6.2 Speckle in Holography 190
  - 6.2.1 Principles of Holography 190
  - 6.2.2 Speckle Suppression in Holographic Images 192
- 6.3 Speckle in Optical Coherence Tomography 195
  - 6.3.1 Overview of the OCT Imaging Technique 195
  - 6.3.2 Analysis of OCT 196
  - 6.3.3 Speckle and Speckle Suppression in OCT 200
- 6.4 Speckle in Optical Projection Displays 203
  - 6.4.1 Anatomies of Projection Displays 204
  - 6.4.2 Speckle Suppression in Projection Displays 208

		6.4.3	Polarization Diversity	208
		6.4.4	A Moving Screen	209
		6.4.5	Wavelength Diversity	211
		6.4.6	Angle Diversity	211
		6.4.7	Overdesign of the Projection Optics	213
		6.4.8	Changing Diffuser Projected onto the Screen	214
		6.4.9	Specially Designed Screens	225
	6.5	Speckle in Projection Microlithography		228
		6.5.1	Coherence Properties of Excimer Lasers	228
		6.5.2	Temporal Speckle	229
		6.5.3	From Exposure Fluctuations to Line Position Fluctuations	231

7 Speckle in Certain Nonimaging Applications ... 235

	7.1	Speckle in Multimode Fibers		235
		7.1.1	Modal Noise in Fibers	237
		7.1.2	Statistics of Constrained Speckle	239
		7.1.3	Frequency Dependence of Modal Noise	243
	7.2	Effects of Speckle on Optical Radar Performance		248
		7.2.1	Spatial Correlation of the Speckle Returned from Distant Targets	250
		7.2.2	Speckle at Low Light Levels	252
		7.2.3	Detection Statistics—Direct Detection	256
		7.2.4	Detection Statistics—Heterodyne Detection	260
		7.2.5	Comparison of Direct Detection and Heterodyne Detection	270
		7.2.6	Reduction of the Effects of Speckle in Optical Radar Detection	273

8 Speckle and Metrology ... 275

	8.1	Speckle Photography		275
		8.1.1	In-Plane Displacement	277
		8.1.2	Simulation	279
		8.1.3	Properties of the Spectra $\mathcal{I}_k(v_X, v_Y)$	281
		8.1.4	Limitations on the Size of the Motion (x_0, y_0)	284
		8.1.5	Analysis with Multiple Specklegram Windows	285
		8.1.6	Object Rotation	286
	8.2	Speckle Interferometry		287
		8.2.1	Systems That Use Photographic Detection	287
		8.2.2	Electronic Speckle Pattern Interferometry (ESPI)	291
		8.2.3	Speckle Shearing Interferometry	294
	8.3	From Fringe Patterns to Phase Maps		296
		8.3.1	The Fourier Transform Method	297
		8.3.2	Phase-Shifting Speckle Interferometry	298
		8.3.3	Phase Unwrapping	299

	8.4	Vibration Measurement Using Speckle	301
	8.5	Speckle and Surface Roughness Measurements	305
		8.5.1 RMS Surface Height and Surface Covariance Area from Speckle Contrast	305
		8.5.2 RMS Surface Height from Two-Wavelength Decorrelation	306
		8.5.3 RMS Surface Height from Two-Angle Decorrelation	307
		8.5.4 Information from Measurement of Angular Power Spectrum	308

9 Speckle in Imaging Through the Atmosphere 311

	9.1	Background	311
		9.1.1 Refractive Index Fluctuations in the Atmosphere	311
	9.2	Point-Spread Functions	313
	9.3	Average Optical Transfer Functions	315
	9.4	Statistical Properties of the Short-Exposure OTF and MTF	316
	9.5	Astronomical Speckle Interferometry	322
		9.5.1 Object Information that Is Retrievable	322
		9.5.2 Results of a More Complete Analysis	325
	9.6	The Cross-Spectrum or Knox–Thompson Technique	327
		9.6.1 The Cross-Spectrum Transfer Function	327
		9.6.2 Recovering Full Object Information from the Cross-Spectrum	329
	9.7	The Bispectrum Technique	331
		9.7.1 The Bispectrum Transfer Function	331
		9.7.2 Recovering Full Object Information from the Bispectrum	332
	9.8	Speckle Correlography	333

A Linear Transformations of Speckle Fields 337

B Contrast of Partially Developed Speckle 341

C Statistics of Derivatives of Speckle 345

	C.1	The Correlation Matrix	345
	C.2	Joint Density Function of the Derivatives of Phase	348
	C.3	Joint Density Function of the Derivatives of Intensity	349

D Wavelength and Angle Dependence 351

	D.1	Free-Space Geometry	351
	D.2	Imaging Geometry	355

E Speckle Contrast with a Projected Diffuser 359

	E.1	Random Phase Diffusers	359
	E.2	Diffuser that Just Fills the Projection Optics	362
	E.3	Diffuser that Overfills the Projection Optics	362

F	**Statistics of Constrained Speckle**	365
G	**Sample *Mathematica* Programs for Simulating Speckle**	369
	G.1 Speckle Simulation With Free Space Propagation	369
	G.2 Speckle Simulation With an Imaging Geometry	369
	Bibliography	373
	Index	383

1 Origins and Manifestations of Speckle

1.1 General Background

In the early 1960s, when continuous-wave lasers first became commercially available, researchers working with these instruments noticed what at the time was regarded as a strange phenomenon. When laser light was reflected from a surface such as paper, or the wall of the laboratory, a high-contrast, fine-scale granular pattern would be seen by an observer looking at the scattering spot. In addition, measurement of the intensity reflected from such a spot showed that such fine scale fluctuations of the intensity exist in space, even though the illumination of the spot was relatively uniform. This type of granularity became known as "speckle."

The origin of these fluctuations was soon recognized to be the "random" roughness of the surfaces from which the light was reflected [133], [121]. In fact, most materials encountered in the real world are rough on the scale of an optical wavelength (notable exceptions being mirrors). Various microscopic facets of the rough scattering surface contribute randomly phased elementary contributions to the total observed field, and those contributions interfere with one another to produce a resultant intensity (the squared magnitude of the field) that may be weak or strong, depending on the particular set of random phases that may be present.

Speckle is also observed when laser light is transmitted through stationary diffusers, for the same basic reason: the optical paths of different light rays passing through the transmissive object vary significantly in length on a scale of a wavelength. Similar effects are observed when light is scattered from particle suspensions. The speckle phenomenon thus appears frequently in optics; it is in fact the rule rather than the exception. Figure 1.1 shows three photos, the first of a rough object illuminated with incoherent light, the second of the same object illuminated with light from a laser, and the third a subregion of the speckled image showing the structure in

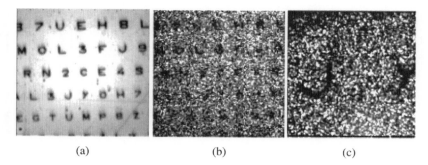

Figure 1.1 Image of a rough object. (a) Image taken with incoherent light; (b) Image taken with coherent light; and (c) a magnified portion of the image shown in (b). (Photos courtesy of P. Chavel and T. Avignon, Institut d'Optique.)

more detail. Speckle clearly has a profound effect on the ability of the human observer to extract the information contained in the image.

Speckle also plays an important role in other fields where radiation is transmitted by or reflected from objects that are rough on the scale of a wavelength. Important cases include synthetic-aperture radar imagery in the microwave region of the spectrum, and ultrasound medical imagery of organs in the human body. Figure 1.2 shows an image from a synthetic-aperture radar in which the speckle is quite visible, and Fig. 1.3 shows an ultrasound image of the human liver in which speckle is also evident.

Exact analogs of the speckle phenomenon appear in many other fields and applications. For example, the squared magnitude of the finite-time Fourier transform of a sample function of almost any random process shows fluctuations in the frequency domain that have the same first-order statistics as speckle.

Previous books treating the subject of speckle are rather few. The most comprehensive and most widely cited is the book edited by J.C. Dainty, first published in 1975, with a second edition in 1984 [30]. Another book worth noting is that of M. Françon, published in 1979 [49]. See also the more recent book by Zel'dovich, Mamaev and Shkunov [182]. Another excellent recent treatment is found in [128] in the chapter by M. Lehmann. Aside from these books, the interested reader must consult diverse scientific and technical journals to learn about the large body of research that has been conducted on the subject of speckle.

1.2 Intuitive Explanation of the Cause of Speckle

Speckle appears in a signal when that signal is composed of a multitude of independently phased additive complex components (i.e. components having both an ampli-

Figure 1.2 Synthetic-aperture radar image of Moffet field, California, dominated by speckle. The runways are clearly visible, as are also industrial buildings and a portion of San Francisco Bay (bottom right). The radar wavelength was 5.67 cm and the ground resolution is approximately 20 m. (Courtesy of Howard Zebker of Stanford University.)

tude and a phase). The components may have both random lengths (amplitudes) and random directions (phases) in the complex plane, or they may have known lengths and random directions. When these components are added together, they constitute what is known as a "random walk." The resultant of the sum may be large or small, depending on the relative phases of the various components of the sum, and in particular whether constructive or destructive interference dominates the sum. The squared length of the resultant is what we usually call the "intensity" of the observed wave.

Figure 1.4 illustrates random walks that generate (a) a large resultant and (b) a small resultant. In the cases illustrated, both lengths and directions are random, and no single contribution dominates the sum. The vector with thicker arrow represents the resultant of the complex sum.

The complex nature of the individual contributions usually arises from the fact that they are phasors representing sinusoidal signal components having both an amplitude and a phase. In other cases, real-valued signal components may be weighted in a signal-processing operation (such as a discrete Fourier transform) by weighting factors that are complex-valued, leading to a sum of corresponding complex

4   CHAPTER 1   Origins and Manifestations of Speckle

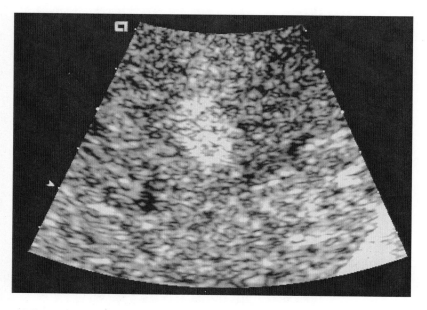

Figure 1.3 Ultrasound image (2.5 MHz) of the human liver showing speckle (courtesy of F. Graham Sommer, Stanford University).

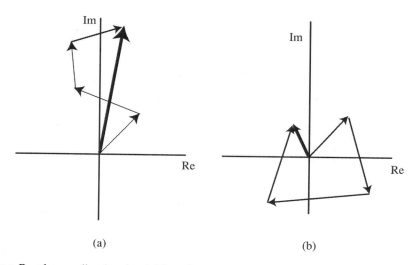

Figure 1.4 Random walks showing (a) largely constructive addition, and (b) largely destructive addition.

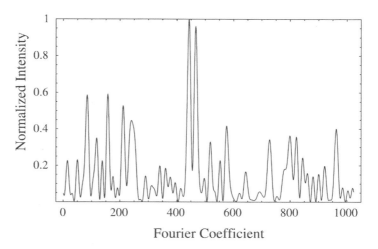

Figure 1.5 Speckle observed in the squared magnitude of the one-dimensional discrete Fourier transform of a random-phase sequence of length 1024.

contributions (again phasors). Figure 1.5 illustrates the squared magnitude of the discrete Fourier transform of a sequence of 1024 complex numbers, each of fixed length and each with an independent random phase drawn with a uniform probability density on the interval $(0, 2\pi)$. Interpolation has been used between the discrete points.

Thus we see that speckle can arise in many different physical situations.

1.3 Some Mathematical Preliminaries

With this brief introduction to the origins and manifestations of speckle, we now turn to a set of simple mathematical representations of the random walks or random phasor sums that underlie the speckle phenomenon.

A typical signal can be represented as a sinusoidal function of time and space,

$$\mathcal{A}(x, y; t) = A(x, y; t) \cos[2\pi v_o t - \theta(x, y; t)], \qquad (1\text{-}1)$$

where $A(x, y; t)$ represents the amplitude or envelope of the signal, and $\theta(x, y; t)$ represents the phase. The quantity v_o is a "carrier" frequency, generally much larger than the bandwidth of either A or θ. In some applications in this book, the amplitude and phase of the resultant may depend on the particular center frequency of the carrier, and in some cases the basic signal will have polarization as a parameter as well, but for the moment the above simplified form will be adequate.

For economy of notation, it is common to create a complex representation of such a signal. This is done by suppressing the positive frequency components in the spectrum of $\mathcal{A}$ and doubling the negative frequency components. The definition of the Fourier transform[1] $\mathbf{G}(v)$ of a signal $g(t)$ used here is[2]

$$\mathbf{G}(v) = \int_{-\infty}^{\infty} g(t) \exp(-j2\pi v t)\, dt. \qquad (1\text{-}2)$$

Hence representing a signal $\cos(2\pi v_o t)$ by a phasor $\exp(-j2\pi v_o t)$ that rotates in the clockwise direction is equivalent to doubling the negative frequency component and eliminating the positive frequency component.

If in addition, the carrier frequency term is suppressed, we have a representation of the form

$$\mathbf{A}(x, y; t) = A(x, y; t)e^{j\theta(x, y; t)}. \qquad (1\text{-}3)$$

Note that the real part of $\mathbf{A}(x, y; t)e^{-j2\pi v_o t}$ is the original real-valued signal we started with.

Such complex representations will be widely used throughout this book. The speckle phenomenon occurs when a resultant complex representation is composed of a superposition (sum) of a multitude of randomly phased "elementary" complex components. Thus, at a single point in space–time,

$$\mathbf{A} = Ae^{j\theta} = \sum_{n=1}^{N} \mathbf{a}_n = \sum_{n=1}^{N} a_n e^{j\phi_n} \qquad (1\text{-}4)$$

where $\mathbf{a}_n$ is the nth complex phasor component of the sum, having length a_n and phase ϕ_n.

In some cases it is convenient to explicitly represent the time or space dependence of the underlying phasors and/or the resultant. In such cases we might write

$$\mathbf{A}(x, y; t) = \sum_{n=1}^{N} a_n(x, y; t)e^{j\phi_n(x, y; t)}. \qquad (1\text{-}5)$$

Finally, there may be cases for which the basic phasor components arise from a set of randomly phased complex orthogonal functions $\boldsymbol{\psi}_n$, such as modes in a waveguide, in which case the sum may take the form

$$\mathbf{A}(x, y; t) = \sum_{n=1}^{N} \boldsymbol{\psi}_n(x, y; t)e^{j\phi_n}. \qquad (1\text{-}6)$$

With this background, we now turn to a detailed study of the statistics of the length and phase of the resultant phasor under a variety of conditions.

---

1. Here and throughout this book, complex quantities will be represented with boldface type, whereas real quantities will be represented with ordinary type. In addition, the symbol j represents the unit length imaginary constant, $j = \sqrt{-1}$.
2. In a few cases, particularly in dealing with characteristic functions of random variables, the form $\mathbf{G}(\omega) = \int_{-\infty}^{\infty} g(t) \exp[-j\omega t]\, dt$ will be used, but which form is being used should be clear from the context.

2 Random Phasor Sums

In this chapter we examine the first order statistical properties of the amplitude and phase of various kinds of random phasor sums. By "first-order" we mean the statistical properties at a point in space or, for time-varying speckle, in space–time. While in optics it is generally the *intensity* of the wave that is ultimately of interest, in both ultrasound and microwave imaging, the amplitude[1] and phase of the field can be detected directly. For this reason, we focus attention in this chapter on the properties of the amplitude and phase of the resultant of a random phasor sum. In the chapter that follows, we examine the corresponding properties of the intensity, as appropriate for speckle in the optical region of the spectrum.

A random phasor sum may be described mathematically as follows:

$$\mathbf{A} = Ae^{j\theta} = \frac{1}{\sqrt{N}} \sum_{n=1}^{N} \mathbf{a}_n = \frac{1}{\sqrt{N}} \sum_{n=1}^{N} a_n e^{j\phi_n}, \qquad (2\text{-}1)$$

where N represents the number of phasor components in the random walk, $\mathbf{A}$ represents the resultant phasor (a complex number), A represents the length (or magnitude) of the complex resultant, θ represents the phase of the resultant, $\mathbf{a}_n$ represents the nth component phasor in the sum (a complex number), a_n is the length of $\mathbf{a}_n$, and ϕ_n is the phase of $\mathbf{a}_n$. The scaling factor $1/\sqrt{N}$ is introduced here and in what follows in order to preserve finite second moments of the sum even when the number of component phasors approaches infinity.

Throughout our discussions of random walks, it will be convenient to adopt certain assumptions about the statistics of the component phasors that make up the

---

1. The term "amplitude" will be used throughout to refer to the modulus of the complex amplitude.

sum. These assumptions are most easily understood by considering the real and imaginary parts of the resultant phasor,

$$\mathcal{R} = \text{Re}\{\mathbf{A}\} = \frac{1}{\sqrt{N}} \sum_{n=1}^{N} a_n \cos \phi_n$$

$$\mathcal{I} = \text{Im}\{\mathbf{A}\} = \frac{1}{\sqrt{N}} \sum_{n=1}^{N} a_n \sin \phi_n,$$

(2-2)

where the symbols Re{ } and Im{ } signify the real and imaginary parts of the complex quantity enclosed in braces. The assumptions are:

1. The amplitudes and phases a_n and ϕ_n are statistically independent of a_m and ϕ_m provided $n \neq m$. That is, knowledge of the values of the amplitude and/or phase of one of the component phasors conveys no knowledge about the amplitude and/or phase of another component phasor.
2. For any n, a_n and ϕ_n are statistically independent of each other. That is, knowledge of the phase of a component phasor conveys no knowledge of the value of the amplitude of that same phasor, and *vice versa*.
3. The phases ϕ_m are uniformly distributed on the interval $(-\pi, \pi)$. That is, all values of phase are equally likely.

Most random walk problems satisfy these assumptions. However, we will later consider at least one case where one of these assumptions is violated. For the moment the assumptions are assumed to be satisfied.

2.1 First and Second Moments of the Real and Imaginary Parts of the Resultant Phasor

Note that these three assumptions have certain implications about the mean (expected value)[2] and variance of $\mathcal{R}$ and $\mathcal{I}$. In particular, the means are easily seen to be

$$E[\mathcal{R}] = E\left[\frac{1}{\sqrt{N}} \sum_{n=1}^{N} a_n \cos \phi_n\right] = \frac{1}{\sqrt{N}} \sum_{n=1}^{N} E[a_n \cos \phi_n]$$

$$= \frac{1}{\sqrt{N}} \sum_{n=1}^{N} E[a_n] E[\cos \phi_n] = 0$$

(2-3)

---

2. Two different but equivalent notations will be used in this book for expected values of random variables. The expected value of a real-valued random variable z will be written either as $E[z]$ or as $\bar{z}$, whichever is notationally more convenient.

2.1 First and Second Moments of the Real and Imaginary Parts of the Resultant Phasor

$$E[\mathcal{I}] = E\left[\frac{1}{\sqrt{N}} \sum_{n=1}^{N} a_n \sin \phi_n\right] = \frac{1}{\sqrt{N}} \sum_{n=1}^{N} E[a_n \sin \phi_n]$$

$$= \frac{1}{\sqrt{N}} \sum_{n=1}^{N} E[a_n] E[\sin \phi_n] = 0, \qquad (2\text{-}4)$$

where the order of averaging and summation were interchanged, the expected value of the product of two independent random variables was replaced by the product of their expected values, and the uniform statistics of ϕ_n imply that the averages of $\cos \phi_n$ and $\sin \phi_n$ are both zero.

In a similar fashion the variances of $\mathcal{R}$ and $\mathcal{I}$, being equal to the second moments because the means are zero, are given by

$$\sigma_{\mathcal{R}}^2 = E[\mathcal{R}^2] = \frac{1}{N} \sum_{n=1}^{N} \sum_{m=1}^{N} E[a_n a_m] E[\cos \phi_n \cos \phi_m]$$

$$\sigma_{\mathcal{I}}^2 = E[\mathcal{I}^2] = \frac{1}{N} \sum_{n=1}^{N} \sum_{m=1}^{N} E[a_n a_m] E[\sin \phi_n \sin \phi_m]. \qquad (2\text{-}5)$$

Now for $n \neq m$, $E[\cos \phi_n \cos \phi_m] = E[\cos \phi_n] E[\cos \phi_m] = 0$ and likewise we have $E[\sin \phi_n \sin \phi_m] = 0$. As a consequence, only the $n = m$ terms in Eq. (2-5) remain. Use of an appropriate trig identity yields

$$\sigma_{\mathcal{R}}^2 = \sum_{n=1}^{N} \frac{1}{N} E[a_n^2] E[\cos^2 \phi_n] = \frac{1}{N} \sum_{n=1}^{N} E[a_n^2] E\left[\frac{1}{2} + \frac{1}{2} \cos 2\phi_n\right]$$

$$= \frac{1}{N} \sum_{n=1}^{N} \frac{E[a_n^2]}{2} \qquad (2\text{-}6)$$

$$\sigma_{\mathcal{I}}^2 = \frac{1}{N} \sum_{n=1}^{N} E[a_n^2] E[\sin^2 \phi_n] = \frac{1}{N} \sum_{n=1}^{N} E[a_n^2] E\left[\frac{1}{2} - \frac{1}{2} \cos 2\phi_n\right]$$

$$= \frac{1}{N} \sum_{n=1}^{N} \frac{E[a_n^2]}{2}, \qquad (2\text{-}7)$$

where we have used the fact that $2\phi_n$ is uniformly distributed on $(-\pi, \pi)$ if ϕ_n is uniformly distributed on that interval. Thus we see that, like the means, the variances of the real and imaginary parts of the resultant phasor are identical.

Finally we consider the correlation between the real and imaginary parts. We have

$$\Gamma_{\mathcal{R},\mathcal{I}} = E[\mathcal{R}\mathcal{I}] = \frac{1}{N} \sum_{n=1}^{N} E[a_n^2] E[\cos \phi_n \sin \phi_n] = 0, \qquad (2\text{-}8)$$

where several steps similar to those used in the previous equations have been skipped. We conclude that there is no correlation between the real and imaginary parts of the resultant phasor under the conditions we have hypothesized.

2.2 Random Walk with a Large Number of Independent Steps

Referring back Eq. (2-1), we consider the number of steps N in the sum to be very large. In this case the real and imaginary parts $\mathcal{R}$ and $\mathcal{I}$ of the resultant phasor **A** are given by sums of large numbers of independent random variables (the $a_n \cos \phi_n$ in the case of R and the $a_n \sin \phi_n$ in the case of I). At this point the Central Limit Theorem ([70], p. 31) can be brought to bear. According to this theorem, under rather general conditions the statistics of the sum of N independent random variables is asymptotically Gaussian as $N \to \infty$. Given the results of the previous section regarding means, variances and correlation, the joint probability density function for the real and imaginary parts of the resultant phasor becomes

$$p_{\mathcal{R},\mathcal{I}}(\mathcal{R},\mathcal{I}) = \frac{1}{2\pi\sigma^2} \exp\left\{-\frac{\mathcal{R}^2 + \mathcal{I}^2}{2\sigma^2}\right\}, \qquad (2\text{-}9)$$

where $\sigma^2 = \sigma_{\mathcal{R}}^2 = \sigma_{\mathcal{I}}^2$.

Contours of constant probability density for this distribution are shown in Fig. 2.1 below. Because of the circular nature of these contours, this resultant complex phasor **A** is said to be a "circular" complex Gaussian variate.

Of equal interest are the statistics of the amplitude (length) A and phase θ of the resultant phasor. The joint probability density function of the amplitude and phase can be found using the rules of probability theory for transformations of variables. We have

$$A = \sqrt{\mathcal{R}^2 + \mathcal{I}^2}$$
$$\theta = \arctan\left\{\frac{\mathcal{I}}{\mathcal{R}}\right\} \qquad (2\text{-}10)$$

and

$$\mathcal{R} = A \cos \theta \qquad (2\text{-}11)$$
$$\mathcal{I} = A \sin \theta. \qquad (2\text{-}12)$$

The joint probability density function of A and θ is related to that of $\mathcal{R}$ and $\mathcal{I}$ through

$$p_{A,\theta}(A,\theta) = p_{\mathcal{R},\mathcal{I}}(A \cos \theta, A \sin \theta)\|J\| \qquad (2\text{-}13)$$

where $\|J\|$ is the magnitude of the Jacobian determinant of the transformation between the two sets of variables,

$$\|J\| = \left\| \begin{matrix} \partial \mathcal{R}/\partial A & \partial \mathcal{R}/\partial \theta \\ \partial \mathcal{I}/\partial A & \partial \mathcal{I}/\partial \theta \end{matrix} \right\| = A. \qquad (2\text{-}14)$$

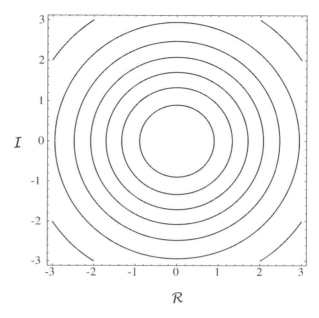

Figure 2.1 Contours of constant probability density for a circular complex Gaussian random variate.

It follows that the joint probability density function of the length and phase of the resultant phasor is given by

$$p_{A,\theta}(A,\theta) = \frac{A}{2\pi\sigma^2} \exp\left\{-\frac{A^2}{2\sigma^2}\right\} \tag{2-15}$$

for $(A \geq 0)$ and $(-\pi \leq \theta < \pi)$, zero otherwise.

Having found the joint statistics of A and θ, we now find the marginal statistics of A and then θ alone. Focusing on the length A first, we obtain

$$p_A(A) = \int_{-\pi}^{\pi} p_{A,\theta}(A,\theta)\, d\theta = \frac{A}{\sigma^2} \exp\left\{-\frac{A^2}{2\sigma^2}\right\} \tag{2-16}$$

for $A \geq 0$, a result known as the Rayleigh density function. Thus we see that the length of the resultant phasor is a Rayleigh variate. Figure 2.2 illustrates the Rayleigh density function. The moments of the amplitude are of some interest. They are found to be[3]

---

3. This result can be easily found with the help of a symbolic manipulation program such as *Mathematica*.

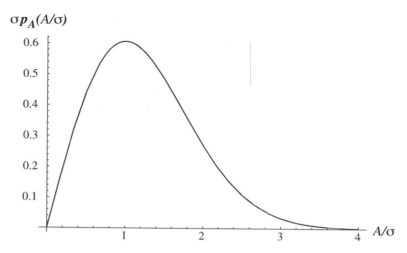

Figure 2.2 Rayleigh probability density function.

$$\overline{A^q} = \int_0^\infty A^q p_A(A)\, dA = 2^{q/2} \sigma^q \Gamma\left(1 + \frac{q}{2}\right), \qquad (2\text{-}17)$$

where $\Gamma(\ )$ is the gamma function. The first moment, the second moment, and the variance are given by

$$\begin{aligned} \overline{A} &= \sqrt{\frac{\pi}{2}}\,\sigma \approx 1.25\sigma \\ \overline{A^2} &= 2\sigma^2 \\ \sigma_A^2 &= \overline{A^2} - (\overline{A})^2 = \left(2 - \frac{\pi}{2}\right)\sigma^2 \approx 0.43\sigma^2. \end{aligned} \qquad (2\text{-}18)$$

The density function of the phase is found either by inspection of Eq. (2-15) or by integrating that equation with respect to amplitude,

$$p_\theta(\theta) = \int_0^\infty \frac{A}{2\pi\sigma^2} \exp\left\{-\frac{A^2}{2\sigma^2}\right\} dA = \frac{1}{2\pi} \qquad (2\text{-}19)$$

for $(-\pi \leq \theta < \pi)$. In reaching this result we have used the fact that the integral of the Rayleigh density function must be unity.

From these results we see that the joint density function for A and θ of Eq. (2-15) factors into a product of $p_A(A)$ and $p_\theta(\theta)$, demonstrating that *the length A and the phase angle θ of the resultant phasor are statistically independent random variables.*

In summary, we have shown that for a random phasor sum in which the component phasors obey the assumptions used here, and for which the number of component phasors is very large, the length of the resultant phasor obeys Rayleigh statistics, and the phase of the resultant is uniformly distributed on the primary interval $(-\pi, \pi)$. For a finite but large number of phasors, these results are approximate. They become exactly correct asymptotically as the number of phasors grows without bound.

2.3 Random Phasor Sum Plus a Known Phasor

In some practical situations, the resultant phasor is the sum of a known constant phasor and a random phasor sum. The geometry is illustrated in Fig. 2.3. Without loss of generality we assume that the known phasor lies along the real axis in the complex plane. The real and imaginary parts of the resultant phasor can be written

$$\mathcal{R} = A_0 + \frac{1}{\sqrt{N}} \sum_{n=1}^{N} a_n \cos \phi_n$$
$$\mathcal{I} = \frac{1}{\sqrt{N}} \sum_{n=1}^{N} a_n \sin \phi_n,$$
(2-20)

where A_0 represents the length of the known phasor.

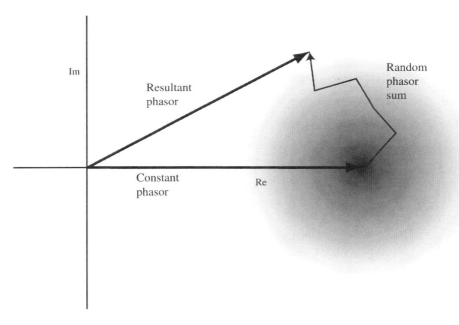

Figure 2.3 Sum of a constant phasor and a random phasor sum.

The effect of the presence of the known phasor is to add a known mean to the real part of the resultant phasor. Then, for a large number of steps in the random phasor sum, the statistics of the real and imaginary parts of the resultant phasor are again asymptotically Gaussian, with joint probability density function given by

$$p_{\mathcal{R},\mathcal{I}}(\mathcal{R},\mathcal{I}) = \frac{1}{2\pi\sigma^2} \exp\left\{-\frac{(\mathcal{R}-A_0)^2 + \mathcal{I}^2}{2\sigma^2}\right\}. \tag{2-21}$$

A transformation of variables, similar to that leading to Eq. (2-15), leads to the joint probability density function of the length and phase of the resultant,

$$p_{A,\theta}(A,\theta) = \frac{A}{2\pi\sigma^2} \exp\left\{-\frac{A^2 + A_0^2 - 2AA_0\cos\theta}{2\sigma^2}\right\} \tag{2-22}$$

valid for $A \geq 0$ and $(-\pi \leq \theta < \pi)$.

To find the marginal statistics of A and θ, integrals of this function must be performed. For the length of the resultant phasor,

$$p_A(A) = \int_{-\pi}^{\pi} p_{A,\theta}(A,\theta)\,d\theta$$

$$= \frac{A}{2\pi\sigma^2} \exp\left\{-\frac{A^2 + A_0^2}{2\sigma^2}\right\} \int_{-\pi}^{\pi} \exp\left\{\frac{2AA_0\cos\theta}{2\sigma^2}\right\} d\theta. \tag{2-23}$$

Using the integral identity

$$\int_{-\pi}^{\pi} e^{b\cos t}\,dt = 2\pi I_0(b), \tag{2-24}$$

where $I_0(\ )$ is a modified Bessel function of the first kind, order zero, we obtain the density function of A,

$$p_A(A) = \frac{A}{\sigma^2} \exp\left\{-\frac{A^2 + A_0^2}{2\sigma^2}\right\} I_0\left(\frac{AA_0}{\sigma^2}\right), \tag{2-25}$$

valid for $A \geq 0$. This density is known as the "Rician" density function, after the individual who first derived it. Figure 2.4 illustrates the Rician density for several values of A_0/σ. Note that for $A_0/\sigma = 0$, the Rician density is the same as the Rayleigh density, as of course it should be. As A_0/σ grows, the Rician density takes on a more symmetrical form, gradually resembling a Gaussian density function. It can in fact be shown that in the limit of ever larger A_0/σ, the result is asymptotically Gaussian, provided the number of steps in the random walk is large.

The qth moment of A is in this case given by

$$\overline{A^q} = (2\sigma^2)^{q/2} e^{-\frac{A_0^2}{2\sigma^2}} \Gamma\left(1+\frac{q}{2}\right) {}_1F_1\left(1+\frac{q}{2}, 1, \frac{A_0^2}{2\sigma^2}\right), \tag{2-26}$$

where ${}_1F_1(\alpha,\beta,x)$ represents the confluent hypergeometric function ([111], p. 1073). The first and second moments are given by

2.3 Random Phasor Sum Plus a Known Phasor

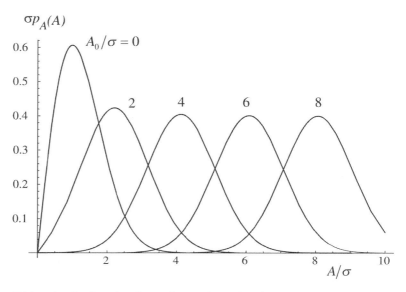

Figure 2.4 Rician density function for various values of A_0/σ.

$$\bar{A} = \frac{1}{2}\sqrt{\frac{\pi}{2\sigma^2}}e^{-\frac{A_0^2}{4\sigma^2}}\left[(A_0^2 + 2\sigma^2)I_0\left(\frac{A_0^2}{4\sigma^2}\right) + A_0^2 I_1\left(\frac{A_0^2}{4\sigma^2}\right)\right] \quad (2\text{-}27)$$

$$\overline{A^2} = A_0^2 + 2\sigma^2.$$

The statistics of the phase are also of interest. Finding the probability density function of θ requires solution of the following integral (cf. Eq. 2-22):

$$p_\theta(\theta) = \frac{e^{-\frac{A_0^2}{2\sigma^2}}}{2\pi\sigma^2}\int_0^\infty A\exp\left[-\frac{A^2 - 2AA_0\cos\theta}{2\sigma^2}\right]dA. \quad (2\text{-}28)$$

The result of the integration can be found in [111] (p. 417) or can be obtained directly with *Mathematica*. One form of the result is

$$p_\theta(\theta) = \frac{e^{-\frac{A_0^2}{2\sigma^2}}}{2\pi} + \sqrt{\frac{1}{2\pi}}\frac{A_0}{\sigma}e^{-\frac{A_0^2}{2\sigma^2}\sin^2\theta}\frac{1 + \operatorname{erf}\left(\frac{A_0\cos\theta}{\sqrt{2}\sigma}\right)}{2}\cos\theta \quad (2\text{-}29)$$

for $(-\pi \leq \theta < \pi)$, zero otherwise. The function $\operatorname{erf}(z)$ is the standard error function,

$$\operatorname{erf}(z) = \frac{2}{\sqrt{\pi}}\int_0^z e^{-t^2}dt. \quad (2\text{-}30)$$

Figure 2.5 shows a 3D plot of $p_\theta(\theta)$ vs. θ for a range of A_0/σ. Another view of the same result is shown in Fig. 2.6, which shows $p_\theta(\theta)$ for several values of A_0/σ.

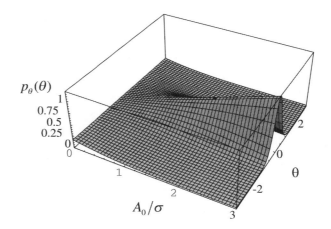

Figure 2.5 Probability density function of the phase θ for a range of A_0/σ.

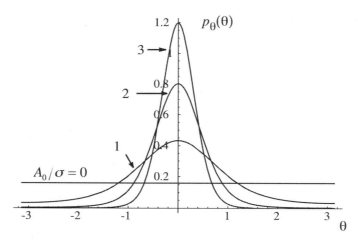

Figure 2.6 Probability density function of the phase θ for several values of A_0/σ.

From these results we see (as we might have suspected in advance) that when $A_0/\sigma = 0$, the phase is uniformly distributed, whereas when A_0/σ grows large, the phase distribution becomes more and more clustered around the phase of the known phasor, $\theta = 0$. When A_0/σ becomes large, it is readily shown that $p_\theta(\theta)$ approaches a Gaussian density function,

$$p_\theta(\theta) \approx \frac{1}{\sqrt{2\pi}\sigma/A_0} e^{-\frac{\theta^2}{2(\sigma/A_0)^2}}. \tag{2-31}$$

The mean of this distribution is zero and the standard deviation is σ/A_0.

2.4 Sums of Random Phasor Sums

In this section we consider the statistics of a random phasor sum that is composed of sums of various independent random phasor sums. The real and imaginary parts of the resultant phasor can in this case be expressed as

$$\mathcal{R} = \frac{1}{\sqrt{N_1}} \sum_{n_1=1}^{N_1} a_{n_1} \cos \phi_{n_1} + \frac{1}{\sqrt{N_2}} \sum_{n_2=1}^{N_2} a_{n_2} \cos \phi_{n_2} + \cdots$$
$$+ \frac{1}{\sqrt{N_M}} \sum_{n_M=1}^{N_M} a_{n_M} \cos \phi_{n_M} \quad (2\text{-}32)$$

$$\mathcal{I} = \frac{1}{\sqrt{N_1}} \sum_{n_1=1}^{N_1} a_{n_1} \sin \phi_{n_1} + \frac{1}{\sqrt{N_2}} \sum_{n_2=1}^{N_2} a_{n_2} \sin \phi_{n_2} + \cdots$$
$$+ \frac{1}{\sqrt{N_M}} \sum_{n=1}^{N_M} a_{n_M} \sin \phi_{n_M} \quad (2\text{-}33)$$

representing a sum of M different random phasor sums. It is quite clear that these sums can be directly combined, giving

$$\mathcal{R} = \frac{1}{\sqrt{N_1 + N_2 + \cdots + N_M}} \sum_{n=1}^{N_1+N_2+\cdots+N_M} b_n \cos \phi_n$$
$$\mathcal{I} = \frac{1}{\sqrt{N_1 + N_2 + \cdots + N_M}} \sum_{n=1}^{N_1+N_2+\cdots+N_M} b_n \sin \phi_n. \quad (2\text{-}34)$$

Thus, when we add independent random phasor sums, we are in effect creating a single new random phasor sum with a number of component phasors equal to the sum of the numbers of components in all the sums.

All of the arguments applied to a single random phasor sum can now be applied to this new sum. In particular, we can again conclude that the resultant phasor has a length that is Rayleigh distributed, and a phase that is uniformly distributed. *The addition of a multitude of sums on an amplitude basis has done nothing to change the functional form of the statistical distributions.*

2.5 Random Phasor Sums with a Finite Number of Equal-Length Components

We will briefly discuss in this section the statistical properties of random phasor sums consisting of a finite number of terms. The simplifications afforded by the Central Limit Theorem for large numbers of contributions are no longer possible in this case. The real and imaginary parts of the resultant phasor are again given by Eqs. (2-2), but this time N is explicitly restricted to being a finite number.

This problem has a long history the interested reader may wish to pursue. Two relevant references are by Lord Rayleigh [130] and Margaret Slack [147].

The ϕ_n are again taken to be uniformly distributed on $(-\pi, \pi)$ and independent of one another. For purposes of simplicity, we will assume here that the a_n are fixed and known quantities, and that the statistical properties of the sum come from the statistics of the phases of the component phasors. These assumptions allow us to write the characteristic function[4] of one term in the sum defining the real part $\mathcal{R}$ of the resultant as

$$\mathbf{M}_n(\omega) = E[e^{j\omega \frac{a_n}{\sqrt{N}} \cos \phi_n}] = \int_{-\pi}^{\pi} p_\phi(\phi_n) e^{j\omega \frac{a_n \cos \phi_n}{\sqrt{N}}} d\phi_n$$

$$= \frac{1}{2\pi} \int_{-\pi}^{\pi} e^{j\omega \frac{a_n \cos \phi_n}{\sqrt{N}}} d\phi_n = J_0\left(\frac{a_n \omega}{\sqrt{N}}\right). \tag{2-35}$$

An identical result is obtained for one term of the imaginary part $\mathcal{I}$.

The characteristic function of the sum of several independent random variables can be found as the product of the characteristic functions of the components of the sum. Therefore

$$\mathbf{M}_\mathcal{R}(\omega) = \mathbf{M}_\mathcal{I}(\omega) = \prod_{n=1}^{N} J_0\left(\frac{a_n \omega}{\sqrt{N}}\right). \tag{2-36}$$

At this point we simplify even further by assuming that the lengths of all the component phasors are identical and equal to $a/\sqrt{N}$. The characteristic function of the real and imaginary parts of the resultant phasor becomes

$$\mathbf{M}_\mathcal{R}(\omega) = \mathbf{M}_\mathcal{I}(\omega) = J_0^N\left(\frac{a\omega}{\sqrt{N}}\right) \tag{2-37}$$

The probability density functions of the real and imaginary parts can be found by Fourier transforming their respective characteristic functions,

$$p_\mathcal{R}(\mathcal{R}) = \mathcal{F}\{\mathbf{M}_\mathcal{R}(\omega)\} = \frac{1}{2\pi} \int_{-\infty}^{\infty} J_0^N\left(\frac{a\omega}{\sqrt{N}}\right) e^{-j\omega \mathcal{R}} d\omega \tag{2-38}$$

and a similar expression for $p_\mathcal{I}(\mathcal{I})$.

At this point we invoke some powerful arguments about symmetry. Since the phase angles of the individual components of the sum are uniformly distributed on $(-\pi, \pi)$, it follows that the real and imaginary parts of the resultant phasor must have a joint probability density function and a joint characteristic function that are

---

4. The 1D characteristic function of a random variable z is by definition the expected value of $e^{j\omega z}$, or equivalently the inverse Fourier transform of the probability density function of z. The 2D characteristic function is the 2D inverse Fourier transform of the joint probability density function of the two random variables in question.

circularly symmetric. It follows that $\mathbf{M}_\mathcal{R}(\omega)$ and $\mathbf{M}_\mathcal{I}(\omega)$ show that the form of the 2D characteristic function $\mathbf{M}_{\mathcal{R},\mathcal{I}}(\omega_\mathcal{R},\omega_\mathcal{I})$ is given by

$$\mathbf{M}_{\mathcal{R},\mathcal{I}}(\omega_\mathcal{R},\omega_\mathcal{I}) = \mathbf{M}_{\mathcal{R},\mathcal{I}}(\omega) = J_0^N\left(\frac{a\omega}{\sqrt{N}}\right), \quad (2\text{-}39)$$

where $\omega = \sqrt{\omega_\mathcal{R}^2 + \omega_\mathcal{I}^2}$.

The circular symmetry of the characteristic function allows the joint probability density function to be recovered by means of a Fourier–Bessel (or Hankel) transform ([71], p. 12). The radial profile $f(A)$ of the joint density function is given by

$$f(A) = 2\pi \int_0^\infty \rho J_0^N\left(\frac{2\pi a \rho}{\sqrt{N}}\right) J_0(2\pi \rho A)\, d\rho. \quad (2\text{-}40)$$

where $\rho = \omega/2\pi$, and $A = \sqrt{\mathcal{R}^2 + \mathcal{I}^2}$ is the length of the resultant phasor.

To find the probability density function of A from the radial profile of the 2D density, we must multiply that profile by $2\pi A$ (the circumference of a circle of radius A). The result is

$$p_A(A) = 4\pi^2 A \int_0^\infty \rho J_0^N\left(\frac{2\pi a \rho}{\sqrt{N}}\right) J_0(2\pi A \rho)\, d\rho \quad (2\text{-}41)$$

for $A \geq 0$.

The integral above is not easily solvable analytically, even for small N. Evaluation numerically is also not easy, due to the highly oscillatory nature of the integrand. Figure 2.7 shows approximate plots for $N = 1, 2, 3, 4, 5$ and ∞ when $a = 1$ (individual phasor lengths $1/\sqrt{N}$). For the case $N = 1$, the density function takes the form of a delta function at $A = 1$. For $N = \infty$ the density function can be shown to become a Rayleigh density.

2.6 Random Phasor Sums with a Nonuniform Distribution of Phases

Next we consider a random phasor sum for which the underlying phasor components have a *nonuniform* distribution of phase. We retain the assumption that all components are identically distributed, as well as all assumptions about the independence of the components and independence of the amplitude and phase of any one component.

We begin with the usual equations for the real and imaginary parts of the resultant phasor,

$$\mathcal{R} = \frac{1}{\sqrt{N}} \sum_{n=1}^{N} a_n \cos \phi_n$$

$$\mathcal{I} = \frac{1}{\sqrt{N}} \sum_{n=1}^{N} a_n \sin \phi_n. \quad (2\text{-}42)$$

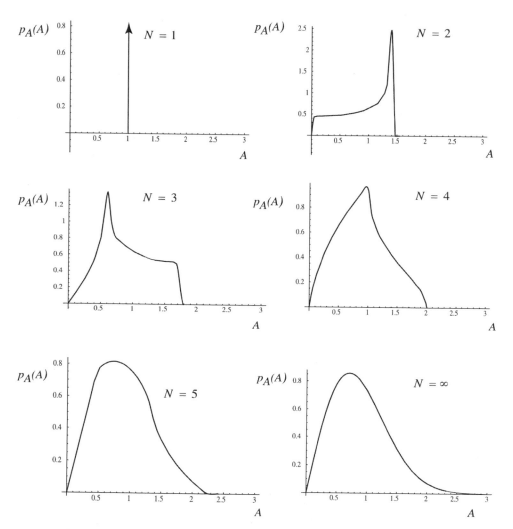

Figure 2.7 Probability density functions of the length A of random phasor sums having N randomly-phased phasors, each with length $1/\sqrt{N}$.

When the number of component phasors is large, the Central Limit Theorem again comes into play, indicating that the joint density function of $\mathcal{R}$ and $\mathcal{I}$ is of the general form (see [111], p. 337)

2.6 Random Phasor Sums with a Nonuniform Distribution of Phases

$$p_{\mathcal{R},\mathcal{I}}(\mathcal{R},\mathcal{I}) = \frac{1}{2\pi\sigma_{\mathcal{R}}\sigma_{\mathcal{I}}\sqrt{1-\rho_{\mathcal{R},\mathcal{I}}^2}}$$

$$\times \exp\left[-\frac{\left(\frac{\mathcal{R}-\bar{\mathcal{R}}}{\sigma_{\mathcal{R}}}\right)^2 + \left(\frac{\mathcal{I}-\bar{\mathcal{I}}}{\sigma_{\mathcal{I}}}\right)^2 - 2\rho_{\mathcal{R},\mathcal{I}}\left(\frac{\mathcal{R}-\bar{\mathcal{R}}}{\sigma_{\mathcal{R}}}\right)\left(\frac{\mathcal{I}-\bar{\mathcal{I}}}{\sigma_{\mathcal{I}}}\right)}{2(1-\rho_{\mathcal{R},\mathcal{I}}^2)}\right], \quad (2\text{-}43)$$

where we have allowed for the possibility of any means, variances and correlation coefficient[5] $\rho_{\mathcal{R},\mathcal{I}}$. A joint density function of length A and phase θ can be found by substituting $\mathcal{R} = A\cos\theta$, $\mathcal{I} = A\sin\theta$, and multiplying the resulting expression by the Jacobian of the transformation, A. However, integration of the resulting expression with respect to θ does not yield a closed form expression for the marginal density function of the length of the resultant.

Turning attention instead to the means of $\mathcal{R}$ and $\mathcal{I}$, we have

$$\bar{\mathcal{R}} = \frac{1}{\sqrt{N}}\sum_{n=1}^{N} \bar{a}_n \overline{\cos\phi_n}$$

$$\bar{\mathcal{I}} = \frac{1}{\sqrt{N}}\sum_{n=1}^{N} \bar{a}_n \overline{\sin\phi_n}. \quad (2\text{-}44)$$

The averages of $\cos\phi_n$ and $\sin\phi_n$ can be expressed in terms of the characteristic function of the random variables ϕ_n,

$$\overline{\cos\phi_n} = \frac{1}{2}\overline{e^{j\phi_n}} + \frac{1}{2}\overline{e^{-j\phi_n}} = \frac{1}{2}[\mathbf{M}_\phi(1) + \mathbf{M}_\phi(-1)]$$

$$\overline{\sin\phi_n} = \frac{1}{2j}\overline{e^{j\phi_n}} - \frac{1}{2j}\overline{e^{-j\phi_n}} = \frac{1}{2j}[\mathbf{M}_\phi(1) - \mathbf{M}_\phi(-1)] \quad (2\text{-}45)$$

yielding expressions for the means of interest,

$$\bar{\mathcal{R}} = \frac{\sqrt{N}\bar{a}}{2}[\mathbf{M}_\phi(1) + \mathbf{M}_\phi(-1)]$$

$$\bar{\mathcal{I}} = \frac{\sqrt{N}\bar{a}}{2j}[\mathbf{M}_\phi(1) - \mathbf{M}_\phi(-1)]. \quad (2\text{-}46)$$

The second moments follow in a similar way, but with more complex algebra (see Ref. [70], Appendix B). The results of interest here are

---

5. The *correlation coefficient* $\rho_{\mathcal{R},\mathcal{I}}$ is defined as the normalized covariance of $\mathcal{R}$ and $\mathcal{I}$, $\rho_{\mathcal{R},\mathcal{I}} = \overline{(\mathcal{R}-\bar{\mathcal{R}})(\mathcal{I}-\bar{\mathcal{I}})}/\sigma_{\mathcal{R}}\sigma_{\mathcal{I}}$.

$$\sigma_{\mathcal{R}}^2 = \frac{\overline{a^2}}{4}[2 + \mathbf{M}_\phi(2) + \mathbf{M}_\phi(-2)]$$

$$- \frac{\bar{a}^2}{4}[2\mathbf{M}_\phi(1)\mathbf{M}_\phi(-1) + \mathbf{M}_\phi^2(1) + \mathbf{M}_\phi^2(-1)]$$

$$\sigma_{\mathcal{I}}^2 = \frac{\overline{a^2}}{4}[2 - \mathbf{M}_\phi(2) - \mathbf{M}_\phi(-2)] \quad (2\text{-}47)$$

$$- \frac{\bar{a}^2}{4}[2\mathbf{M}_\phi(1)\mathbf{M}_\phi(-1) - \mathbf{M}_\phi^2(1) - \mathbf{M}_\phi^2(-1)]$$

$$C_{\mathcal{R},\mathcal{I}} = \frac{\overline{a^2}}{4j}[\mathbf{M}_\phi(2) - \mathbf{M}_\phi(-2)] - \frac{\bar{a}^2}{4j}[\mathbf{M}_\phi^2(1) - \mathbf{M}_\phi^2(-1)]$$

where $C_{\mathcal{R},\mathcal{I}}$ signifies the *covariance* of $\mathcal{R}$ and $\mathcal{I}$,

$$C_{\mathcal{R},\mathcal{I}} = \overline{(\mathcal{R} - \bar{\mathcal{R}})(\mathcal{I} - \bar{\mathcal{I}})}. \quad (2\text{-}48)$$

As a special case, consider phases ϕ_n that obey zero-mean Gaussian statistics. The density function and characteristic function of the phase in this case are

$$p_\phi(\phi) = \frac{1}{\sqrt{2\pi}\sigma_\phi} \exp\left(-\frac{\phi^2}{2\sigma_\phi^2}\right)$$

$$\mathbf{M}_\phi(\omega) = \exp\left(-\frac{\sigma_\phi^2 \omega^2}{2}\right). \quad (2\text{-}49)$$

Substitution in Eq. (2-47) yields

$$\bar{\mathcal{R}} = \sqrt{N}\bar{a}e^{-\sigma_\phi^2/2}$$

$$\bar{\mathcal{I}} = 0$$

$$\sigma_{\mathcal{R}}^2 = \frac{\overline{a^2}}{2}[1 + e^{-2\sigma_\phi^2}] - \bar{a}^2 e^{-\sigma_\phi^2} \quad (2\text{-}50)$$

$$\sigma_{\mathcal{I}}^2 = \frac{\overline{a^2}}{2}[1 - e^{-2\sigma_\phi^2}]$$

$$C_{\mathcal{R},\mathcal{I}} = 0.$$

As an aside we note that both $\bar{\mathcal{I}}$ and $C_{\mathcal{R},\mathcal{I}}$ will always vanish for any *even* probability density function for the ϕ_n.

Figure 2.8 shows a contour plot of the (approximate) joint density function of $\mathcal{R}$ and $\mathcal{I}$ when $N = 100$ and $\sigma_\phi = 1$ radian. In this plot we have assumed that all component phasors have length 1. Note that the contours are elliptical in this case (i.e. the resultant is *not* a circular complex random variable). In addition, the distribution is seen to be centered on a point on the real axis other than the origin.

2.6 Random Phasor Sums with a Nonuniform Distribution of Phases 23

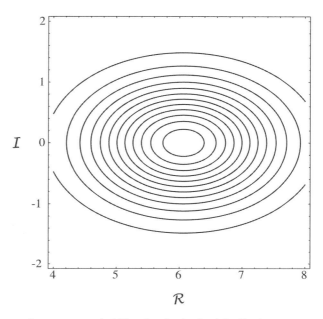

Figure 2.8 Contours of constant probability density in the $(\mathcal{R}, \mathcal{I})$ plane.

Finally, consider the case in which the variance σ_ϕ^2 of the phases becomes arbitrarily large. The moments above become

$$\begin{align}
\bar{\mathcal{R}} &= 0 \\
\bar{\mathcal{I}} &= 0 \\
\sigma_\mathcal{R}^2 &= \overline{a^2}/2 \\
\sigma_\mathcal{I}^2 &= \overline{a^2}/2 \\
C_{\mathcal{R},\mathcal{I}} &= 0,
\end{align} \tag{2-51}$$

which agree with the results found in Section 2.1 for the case of a uniform phase distribution (provided the lengths are identically distributed). Indeed, as the variance of the phase increases without bound, the Gaussian distribution, taken modulo 2π, approaches a uniform distribution asymptotically.

3 First-Order Statistical Properties of Optical Speckle

In this chapter we consider the first-order statistical properties of optical speckle patterns. By "first-order" we again mean the properties at a single point in space and time. The distinguishing characteristic of optical speckle is that we can measure only the *intensity* of the wavefield, rather than the complex amplitude. In other words, detectors in the optical wavelength regime respond to incident power rather than to the amplitude of the electromagnetic field. Interferometry can be used to determine amplitude, but this is an indirect measurement rather than a direct measurement.

Again the origin of speckle lies in the addition of a number of randomly phased complex contributions, usually complex amplitudes of optical wavefields. Thus the random-phasor-sum results of the previous chapter will be of great help to us here. We will assume that the statistical conditions outlined at the beginning of Chapter 2 hold here for the component phasors.

3.1 Definition of Intensity

Since intensity is of critical interest in the optical regime, we provide a brief definition of this quantity. We begin with a definition of the Poynting vector $\vec{\mathcal{P}}$, of an electromagnetic disturbance,

$$\vec{\mathcal{P}} = \vec{\mathcal{E}} \times \vec{\mathcal{H}} \tag{3-1}$$

where $\vec{\mathcal{E}}$ is the time-varying vector electric field, $\vec{\mathcal{H}}$ is the time-varying vector magnetic field, and $\times$ signifies a vector cross-product operation. The time-averaged intensity of the wavefield is usually defined as being proportional to the magnitude of the time-averaged Poynting vector[1]

$$I \propto |\langle \vec{\mathcal{P}} \rangle| \tag{3-2}$$

where $\langle \ldots \rangle$ indicates an infinite time average. The constant of proportionality is chosen for simplicity. For the case of a locally transverse, monochromatic electromagnetic plane wave in an isotropic medium,

$$\vec{\mathcal{E}} = \text{Re}\{\vec{\mathbf{E}}_0 \exp[-j(2\pi \nu t - \vec{k} \cdot \vec{r})]\}$$
$$\vec{\mathcal{H}} = \text{Re}\{\vec{\mathbf{H}}_0 \exp[-j(2\pi \nu t - \vec{k} \cdot \vec{r})]\}, \tag{3-3}$$

where ν is the optical frequency of the wave, while $\vec{k}$ is the wave vector (length $2\pi/\lambda$, direction orthogonal to $\vec{\mathcal{E}}$ and $\vec{\mathcal{H}}$). The quantities $\vec{\mathbf{E}}_0$ and $\vec{\mathbf{H}}_0$ represent the complex vector amplitudes of the electric and magnetic fields, respectively. The vector $\vec{r}$ represents a position in three-dimensional space. The magnitude of the time-averaged Poynting vector can now be expressed as

$$|\langle \vec{\mathcal{P}} \rangle| = \frac{\vec{\mathbf{E}}_0 \cdot \vec{\mathbf{E}}_0^*}{2\eta}, \tag{3-4}$$

where we have used the fact that $\vec{\mathbf{H}}_0 = \vec{\mathbf{E}}_0/2\eta$, η being the characteristic impedance of the medium.

From Eq. (3-4) we arrive at a representation for the (time-averaged) intensity of the wavefield of the form (suppressing an unnecessary constant)

$$I = |\mathbf{E}_{0_x}|^2 + |\mathbf{E}_{0_y}|^2 + |\mathbf{E}_{0_z}|^2, \tag{3-5}$$

where $\mathbf{E}_{0_x}$, $\mathbf{E}_{0_y}$, $\mathbf{E}_{0_z}$ are the three cartesian components of the complex vector $\vec{\mathbf{E}}_0$.

For *paraxial* waves, that is, waves traveling with local $\vec{k}$ vectors that always subtend small angles with respect to the z axis, the component $\mathbf{E}_{0_z}$ is small and is generally neglected. This leaves the intensity expressed as a sum of the squared magnitudes of two complex scalar functions, one for the x component of the field and one for the y component of the field.

In the future we shall suppress the fact that the scalar quantities involved here are electric fields (an entirely equivalent development could express them in terms of magnetic fields), and represent the intensity in the general case as

$$I = \begin{cases} |\mathbf{A}_x|^2 + |\mathbf{A}_y|^2 & \text{for an unpolarized wave} \\ |\mathbf{A}|^2 & \text{for a polarized wave,} \end{cases} \tag{3-6}$$

where the scalar quantity $\mathbf{A}$ in the polarized case corresponds to the complex amplitude of the field along the direction of polarization.

---

1. While this definition of intensity is generally satisfactory, there do occasionally arise some difficulties with its application. See, for example, Ref. [42], pp. 200–203. Note also that we have defined the basic phasors $\exp(-j2\pi \nu t)$ to rotate in the clockwise direction.

The discussion above has assumed an infinite time average for the definition of intensity. It is also useful to define an *instantaneous* intensity as well. If the wave has nonzero but narrow bandwidth (i.e. the bandwidth Δv is much smaller than the center frequency v_o), then the phasor amplitudes $\mathbf{A}_x$ and $\mathbf{A}_y$ are functions of time, with variations taking place on time scales of the order of $1/\Delta v$. In such cases we represent the instantaneous intensity by

$$I(t) = \begin{cases} |\mathbf{A}_x(t)|^2 + |\mathbf{A}_y(t)|^2 & \text{for an unpolarized wave} \\ |\mathbf{A}(t)|^2 & \text{for a polarized wave.} \end{cases} \quad (3\text{-}7)$$

This concludes our introduction to the concept of intensity.

3.2 First-Order Statistics of the Intensity and Phase

Many of the statistical properties of optical speckle can be deduced directly from the information we have already developed in the previous chapter. A critical relationship that will help us in that regard is the rule governing the change of probability density functions that results from a monotonic transformation of an underlying random variable. Let v be a random variable that is related to a random variable u through a monotonic transformation $v = f(u)$. Then it is a fundamental result of probability theory that the probability density function $p_V(v)$ of v can be found from the probability density function $p_U(u)$ of u through (see, for example, [70], Section 2.5.2)

$$p_V(v) = p_U(f^{-1}(v)) \left| \frac{du}{dv} \right|. \quad (3\text{-}8)$$

For the particular case of interest, $v = I$, the intensity, $u = |\mathbf{A}| = A$, the amplitude, and

$$I = f(A) = A^2. \quad (3\text{-}9)$$

It follows that, knowing the probability density function $p_A(A)$ we can find the corresponding probability density function $p_I(I)$ through

$$p_I(I) = p_A(\sqrt{I}) \left| \frac{dA}{dI} \right| = \frac{1}{2\sqrt{I}} p_A(\sqrt{I}). \quad (3\text{-}10)$$

This result then allows us to specify the probability density function of intensity for every case in which the probability density function of amplitude A is known.

3.2.1 Large Number of Random Phasors

When the number of contributing random phasors (each with phase uniformly distributed on $(-\pi, \pi)$) is large, the conditions of Section 2.2 apply, and the probability density function of the amplitude is Rayleigh distributed,

$$p_A(A) = \frac{A}{\sigma^2} \exp\left(-\frac{A^2}{2\sigma^2}\right) \qquad (3\text{-}11)$$

for $A \geq 0$. Applying the transformation law of Eq. (3-10), we find that the intensity is distributed according to an *exponential* probability density,

$$p_I(I) = \frac{\sqrt{I}}{\sigma^2} \exp\left(-\frac{I}{2\sigma^2}\right) \cdot \frac{1}{2\sqrt{I}} = \frac{1}{2\sigma^2} \exp\left(-\frac{I}{2\sigma^2}\right) \qquad (3\text{-}12)$$

for $I \geq 0$. The moments of this distribution are easily found by direct integration, with the result

$$\overline{I^q} = (2\sigma^2)^q q!. \qquad (3\text{-}13)$$

From this result we see that the mean intensity $\bar{I}$ is $2\sigma^2$, so that an equivalent statement is that the qth moment is given by

$$\overline{I^q} = \bar{I}^q q!, \qquad (3\text{-}14)$$

and the probability density function can be rewritten

$$p_I(I) = (1/\bar{I}) \exp(-I/\bar{I}). \qquad (3\text{-}15)$$

Speckle with this intensity distribution is often referred to as *fully developed* speckle, contrasting the result with cases in which the phases of the contributing phasors may not be uniformly distributed. The second moment, variance, and standard deviation of the intensity for the current case are given by

$$\overline{I^2} = 2\bar{I}^2$$
$$\sigma_I^2 = \bar{I}^2 \qquad (3\text{-}16)$$
$$\sigma_I = \bar{I}.$$

Figure 3.1 shows a plot of the negative exponential distribution.

Two quantities that will appear frequently in future discussions of applications are the *contrast* C and the *signal-to-noise ratio* S/N of a speckle pattern, defined by

$$C = \frac{\sigma_I}{\bar{I}} \qquad (3\text{-}17)$$

$$S/N = 1/C = \frac{\bar{I}}{\sigma_I}. \qquad (3\text{-}18)$$

Contrast is a measure of how strong the fluctuations of intensity are in a speckle pattern compared with the average intensity. Signal-to-noise ratio is the reciprocal concept, that is, the strength of the average intensity with respect to the strength of the fluctuations of intensity.

For fully developed speckle, as considered in this section, Eq. (3-16) implies that the contrast and signal-to-noise ratio for such speckle are

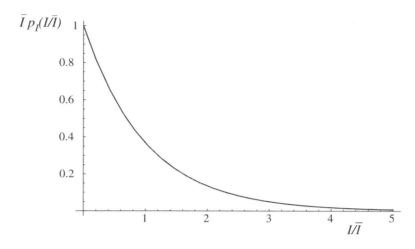

Figure 3.1 The negative exponential density function.

$$C = 1$$
$$S/N = 1. \quad (3\text{-}19)$$

Thus the fluctuations of fully developed speckle are of the same order as the average, making this type of noise potentially very severe indeed.

In some applications (for example, when possible eye damage due to excess optical power incident on the retina is of concern) it is of interest to know the probability that the intensity will exceed a certain threshold level I_t. This quantity can be calculated as follows:

$$P(I \geq I_t) = \int_{I_t}^{\infty} p_I(I)\, dI = \frac{1}{\bar{I}} \int_{I_t}^{\infty} e^{-I/\bar{I}}\, dI = e^{-I_t/\bar{I}}. \quad (3\text{-}20)$$

Thus the probability of exceeding the threshold I_t falls exponentially with the ratio of I_t to the average intensity $\bar{I}$.

For future use, we also present the characteristic function of the intensity fluctuations. The required Fourier transform is easily performed, as indicated below:

$$\mathbf{M}_\mathbf{I}(\omega) = \int_0^{\infty} e^{j\omega I} p_I(I)\, dI = \int_0^{\infty} e^{j\omega I} \left[\frac{1}{\bar{I}} e^{-I/\bar{I}} \right] dI = \frac{1}{1 - j\omega \bar{I}}. \quad (3\text{-}21)$$

While our primary interest here is the statistics of the intensity, we remark in passing that the statistics of the phase that correspond to this intensity are exactly the same as the statistics of the phase given in Eq. (2-19), namely the probability density function of the resultant phase is uniformly distributed on $(-\pi, \pi)$. The statistics of the phase are not changed by the fact that our interest here is in the squared length of the resultant phasor rather than the length of that resultant.

3.2.2 Constant Phasor plus a Random Phasor Sum

In a variety of applications, the field in a speckle pattern consists of a sum of a single "constant" or known phasor, plus a random phasor sum with uniformly distributed phase. Such is the case in holography, where the known phasor corresponds to the reference beam, and the random phasor sum corresponds to the field scattered from a rough object. In addition, when the speckle pattern results from scattering from a surface with a roughness that is less than a wavelength, the scattered wave often consists of a constant specular component and a diffuse scattered component. In all such cases, knowledge about the statistical properties of the resultant *intensity* is of considerable interest.

The resultant phasor $\mathbf{A}$ in this case can be written

$$\mathbf{A} = \mathbf{A}_0 + \mathbf{A}_n = A_0 + A_n e^{j\theta_n} \tag{3-22}$$

where $\mathbf{A}_0$ is the known phasor, $\mathbf{A}_n$ is the random phasor sum; without loss of generality, the known phasor can be taken to have zero phase, allowing us to write $\mathbf{A}_0 = A_0$. The random phasor sum, $\mathbf{A}_n$, is assumed to obey circular Gaussian statistics, and to have length A_n and phase θ_n.

The intensity of the field is easily calculated to be given by

$$I = |\mathbf{A}|^2 = A_0^2 + A_n^2 + 2A_0 A_n \cos\theta_n. \tag{3-23}$$

The first term in this expression is the intensity of the known phasor alone, and the second term is the intensity of the random phasor sum alone. The third term represents *interference* between the constant phasor and the random phasor sum. While this interference term is readily shown to have a mean of zero (due to the uniform distribution of θ_n), nonetheless it has a major effect on the statistical distribution of the resultant intensity.

The probability density function of the intensity in such a speckle pattern can be found with the help of Eq. (2-25) expressing the probability density function of the amplitude, and the transformation rule of Eq. (3-10) relating the density functions of amplitude and intensity. The result is a probability density function

$$p_I(I) = \frac{1}{2\sigma^2} \exp\left\{-\frac{I + A_0^2}{2\sigma^2}\right\} I_0\left(\frac{\sqrt{I} A_0}{\sigma^2}\right) \tag{3-24}$$

for $I \geq 0$.

It is convenient to express this result in terms of the following parameters:

- $\bar{I}_n = \overline{A_n^2} = 2\sigma^2$, representing the average intensity of the random phasor sum alone,
- $I_0 = A_0^2$, representing the intensity of the known phasor alone, and
- $r = I_0/\bar{I}_n$, which is the ratio of the intensity of the known phasor to the average intensity of the random phasor sum (known as the "beam ratio" in holography).

The expression for the probability density function of intensity can now be written

3.2 First-Order Statistics of the Intensity and Phase

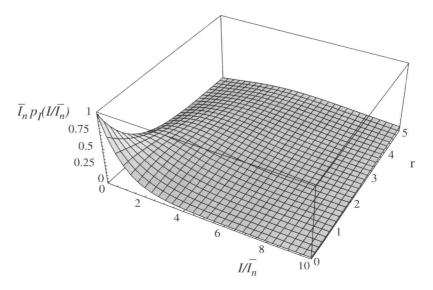

Figure 3.2 The modified Rician density function as a function of $I/\bar{I}_n$ and r.

$$p_I(I) = \frac{1}{\bar{I}_n} \exp\left\{-\frac{I + I_0}{\bar{I}_n}\right\} I_0\left(2\frac{\sqrt{II_0}}{\bar{I}_n}\right)$$

$$= \frac{1}{\bar{I}_n} \exp\left\{-\left(\frac{I}{\bar{I}_n} + r\right)\right\} I_0\left(2\sqrt{\frac{I}{\bar{I}_n}r}\right) \quad (3\text{-}25)$$

(for $I \geq 0$), which is often referred to as the *modified* Rician density function.

Figure 3.2 shows a plot of $\bar{I}_n p_I(I/\bar{I}_n)$ vs. $I/\bar{I}_n$ and r. Figure 3.3 shows plots of $\bar{I}_n p_I(I/\bar{I}_n)$ vs. $I/\bar{I}_n$ for $r = 0, 2$ and 4.

Note that for $r = 0$, the distribution is negative exponential, as expected when the constant phasor vanishes. We also notice from these plots that (for $r > 1$) the center of the distribution is found roughly at $I/\bar{I}_n = r$, but that the distributions do not bunch about the center as strongly as they did in the case of the amplitude distributions. This latter behavior is due to the interference of the strong constant phasor with the weaker random phasor sum, which causes the distributions to spread by significant amounts.

It is also of interest to express the contrast and signal-to-noise ratio for this type of speckle. To do so, we must first find a general expression for the qth moment of the intensity,

$$\overline{I^q} = \int_0^\infty I^q p_I(I)\, dI. \quad (3\text{-}26)$$

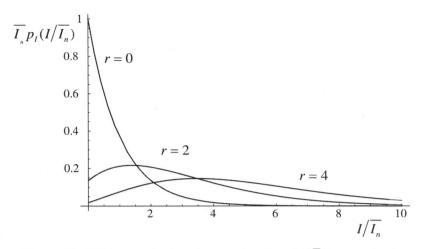

Figure 3.3 The modified Rician density function as a function of $I/\bar{I}_n$ for several values of r.

where $p_I(I)$ is given by Eq. (3-24). With the help of a symbolic manipulation program, it is easily shown that

$$\overline{I^q} = \bar{I}_n^q e^{-r} q!\,_1F_1(q+1,1,r), \tag{3-27}$$

where $_1F_1$ is a confluent hypergeometric function. For the first and second moments, this equation reduces to

$$\begin{aligned}\bar{I} &= (1+r)\bar{I}_n \\ \overline{I^2} &= (2+4r+r^2)\bar{I}_n^2.\end{aligned} \tag{3-28}$$

It follows that the standard deviation of intensity, σ_I, is given by

$$\sigma_I = \bar{I}_n\sqrt{1+2r}. \tag{3-29}$$

From this result the contrast and signal-to-noise ratio are given by

$$\begin{aligned}C &= \frac{\sigma_I}{\bar{I}} = \frac{\sqrt{1+2r}}{1+r} \\ S/N &= \frac{\bar{I}}{\sigma_I} = \frac{1+r}{\sqrt{1+2r}}.\end{aligned} \tag{3-30}$$

In Figure 3.4 we show plots of contrast and signal-to-noise ratio as a function of beam ratio r. Note that the contrast and the signal-to-noise ratio both begin at unity for $r = 0$, as appropriate for a negative exponential distribution, and then drop and increase, respectively, slowly as a function of beam ratio.

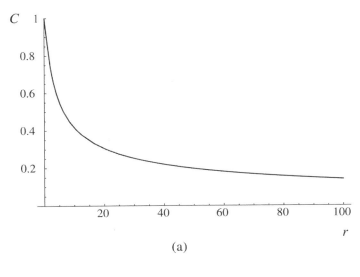

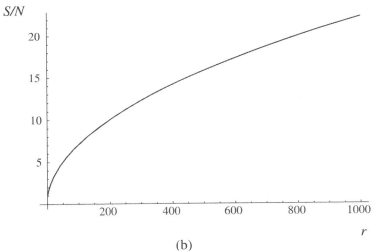

Figure 3.4 (a) Contrast and (b) signal-to-noise ratio of speckle produced by the sum of a known phasor and a random phasor sum, as a function of beam ratio r.

In applications for which the probability of total intensity exceeding a threshold I_t is of concern, we can calculate the desired probability as follows:

$$P(I \geq I_t) = \int_{I_t}^{\infty} \frac{1}{\bar{I}_n} \exp\left\{-\left(\frac{I}{\bar{I}_n}+r\right)\right\} I_0\left(2\sqrt{\frac{I}{\bar{I}_n}r}\right) dI. \quad (3\text{-}31)$$

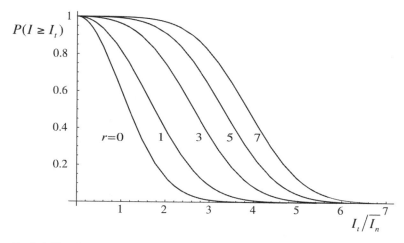

Figure 3.5 Probability that the intensity I exceeds a threshold I_t vs. normalized threshold $I_t/\overline{I_n}$ for various values of the beam ratio r.

With a change of variables $x = \sqrt{2y}$, the integral becomes

$$P(I \geq I_t) = e^{-r} \int_{\sqrt{2\beta}}^{\infty} x e^{-\frac{1}{2}x^2} I_0(\sqrt{2r}x) \, dx, \tag{3-32}$$

where $\beta = I_t/\overline{I_n}$. This integral can be expressed in terms of the so-called Marcum Q-functions [107], which are tabulated,

$$P(I \geq I_t) = Q(\sqrt{2r}, \sqrt{2\beta}). \tag{3-33}$$

Alternatively the integral can be directly performed with a numerical integration routine.

Figure 3.5 shows plots of $P(I \geq I_t)$ vs. $I_t/\overline{I_n}$ for several values of beam ratio r.

3.2.3 Finite Number of Equal-Length Phasors

We look briefly now at the statistical properties of the intensity of random phasor sums with a finite number of equal-length but randomly phased steps. The problem is similar to that studied in Section 2.5, the only difference being that we are now interested in the statistics of the intensity, rather than the amplitude, of the sum. For a useful reference, consult [170].

Application of the probability transformation of Eq. (3-10) to the previously derived expressions for the probability density functions of the amplitudes, as given in Eq. (2-41), yields the following expression for the probability density function of intensity with the number N of equal-length phasors as a parameter:

3.2 First-Order Statistics of the Intensity and Phase

$$p_I(I) = 2\pi^2 \int_0^\infty \rho J_0^N\left(\frac{2\pi a \rho}{\sqrt{N}}\right) J_0(2\pi\sqrt{I}\rho)\,d\rho, \qquad (3\text{-}34)$$

where $a/\sqrt{N}$ is the *amplitude* of any one of the component phasors in the sum, and N is the number of such phasors. As before, the phases of all component phasors have been assumed uniformly distributed on $(-\pi, \pi)$ and independent.

Figure 3.6 shows the calculated probability density functions of the total intensity for various values of N, under the assumption that $a = 1$. Note that for a single phasor, the intensity must be unity, as represented by the delta function. For two phasors, an analytical solution has been found previously (see Fig. 4-12 of [70]). Note also that for $N = \infty$, the probability density function is a negative exponential distribution, as appropriate for the fully developed speckle pattern with a very large number of individual phasors.

While we have been forced to resort to numerical evaluation in order to find the probability density functions of the resultant intensity, it is possible to find analytical expressions for the most important moments of the intensity. For example, using Eq. (2-1), the mean of the resultant intensity is given by

$$E[I] = \frac{1}{N} E\left[\left|\sum_{n=1}^N a_n e^{j\phi_n}\right|^2\right] = \frac{a^2}{N}\sum_{n=1}^N \sum_{m=1}^N E[e^{j(\phi_n-\phi_m)}] = a^2, \qquad (3\text{-}35)$$

where we have assumed again that $a_n = a_m = a$, and that ϕ_n and ϕ_m are uniformly distributed on $(-\pi, \pi)$ and uncorrelated for $n \neq m$. The second moment similarly takes the form

$$E[I^2] = \frac{a^4}{N^2}\sum_{n=1}^N \sum_{m=1}^N \sum_{p=1}^N \sum_{q=1}^N E[e^{j(\phi_n-\phi_m-\phi_p+\phi_q)}]. \qquad (3\text{-}36)$$

Due to the uniform distribution of the phases, and their lack of correlation, only the following terms survive the averaging operation:

N terms with $n = m = p = q$, each of magnitude a^4/N^2

$N(N-1)$ terms with $n = m$, $p = q$, $n \neq p$, each of magnitude a^4/N^2 $\qquad (3\text{-}37)$

$N(N-1)$ terms with $n = p$, $m = q$, $n \neq m$, each of magnitude a^4/N^2.

The result is a second moment given by

$$E[I^2] = \frac{a^4}{N^2}[N + 2N(N-1)] = \left(2 - \frac{1}{N}\right)a^4. \qquad (3\text{-}38)$$

It follows that the variance of the speckle is

$$\sigma_I^2 = E[I^2] - E[I]^2 = \left(1 - \frac{1}{N}\right)a^4. \qquad (3\text{-}39)$$

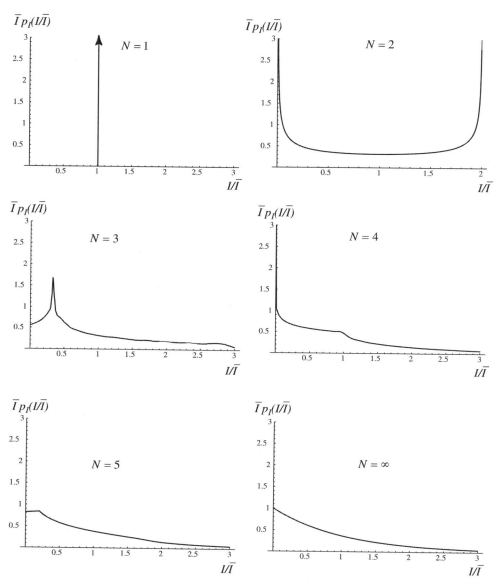

Figure 3.6 Probability density functions of intensity for random phasor sums with $N = 1, 2, 3, 4, 5$ and ∞ steps, each with amplitude of length $1/\sqrt{N}$. For $N = \infty$ the density function is a negative exponential.

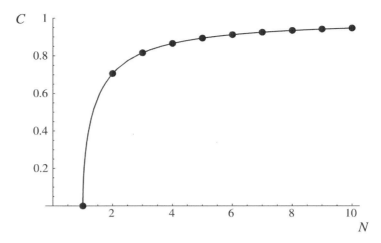

Figure 3.7 Contrast of a speckle pattern as a function of the number of independent contributions N.

Especially important in applications are the contrast and signal-to-noise ratio of the speckle, which are easily seen to be

$$C = \frac{\sigma_I}{\bar{I}} = \sqrt{1 - \frac{1}{N}}$$
$$\frac{S}{N} = \frac{\bar{I}}{\sigma_I} = \sqrt{\frac{N}{N-1}}.$$
(3-40)

Figure 3.7 shows a plot of contrast C vs. the number of contributions N. Note that when $N = 1$ the contrast is zero (meaning there are no fluctuations of the intensity) and when $N \to \infty$ the contrast asymptotically approaches unity, the result for a sum with an infinite number of independent contributions.

3.3 Sums of Speckle Patterns

In this section, we explore the statistics of the intensity of sums of fully developed speckle patterns. We first consider sums on an amplitude basis, and then consider sums on an intensity basis for both uncorrelated and correlated patterns. The results of the latter cases are fundamental to an understanding of various methods for suppressing speckle, to be discussed in a later chapter.

3.3.1 Sums on an Amplitude Basis

In Section 2.4 we considered the amplitude statistics of a sum of several random phasor sums, and argued that summations on an amplitude basis have no effect on the form of the statistical distribution of amplitude. The same holds true for the intensity statistics when two or more speckle patterns are added on an amplitude basis. In particular, we can not emphasize enough the following: *the addition of speckle patterns on an amplitude basis does not reduce the contrast of speckle and likewise has no effect on the signal-to-noise ratio.* Speckle patterns must be added on an intensity basis if any suppression of fluctuations is to be expected, and even then, as we will soon discuss, such suppression will occur only when there is some degree of decorrelation between the patterns being added.

3.3.2 Sum of Two Independent Speckle Intensities

We now consider the sum of two speckle patterns on an intensity basis when those speckle patterns are known to be statistically independent of one another. As an aside, we note that, for fully developed speckle, the terms *independent* and *uncorrelated* can be used interchangeably in this discussion, due to the fact that the underlying amplitude statistics are circular complex Gaussian, and for such statistics, lack of correlation implies statistical independence ([70], Section 2.8.2).

Independent speckle patterns can arise in a multitude of circumstances, as will be discussed in later sections. As one of many possible examples, if two speckle patterns arise by reflection of light from non-overlapping sections of a rough surface, the resulting individual speckle patterns can be expected to have intensities that are statistically independent. Independent speckle intensities can also be generated under the proper circumstances (to be discussed in later sections) from orthogonal polarization components, from different wavelengths of light, and from different angles of illumination.

If a single detector integrates two independent speckle patterns sequentially, then the total detector response is proportional to the sum of those two intensity patterns. Thus the total detected intensity I_s can be expressed as the sum of component independent intensities I_1 and I_2,

$$I_s = I_1 + I_2. \tag{3-41}$$

Both I_1 and I_2 are assumed to be fully developed speckle patterns, and therefore their intensities obey negative exponential probability density functions,

$$p_1(I_1) = \frac{1}{\overline{I_1}} \exp(-I_1/\overline{I_1})$$
$$p_2(I_2) = \frac{1}{\overline{I_2}} \exp(-I_2/\overline{I_2}). \tag{3-42}$$

A fundamental result from probability theory states that the probability density function of the sum of independent random variables is the *convolution* of the probability density functions of components of the sum ([70], Section 2.6.2). Equivalently, according to the convolution theorem, the characteristic function of the sum is the product of the characteristic functions of the components of the sum. If $\mathbf{M}_s$, $\mathbf{M}_1$, and $\mathbf{M}_2$ represent the characteristic functions of I_s, I_1 and I_2, respectively, then

$$\mathbf{M}_s = \mathbf{M}_1 \mathbf{M}_2. \quad (3\text{-}43)$$

Now Eq. (3-21) implies that

$$\mathbf{M}_1 = \frac{1}{1 - j\omega \bar{I}_1}$$
$$\mathbf{M}_2 = \frac{1}{1 - j\omega \bar{I}_2}, \quad (3\text{-}44)$$

yielding

$$\mathbf{M}_s = \frac{1}{1 - j\omega \bar{I}_1} \frac{1}{1 - j\omega \bar{I}_2}. \quad (3\text{-}45)$$

An inverse Fourier transform of this expression yields

$$p_s(I_s) = \frac{1}{\bar{I}_1 - \bar{I}_2} \left[\exp\left(-\frac{I_s}{\bar{I}_1}\right) - \exp\left(-\frac{I_s}{\bar{I}_2}\right) \right] \quad \text{when } \bar{I}_1 > \bar{I}_2,$$
$$p_s(I_s) = \frac{I_s}{\bar{I}^2} \exp\left(-\frac{I_s}{\bar{I}}\right) \quad \text{when } \bar{I}_1 = \bar{I}_2 = \bar{I}. \quad (3\text{-}46)$$

Figure 3.8 shows plots of this probability density function for various values of the ratio $r = \bar{I}_2/\bar{I}_1$ between 0 and 1. The quantity $\bar{I}_s$ represents the average total intensity,

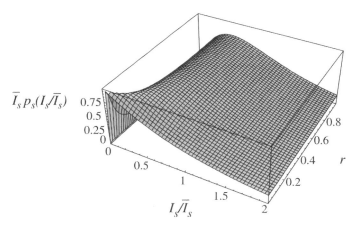

Figure 3.8 Probability density function of the sum of two independent speckle patterns as a function of $I_s/\bar{I}_s$ and $r = \bar{I}_2/\bar{I}_1$.

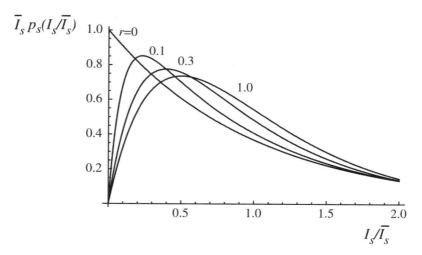

Figure 3.9 Probability density function of the sum of two independent speckle patterns as a function of $I_s/\bar{I}_s$ for $r = 0, 0.1, 0.3,$ and 1.0.

and is given by $\bar{I}_1 + \bar{I}_2$. Slices through this three-dimensional plot are shown in Fig. 3.9. Note that the curve for $r = 0$ corresponds to the probability density function of a *single* speckle pattern and is negative exponential, while the curve for $r = 1.0$ corresponds to the second line of Eq. (3-46) and applies for the sum of two independent speckle patterns with equal average intensities.

The contrast of the sum intensity is easily found without referring directly to the probability density functions just found. The first and second moments of the sum intensity are given by

$$\bar{I}_s = \overline{I_1 + I_2} = \bar{I}_1 + \bar{I}_2$$
$$\overline{I_s^2} = \overline{(I_1 + I_2)^2} = \overline{I_1^2} + \overline{I_2^2} + 2\bar{I}_1\bar{I}_2 \qquad (3\text{-}47)$$

where we have used the assumed independence of I_1 and I_2 to replace $\overline{2I_1I_2}$ with $2\bar{I}_1\bar{I}_2$. We can also make use of the known exponential statistics of I_1 and I_2 to write $\overline{I_1^2} = 2\bar{I}_1^{\,2}$, $\overline{I_2^2} = 2\bar{I}_2^{\,2}$, with the result

$$\sigma_s^2 = \overline{I_s^2} - \bar{I}_s^{\,2} = \bar{I}_1^{\,2} + \bar{I}_2^{\,2}. \qquad (3\text{-}48)$$

It follows that the contrast is given by

$$C = \frac{\sigma_s}{\bar{I}_s} = \frac{\sqrt{\bar{I}_1^{\,2} + \bar{I}_2^{\,2}}}{\bar{I}_1 + \bar{I}_2} = \frac{\sqrt{1 + r^2}}{1 + r}, \qquad (3\text{-}49)$$

where again $r = \bar{I}_2/\bar{I}_1$. This expression for contrast is plotted vs. r in Fig. 3.10. Note that a minimum contrast of $1/\sqrt{2}$ occurs for $r = 1$, that is, when the two speckle patterns have equal average intensities.

3.3 Sums of Speckle Patterns   41

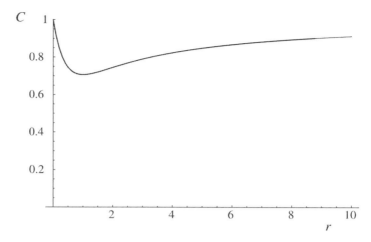

Figure 3.10 Contrast C for the sum of two independent speckle patterns with a ratio of average intensities given by r.

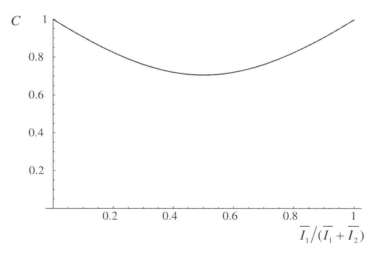

Figure 3.11 Contrast C for the sum of two independent speckle patterns as a function of the fraction of average intensity contributed by one of the components.

An alternate presentation of the same result is a plot of contrast C vs. the fraction of total average intensity contributed by one of the two components, e.g. $\bar{I}_1/(\bar{I}_1 + \bar{I}_2)$, as shown in Fig. 3.11.

3.3.3 Sum of *N* Independent Speckle Intensities

As a generalization of the case considered in the previous section, we now study the properties of a sum of N independent speckle patterns. Thus the total intensity of interest is now given by the sum

$$I_s = \sum_{n=1}^{N} I_n. \tag{3-50}$$

Following the reasoning used in the previous sections, the assumed statistical independence of the individual speckle patterns allows us to write the characteristic function of the sum as the product of the characteristic functions of the component intensities,

$$\mathbf{M}_s(\omega) = \prod_{n=1}^{N} \mathbf{M}_n(\omega), \tag{3-51}$$

where $\mathbf{M}_n(\omega)$ is the characteristic function of I_n. Again using the characteristic function of Eq. (3-21) for an individual speckle pattern we obtain

$$\mathbf{M}_s(\omega) = \prod_{n=1}^{N} \frac{1}{1 - j\omega \overline{I}_n}, \tag{3-52}$$

where $\overline{I}_n$ is the mean of the nth component speckle pattern.

A Fourier inversion of this characteristic function will yield the probability density function $p_s(I_s)$ for the total intensity. The form of this density function depends on the relationship between the values of the various average intensities $\overline{I}_n$. Results for two important cases follow.

When the various values of average intensity $\overline{I}_n$ are all nonzero and distinct (i.e. no two alike), we find (for $I_s \geq 0$)

$$p_s(I_s) = \sum_{n=1}^{N} \frac{\overline{I}_n^{(N-2)}}{\prod_{p=1, p \neq n}^{N}(\overline{I}_n - \overline{I}_p)} \exp\left(-\frac{I_s}{\overline{I}_n}\right). \tag{3-53}$$

On the other hand, when all $\overline{I}_n$ are identical and equal to I_0, the corresponding result is

$$p_s(I_s) = \frac{I_s^{N-1}}{\Gamma(N)I_0^N} \exp\left(-\frac{I_s}{I_0}\right) = \frac{N^N I_s^{N-1}}{\Gamma(N)\overline{I}^N} \exp\left(-N\frac{I_s}{\overline{I}}\right), \tag{3-54}$$

where $\overline{I} = NI_0$ represents the total mean intensity. Such a density function is known as a *gamma* density function of order N.

In more complicated situations where some of the average intensities are identical and some are different, standard techniques of partial fraction expansions applied to $\mathbf{M}_s(\omega)$ can be used to obtain expressions for $p_s(I_s)$ that are combinations of the two forms above.

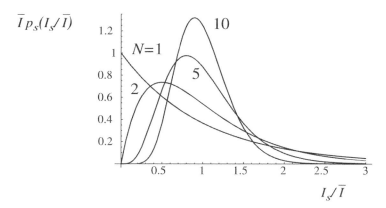

Figure 3.12 Probability density functions of the sum of N equal-strength, independent speckle patterns on an intensity basis, with the average intensity of the sum held constant.

Figure 3.12 shows plots of the probability density functions of the intensity of N equal-strength, independent speckle patterns, summed on an intensity basis, for various values of N. The average intensity of the sum has been kept constant as N is changed. The probability density function is seen to change from negative exponential ($N = 1$) to a curve that begins to resemble a Gaussian density function (large N), in agreement with the predictions of the Central Limit Theorem.

Finally we will calculate the contrast and signal-to-noise ratio of the sum of N independent speckle pattern intensities for arbitrary component means. The total intensity is written

$$I_s = \sum_{n=1}^{N} I_n, \tag{3-55}$$

where each I_n has mean value $\overline{I_n}$. The mean value of the total intensity is clearly

$$\overline{I_s} = \sum_{n=1}^{N} \overline{I_n}. \tag{3-56}$$

The second moment of the total intensity is

$$\overline{I_s^2} = \sum_{n=1}^{N} \sum_{m=1}^{N} \overline{I_n I_m} = \sum_{n=1}^{N} \overline{I_n^2} + \sum_{n=1}^{N} \sum_{m=1, m \neq n}^{N} \overline{I_n}\, \overline{I_m} \tag{3-57}$$

where the independence of I_n and I_m for $n \neq m$ has been used. At this point we use the fact that the individual component intensities are speckle patterns, and therefore they obey negative exponential probability distributions, for which $\overline{I_n^2} = 2\overline{I_n}^2$. The expression for $\overline{I_s^2}$ becomes

$$\overline{I_s^2} = 2\sum_{n=1}^{N}\overline{I}_n^2 + \sum_{n=1}^{N}\sum_{m=1,m\neq n}^{N}\overline{I}_n\overline{I}_m = \sum_{n=1}^{N}\overline{I}_n^2 + \left(\sum_{n=1}^{N}\overline{I}_n\right)^2 = \sum_{n=1}^{N}\overline{I}_n^2 + \overline{I}_s^2. \quad (3\text{-}58)$$

It follows that the variance of the total intensity is

$$\sigma_s^2 = \sum_{n=1}^{N}\overline{I}_n^2, \quad (3\text{-}59)$$

and that the contrast and signal-to-noise ratio of the total intensity are

$$C = \frac{\sigma_s}{\overline{I}_s} = \frac{\sqrt{\sum_{n=1}^{N}\overline{I}_n^2}}{\sum_{n=1}^{N}\overline{I}_n}$$

$$S/N = 1/C = \frac{\sum_{n=1}^{N}\overline{I}_n}{\sqrt{\sum_{n=1}^{N}\overline{I}_n^2}}. \quad (3\text{-}60)$$

For the special case of components with equal mean intensities ($\overline{I}_n = I_0$, all n), this result reduces to the important expression

$$C = \frac{1}{\sqrt{N}}$$

$$S/N = \sqrt{N}. \quad (3\text{-}61)$$

Thus the contrast falls in proportion to $1/\sqrt{N}$ as the number of independent patterns N increases, and similarly the signal-to-noise ratio increases as $\sqrt{N}$.

We note in closing this section that the results have depended on the assumption that the N individual speckle patterns are statistically independent. In the next section we consider the case of a sum of partially correlated speckle patterns.

3.3.4 Sums of Correlated Speckle Intensities

Correlations between speckle patterns can arise in many different ways. For example, suppose that a random reflecting surface is illuminated with coherent light, and a speckle pattern is recorded at some distance from the surface. Now suppose that the random surface is translated by a small amount, illuminated again with the same coherent light, and a second speckle pattern is recorded. The two speckle patterns recorded in this fashion will in general be partially correlated. The degree of correlation will depend on several factors, including how far the surface has been translated, the exact geometry, and the size of the speckles.

The sum of N speckle patterns is represented by

$$I_s = \sum_{n=1}^{N} I_n = \sum_{n=1}^{N} |\mathbf{A}_n|^2. \quad (3\text{-}62)$$

The normalized correlation between the nth and mth speckle pattern intensities is represented by

$$\rho_{n,m} = \frac{\overline{I_n I_m} - \overline{I_n}\,\overline{I_m}}{[\overline{(I_n - \overline{I_n})^2}\,\overline{(I_m - \overline{I_m})^2}]^{1/2}}. \qquad (3\text{-}63)$$

Such intensity correlations can be present only if the underlying fields are correlated. We represent such field correlations by

$$\mu_{n,m} = \frac{\overline{\mathbf{A}_n \mathbf{A}_m^*}}{[\overline{|\mathbf{A}_n|^2}\,\overline{|\mathbf{A}_m|^2}]^{1/2}}. \qquad (3\text{-}64)$$

For fully developed speckle, the circular complex Gaussian statistics of the fields imply ([70], p. 44) that the correlations of the intensities are related to the correlations of the fields through (cf. Eq. 4-43)

$$\overline{I_n I_m} = \overline{I_n}\,\overline{I_m}[1 + |\mu_{n,m}|^2], \qquad (3\text{-}65)$$

from which it follows that

$$\rho_{n,m} = |\mu_{n,m}|^2 \qquad (3\text{-}66)$$

and

$$\mu_{n,m} = \sqrt{\rho_{n,m}}\,\exp j\psi_{n,m}, \qquad (3\text{-}67)$$

where $\psi_{n,m}$ is a phase factor associated with the correlation of the nth and mth amplitude patterns $\mathbf{A}_n$ and $\mathbf{A}_m$.

At this point a fundamental question arises, namely, does there exist a transformation of the N fields $\mathbf{A}_n$ which will preserve the total intensity I, but eliminate the correlation between the various field components, possibly changing their individual intensities I_n in the process. To help answer this question, we first define a column vector $\underline{\mathcal{A}}$ of the N different complex speckle fields, as shown below:

$$\underline{\mathcal{A}} = \begin{bmatrix} \mathbf{A}_1 \\ \mathbf{A}_2 \\ \vdots \\ \mathbf{A}_N \end{bmatrix}. \qquad (3\text{-}68)$$

We then define a *coherency matrix* $\underline{\mathcal{J}}$ by

$$\underline{\mathcal{J}} = \overline{\underline{\mathcal{A}}\underline{\mathcal{A}}^\dagger}, \qquad (3\text{-}69)$$

where the superscript † represents the Hermitian transpose operation (i.e. a transpose of the conjugate of the matrix in question). Thus the element $\mathbf{J}_{n,m}$ of $\underline{\mathcal{J}}$ is

$$\mathbf{J}_{n,m} = \overline{\mathbf{A}_n \mathbf{A}_m^*} = \sqrt{\overline{I_n}\,\overline{I_m}}\,\mu_{n,m}, \qquad (3\text{-}70)$$

and the coherency matrix is given by

$$\underline{\mathcal{J}} = \begin{bmatrix} \overline{I_1} & \sqrt{\overline{I_1}\overline{I_2}}\mu_{1,2} & \cdots & \sqrt{\overline{I_1}\overline{I_N}}\mu_{1,N} \\ \sqrt{\overline{I_1}\overline{I_2}}\mu_{1,2}^* & \overline{I_2} & \cdots & \sqrt{\overline{I_2}\overline{I_N}}\mu_{2,N} \\ \vdots & \vdots & \vdots & \vdots \\ \sqrt{\overline{I_1}\overline{I_N}}\mu_{1,N}^* & \sqrt{\overline{I_2}\overline{I_N}}\mu_{2,N}^* & \cdots & \overline{I_N} \end{bmatrix}. \qquad (3\text{-}71)$$

The conjugate symmetry above and below the diagonal establishes that $\underline{\mathcal{J}}$ is what is called an *Hermitian* matrix.

We seek a linear transformation of the field matrix $\underline{\mathcal{A}}$ that will result in decorrelation of the fields, without loss of total intensity (as summed along the diagonal of the coherency matrix). Let $\underline{\mathcal{L}}$ represent an $N \times N$ transformation matrix, which transforms the fields from $\underline{\mathcal{A}}$ to a new field matrix $\underline{\mathcal{A}}'$ according to

$$\underline{\mathcal{A}}' = \underline{\mathcal{L}}\underline{\mathcal{A}}. \qquad (3\text{-}72)$$

After such a transformation, the coherency matrix becomes

$$\underline{\mathcal{J}}' = \overline{\underline{\mathcal{A}}'\underline{\mathcal{A}}'^\dagger} = \overline{\underline{\mathcal{L}}\underline{\mathcal{A}}\underline{\mathcal{A}}^\dagger\underline{\mathcal{L}}^\dagger} = \underline{\mathcal{L}}\underline{\mathcal{J}}\underline{\mathcal{L}}^\dagger. \qquad (3\text{-}73)$$

We now invoke a well-known result of matrix theory [153]: for every Hermitian matrix $\underline{\mathcal{J}}$, there exists a unitary linear transformation $\underline{\mathcal{L}}_0$ that will diagonalize $\underline{\mathcal{J}}$, i.e.

$$\underline{\mathcal{J}}' = \underline{\mathcal{L}}_0 \underline{\mathcal{J}} \underline{\mathcal{L}}_0^\dagger = \begin{bmatrix} \lambda_1 & 0 & \cdots & 0 \\ 0 & \lambda_2 & \cdots & 0 \\ \vdots & \vdots & \vdots & \vdots \\ 0 & 0 & \cdots & \lambda_N \end{bmatrix}, \qquad (3\text{-}74)$$

where the λ_n represent the eigenvalues of the matrix $\underline{\mathcal{J}}$. Furthermore, because the transformation achieving diagonalization is unitary, i.e. $\underline{\mathcal{L}}_0 \underline{\mathcal{L}}_0'$ equals the identity matrix, this transformation is lossless, and the sum of the diagonal elements of the coherency matrix remains unchanged. The coherency matrix, like all correlation matrices, has the property that it is nonnegative definite, that is, its eigenvalues are nonnegative. Since they lie along the diagonal of the new coherency matrix, the λ_n represent the intensities of the new field components after the linear transformation.

We would like to say that we have succeeded in transforming the N correlated circular Gaussian complex fields of the original speckle patterns into N uncorrelated circular Gaussian complex fields representing new speckle patterns having the same summed average intensity. The fact that the new fields have been produced by a linear transformation guarantees that they are complex Gaussian. However, it is not necessarily obvious that the transformed variables are also *circular*. Appendix A explores this question, and demonstrates that circularity holds for any linear transformation of circular random variables.

To summarize the results of this section, the sum of N correlated speckle patterns can be transformed to an equivalent sum of N uncorrelated speckle patterns.

While the total average intensity of the sum remains unchanged under this transformation, the particular average intensities composing the sum do in general change. Once the eigenvalues of the original coherency matrix are known, the average intensities composing the sum are known, and the results of the previous section regarding sums of uncorrelated speckle patterns can be directly applied.

3.4 Partially Polarized Speckle

The polarization properties of linearly polarized incident light after reflection from different types of surfaces can be quite different. For example, when light is reflected from a rough dielectric surface such as paper, multiple scattering usually takes place and the light emerges in a state we could reasonably call unpolarized. If the intensity of the light is observed through a polarization analyzer, oriented first in the x-direction and then in the y-direction, the two speckle patterns observed will bear little resemblance to one another. On the other hand, when linearly polarized incident light is reflected from a rough metallic surface, the reflected light often has a polarization state we could reasonably call polarized; that is, often the speckle patterns observed in two orthogonal polarizations will be highly correlated. Precise definitions of what we mean by the terms "polarized" and "unpolarized" will be presented later in this section.

We assume in what follows that the light amplitude incident on a rough surface is linearly polarized in the x direction, and thus describable by

$$\vec{A}_i = \sqrt{I_i}\hat{x}, \tag{3-75}$$

where I_i is the incident intensity and $\hat{x}$ is a unit vector in the x-direction. The complex amplitude of the reflected light can then be written as

$$\vec{A}_r = \mathbf{A}_x\hat{x} + \mathbf{A}_y\hat{y}. \tag{3-76}$$

The total intensity observed is given by the sum of the intensities carried by the x- and y-components of the light,

$$I = I_x + I_y = |\mathbf{A}_x|^2 + |\mathbf{A}_y|^2. \tag{3-77}$$

When the surface from which the light has been reflected is very rough on the scale of an optical wavelength, both I_x and I_y are fully developed speckle patterns. As we know from Section 3.3.4, the statistics of the total intensity will depend on the *correlation* between the intensities I_x and I_y. The correlation between I_x and I_y in turn depends on the correlation between the fields $\mathbf{A}_x$ and $\mathbf{A}_y$, as described in Eq. (3-66), which we modify here to apply to the current situation,

$$\rho_{x,y} = |\boldsymbol{\mu}_{x,y}|^2. \tag{3-78}$$

The polarization properties of the two orthogonal field components are described by a 2×2 coherency matrix,

$$\underline{\mathcal{J}} = \begin{bmatrix} \overline{I}_x & \sqrt{\overline{I}_x \overline{I}_y}\,\mu_{x,y} \\ \sqrt{\overline{I}_x \overline{I}_y}\,\mu_{x,y}^* & \overline{I}_y \end{bmatrix}. \tag{3-79}$$

As in the case of N correlated speckle patterns considered in the previous section, there exists a unitary matrix $\underline{\mathcal{L}}_0$ that diagonalizes the coherency matrix. In this case the transformation matrix can be interpreted as a Jones matrix [87] representation of a combination of a coordinate rotation and a relative retardation of the two polarization components. After transformation, the coherency matrix becomes

$$\underline{\mathcal{J}}' = \begin{bmatrix} \lambda_1 & 0 \\ 0 & \lambda_2 \end{bmatrix} \tag{3-80}$$

where λ_1 and λ_2 are the (real and nonnegative) eigenvalues of the original coherency matrix. The two eigenvalues are given explicitly by

$$\lambda_{1,2} = \frac{1}{2}\,\mathrm{tr}(\underline{\mathcal{J}})\left[1 \pm \sqrt{1 - 4\frac{\det(\underline{\mathcal{J}})}{(\mathrm{tr}(\underline{\mathcal{J}}))^2}}\right] \tag{3-81}$$

where "tr" signifies the trace operation and "det" signifies the determinant.

Thus we see that a wave with correlated x- and y-polarization components is equivalent to a wave with uncorrelated polarization components having different values of intensity in the two components, but the same sum of the two intensities, i.e.

$$I_x + I_y = \lambda_1 + \lambda_2 = \overline{I}. \tag{3-82}$$

Wolf [178] recognized that a general partially polarized wave can be regarded as the sum of two components, one linearly polarized and the other completely unpolarized (a *completely unpolarized* wave has uncorrelated intensity components of equal average intensity). Such a decomposition is accomplished by rewriting the diagonalized coherency matrix in the form (for $\lambda_1 > \lambda_2$)

$$\underline{\mathcal{J}}' = \begin{bmatrix} \lambda_2 & 0 \\ 0 & \lambda_2 \end{bmatrix} + \begin{bmatrix} \lambda_1 - \lambda_2 & 0 \\ 0 & 0 \end{bmatrix}. \tag{3-83}$$

The first matrix represents a completely unpolarized wave, while the second represents a completely polarized wave (all power in the x-component of polarization). The degree of polarization can reasonably be defined as the ratio of the intensity in the completely polarized wave component to the total intensity in the wave,

$$\mathcal{P} = \left|\frac{\lambda_1 - \lambda_2}{\lambda_1 + \lambda_2}\right|. \tag{3-84}$$

Note that the degree of polarization always lies between 1 (for a perfectly polarized wave) and 0 (for a completely unpolarized wave). With the help of Eq. (3-81), we can express the degree of polarization in terms of the original coherency matrix through

3.4 Partially Polarized Speckle

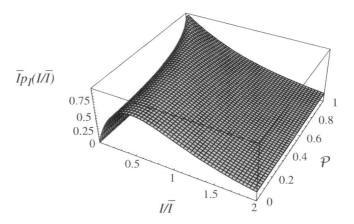

Figure 3.13 Probability density function of the intensity of a partially polarized speckle pattern as a function of $I/\bar{I}$ and $\mathcal{P}$.

$$\mathcal{P} = \left[1 - 4\frac{\det(\underline{\mathcal{J}})}{(\operatorname{tr}(\underline{\mathcal{J}}))^2}\right]^{\frac{1}{2}}, \quad (3\text{-}85)$$

and with the help of Eq. (3-84) we can express the eigenvalues in terms of the degree of polarization,

$$\lambda_1 = \frac{1}{2}\bar{I}(1 + \mathcal{P})$$
$$\lambda_2 = \frac{1}{2}\bar{I}(1 - \mathcal{P}). \quad (3\text{-}86)$$

From Eq. (3-46), the probability density function of the intensity of a partially polarized wave is given by

$$p_I(I) = \frac{1}{\mathcal{P}\bar{I}}\left[\exp\left(-\frac{2}{1+\mathcal{P}}\frac{I}{\bar{I}}\right) - \exp\left(-\frac{2}{1-\mathcal{P}}\frac{I}{\bar{I}}\right)\right]. \quad (3\text{-}87)$$

Figure 3.13 shows this probability density as a function of the degree of polarization $\mathcal{P}$ and the intensity relative to the mean intensity, $I/\bar{I}$. Note that the probability density function varies from a negative exponential distribution ($\mathcal{P} = 1$) to a function of the form $x \exp(-x)$ for $\mathcal{P} = 0$.

It is straightforward to prove that the contrast of partially polarized speckle is given by

$$C = \sqrt{\frac{1 + \mathcal{P}^2}{2}}, \quad (3\text{-}88)$$

which is plotted in Fig. 3.14.

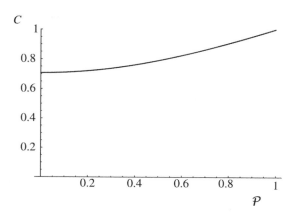

Figure 3.14 Contrast C of partially polarized speckle vs. the degree of polarization $\mathcal{P}$.

3.5 Partially Developed Speckle

When a large number of phasors having nonuniform phase statistics are added, the result can be what is referred to as "partially developed" speckle. We have touched upon such random phasor sums in Section 2.6. We consider in this section the *intensity* statistics for such sums.

In Section 2.6 we commented that a closed form solution for the statistics of amplitude is not available in this general case. It is therefore not possible to simply transform known amplitude statistics to obtain intensity statistics. Instead, we satisfy ourselves with finding expressions for the mean intensity and the standard deviation of intensity, which in turn will allow us to find an expression for the speckle contrast.

We begin with an explicit expression for the random sum of interest. Given that the amplitude of the random walk is expressible as

$$\mathbf{A} = \frac{1}{\sqrt{N}} \sum_{n=1}^{N} a_n e^{j\phi_n}, \qquad (3\text{-}89)$$

the intensity must be given by

$$I = \mathbf{A}\mathbf{A}^* = \frac{1}{N} \sum_{n=1}^{N} \sum_{m=1}^{N} a_n a_m e^{j(\phi_n - \phi_m)}. \qquad (3\text{-}90)$$

Allowing, for the moment, arbitrary statistics for both the amplitudes a_n and the phases ϕ_n, but assuming that amplitudes and phases are independent, we can express the mean intensity as

3.5 Partially Developed Speckle

$$\bar{I} = \frac{1}{N}\sum_{n=1}^{N}\sum_{m=1}^{N}\overline{a_n a_m}\overline{e^{j(\phi_n-\phi_m)}}$$

$$= \frac{1}{N}\sum_{n=1}^{N}\overline{a_n^2} + \frac{1}{N}\sum_{n=1}^{N}\sum_{\substack{m=1\\m\neq n}}^{N}\overline{a_n a_m}\mathbf{M}_\phi(1)\mathbf{M}_\phi(-1)$$

$$= \overline{a^2} + (N-1)(\bar{a})^2 \mathbf{M}_\phi(1)\mathbf{M}_\phi(-1) \tag{3-91}$$

where we have assumed that all a_n are identically distributed with mean $\bar{a}$ and second moment $\overline{a^2}$. We have also assumed that all ϕ_n are identically distributed and therefore have a common characteristic function $\mathbf{M}_\phi(\omega)$.

We now turn attention to the more difficult problem of calculating the second moment of the intensity. The details of this calculation are presented in Appendix B. The result is

$$\overline{I^2} = \frac{1}{N}\overline{a^4} + 2\left(1-\frac{1}{N}\right)(\overline{a^2})^2 + 4\left(1-\frac{1}{N}\right)(N-2)\overline{a^2}(\bar{a})^2|\mathbf{M}_\phi(1)|^2$$

$$+ 4\left(1-\frac{1}{N}\right)\overline{a^3}\bar{a}|\mathbf{M}_\phi(1)|^2 + \left(1-\frac{1}{N}\right)(N-2)\overline{a^2}(\bar{a})^2\mathbf{M}_\phi^2(-1)|\mathbf{M}_\phi(2)$$

$$+ \left(1-\frac{1}{N}\right)(N-2)\overline{a^2}(\bar{a})^2\mathbf{M}_\phi^2(1)\mathbf{M}_\phi(-2)$$

$$+ \left(1-\frac{1}{N}\right)(N-2)(N-3)(\bar{a})^4|\mathbf{M}_\phi(1)|^4 + \left(1-\frac{1}{N}\right)(\overline{a^2})^2|\mathbf{M}_\phi(2)|^2. \tag{3-92}$$

Subtracting the square of the mean of I, as given by Eq. (3-91), yields the variance. The standard deviation is the square root of the variance, and the contrast is the ratio of the standard deviation of I to the mean of I.

The expressions for these quantities are complex (see Appendix B), but can be simplified with some assumptions. The assumptions adopted here will be:

1. The lengths of all random phasors are unity, and hence $\overline{a^4} = \overline{a^3} = \overline{a^2} = \bar{a} = 1$; and
2. The phase ϕ is a zero mean Gaussian random variable with standard deviation σ_ϕ, yielding a characteristic function of the form

$$\mathbf{M}_\phi(\omega) = \exp\left(-\frac{\omega^2 \sigma_\phi^2}{2}\right). \tag{3-93}$$

The resulting expression for the speckle contrast is (again, see Appendix B)

$$C = \sqrt{\frac{8(N-1)[N-1+\cosh(\sigma_\phi^2)]\sinh^2(\sigma_\phi^2/2)}{N(N-1+e^{\sigma_\phi^2})^2}} \tag{3-94}$$

Figure 3.15 shows plots of (a) contrast C vs. σ_ϕ for various values of N, and (b) C vs. N for various values of σ_ϕ. In part (a), note that for $N = 1$, the contrast is always zero. The plot shows curves for $N = 2, 5, 10, 100,$ and 1000.

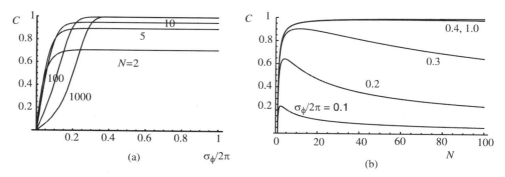

Figure 3.15 Case of a Gaussian phase distribution: (a) contrast C of partially developed speckle vs. phase standard deviation σ_ϕ for various values of N, and (b) C vs. N for various values of σ_ϕ.

The behavior of these curves requires some explanation. In particular, the reader may wonder why for some σ_ϕ they show a pronounced maximum first rising with N, and then falling towards zero. The surprising fact, not visible in the figure, is that *all* the curves for C eventually fall towards zero, regardless of the value of σ_ϕ, but for large σ_ϕ the value of N where this begins to happen is extremely large. An explanation for this phenomenon rests on the different dependencies of $\bar{I}$ and σ_I on N. For the normalization chosen in Eq. (3-89), the dependence of average intensity on N (for large N) is $\bar{I} \propto N$, while the dependence of σ_I is $\sigma_I \propto \sqrt{N}$. Thus for any σ_ϕ, eventually for large enough N the denominator of the contrast will begin to dominate and will cause the contrast to fall, although for large σ_ϕ the value of N required to see this effect may be extremely large. Note that this property is a consequence of the assumption that the phase statistics are Gaussian (in which case a nonzero average intensity always exists) and does not occur when the phase statistics are uniform on $(-\pi, \pi)$.

As a matter of curiosity, we show similar results in Fig. 3.16 for the case of a uniform probability density function of phase on the interval $(-\sqrt{3}\sigma_\phi, \sqrt{3}\sigma_\phi)$ for the same cases shown in the previous figure. In this case, $M_\phi(\omega) = \frac{\sin(\sqrt{3}\sigma_\phi \omega)}{\sqrt{3}\sigma_\phi \omega}$. The discussion following the previous figure also applies here, but we have the added phenomenon of a quasi-oscillatory behavior of C vs. σ_ϕ for large values of N. The oscillations are caused by folding of the uniform distribution on $(-\sqrt{3}\sigma_\phi, \sqrt{3}\sigma_\phi)$ into the primary interval $(-\pi, \pi)$. For certain values of σ_ϕ the folded distribution is perfectly uniform on the primary interval, in which case the contrast rises to unity. The oscillations for large N are caused by oscillations of both $\bar{I}$ and σ_ϕ as the phase distribution varies from uniform to nonuniform on the primary interval.

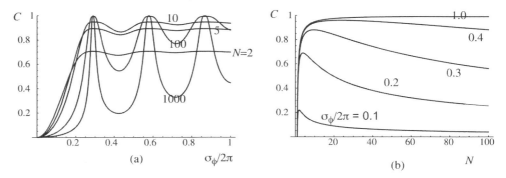

Figure 3.16 Case of a uniform phase distribution: (a) contrast C of partially developed speckle vs. $\sigma_\phi/2\pi$ for various values of N, and (b) C vs. N for various values of $\sigma_\phi/2\pi$.

3.6 Speckled Speckle, or Compound Speckle Statistics

We shall see examples in later chapters of situations in which the speckle process is "driven" by another random process. For example, when light passes through one diffuser, and the speckled light from that diffuser falls on a finite-sized second diffuser, there can be a compounding of the speckle statistics. A similar process, but with different statistics, results when light propagates through the Earth's atmosphere and then falls on a a rough surface, resulting in speckle reflected from the rough surface that is a compounded effect of the atmospherically induced intensity fluctuations and the speckle fluctuations.

Such problems can be analyzed by the application of conditional speckle statistics. Normally, the probability density function of speckle is a negative-exponential distribution. But we can regard this distribution to be a *conditional* density function, conditioned on knowledge of the mean intensity, which we represent here by the variable x,

$$p_{I|x}(I|x) = \frac{1}{x} \exp\left[-\frac{I}{x}\right]. \qquad (3\text{-}95)$$

The unconditional probability density function of the intensity is found by averaging the above density function with respect to the statistics of the conditional mean intensity x,

$$p_I(I) = \int_0^\infty p_{I|x}(I|x) p_x(x)\, dx = \int_0^\infty \frac{1}{x} \exp\left[-\frac{I}{x}\right] p_x(x)\, dx. \qquad (3\text{-}96)$$

3.6.1 Speckle Driven by a Negative Exponential Intensity Distribution

Suppose that we illuminate a small, optically rough reflecting disk with coherent light that has first passed through a diffuser. We assume that the disk is sufficiently small that only a single speckle cell from the diffuser strikes it (we consider the spatial extent of a speckle cell in Chapter 4). The disk is thus illuminated by an approximately constant intensity that, over an ensemble of illuminating diffusers, obeys negative exponential statistics. The intensity of the light reflected from the disk would, under known illumination intensity, also generate a speckle pattern with negative exponential statistics. However, in this case the illumination intensity is random, and we wish to calculate the statistics of the speckle intensity at a point, reflected from the disk.

We make use of Eq. (3-96), with

$$p_x(x) = \frac{1}{\bar{I}} \exp\left[-\frac{x}{\bar{I}}\right] \tag{3-97}$$

to yield

$$p_I(I) = \frac{1}{\bar{I}} \int_0^\infty \frac{1}{x} \exp\left[-\left(\frac{I}{x} + \frac{x}{\bar{I}}\right)\right] dx = \frac{2}{\bar{I}} K_0\left(2\sqrt{\frac{I}{\bar{I}}}\right) \tag{3-98}$$

where $K_0(x)$ is a modified Bessel function of the second kind, order 0.

This density function is an example of a more general class of density functions that have become known as "K distributions." We will see more general forms of such density functions in the two subsections that follow. K distributions were first introduced in scattering problems by E. Jakeman and P.N. Pusey in 1976 [84], but the model that led to such distributions was different than used here.[2] Figure 3.17 shows a plot of $\bar{I} p_I(I/\bar{I})$ vs. $I/\bar{I}$ for the density function above.

The qth moment of the observed speckle pattern can be calculated as follows:

$$\overline{I^q} = \int_0^\infty I^q p_I(I) \, dI = \int_0^\infty I^q \frac{2}{\bar{I}} K_0\left(2\sqrt{\frac{I}{\bar{I}}}\right) dI = \bar{I}^q (q!)^2. \tag{3-99}$$

Of special interest is the ratio of the standard deviation of the speckled speckle to its mean (the contrast). In this case we find

$$\frac{\sigma_I}{\bar{I}} = \sqrt{3} \approx 1.73. \tag{3-100}$$

While we normally think of optical contrast as a quantity that lies between 0 and 1, in this case what we have defined to be contrast, a ratio of two averaged quantities, has no such constraint. The result greater than unity is a consequence of the compounding of two sources of fluctuation, the illuminating intensity and the speckle.

---

2. Jakeman and Pusey derived the K distribution by assuming a random walk with finite number of terms, in which the number of scattered contributions obeyed negative binomial statistics. See Refs. [85], [82] and [83].

3.6 Speckled Speckle, or Compound Speckle Statistics

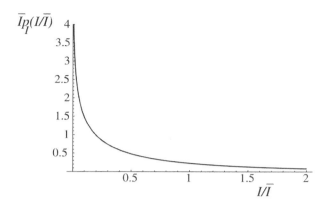

Figure 3.17 Probability density function of intensity for speckle driven by a negative exponential distribution.

3.6.2 Speckle Driven by a Gamma Intensity Distribution

A generalization of the above problem is found when we consider the small reflecting disk to be illuminated not by a single speckle pattern, but instead by the sum of N independent speckle patterns, each of the same average intensity I_0, as might be generated when several equal-strength independent lasers[3] illuminate separate diffusers, and the diffused light falls on the small reflecting disk. Again, the disk is assumed to be sufficiently small that only a single speckle from each of the speckle patterns strikes the disk. From Eq. 3-54 the intensity illuminating the disk in this case obeys a gamma density function, and use of Eq. (3-96) yields

$$p_I(I) = \int_0^\infty \frac{1}{x} \exp\left[-\frac{I}{x}\right] \frac{x^{N-1}}{\Gamma(N) I_0^N} \exp\left[-\frac{x}{I_0}\right] dx$$

$$= \int_0^\infty \frac{1}{x} \exp\left[-\frac{I}{x}\right] \frac{N^N x^{N-1}}{\Gamma(N) \bar{I}^N} \exp\left[-N\frac{x}{\bar{I}}\right] dx \qquad (3\text{-}101)$$

where the total average intensity $\bar{I}$ is the sum of the intensities contributed by the N sources, NI_0. The result of this integration can be reduced to

$$p_I(I) = \frac{2N^{\frac{N+1}{2}}}{\bar{I}\Gamma(N)} \left(\frac{I}{\bar{I}}\right)^{\frac{N-1}{2}} K_{N-1}\left(2\sqrt{\frac{NI}{\bar{I}}}\right), \qquad (3\text{-}102)$$

which is a more general type of K-distribution than dealt with in the previous section. Here $K_N(x)$ is a modified Bessel function of the second kind, order N. Note

---

[3]. The fact that the lasers are oscillating independently, with no phase locking between them, allows us to add the intensities, rather than the amplitudes, at the small reflecting disk.

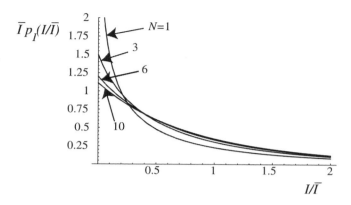

Figure 3.18 Probability density function of intensity for speckle driven by a gamma distribution of order N.

that for $N = 1$ the result reduces to that found in the previous subsection. Figure 3.18 shows plots of this density function for $N = 1, 3, 6$ and 10. For $N \geq 10$, the curves are indistinguishable from a negative exponential distribution, a consequence of the fact that the gamma density function approaches a δ-function as N grows large. Substitution of a δ-function for $p_x(x)$ in Eq. (3-96) clearly yields a negative exponential.

The qth moment of the more general K-distribution of Eq. (3-102) is found to be

$$\overline{I^q} = \left(\frac{\bar{I}}{N}\right)^q \frac{q!\Gamma(q+N)}{\Gamma(N)}. \tag{3-103}$$

The ratio of the intensity standard deviation to mean is found to be

$$\frac{\sigma_I}{\bar{I}} = \sqrt{1 + \frac{2}{N}}. \tag{3-104}$$

Note that for $N = 1$ this ratio is $\sqrt{3}$, as found in the previous subsection, and that as $N \to \infty$, the ratio approaches unity, this result corresponding to a negative-exponential distribution of intensity. A plot of $\sigma_I/\bar{I}$ vs. $I/\bar{I}$ is shown in Fig. 3.19.

Finally, the *cumulative* probability that the intensity exceeds a threshold I_t in this case is given by

$$P(I \geq I_t) = 1 - \int_0^{I_t} p_I(I)\, dI = \frac{2(NI_t/\bar{I})^{N/2}}{\Gamma(N)} K_N(2\sqrt{NI_t/\bar{I}}). \tag{3-105}$$

Figure 3.20 shows the probability that $I > I_t$ as a function of $I_t/\bar{I}$ for $N = 1$ and $N = 10$.

3.6 Speckled Speckle, or Compound Speckle Statistics

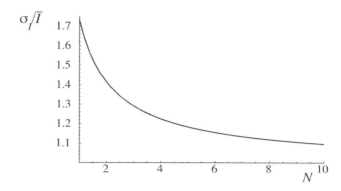

Figure 3.19 "Contrast" of the intensity fluctuations for speckle driven by a gamma distribution of order N.

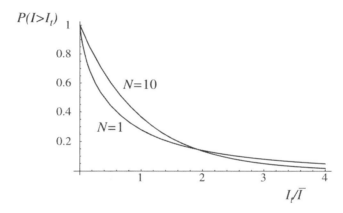

Figure 3.20 Cumulative probability that I exceeds a threshold I_t vs. $I_t/\bar{I}$ for K-distributed speckle.

3.6.3 Sums of Independent Speckle Patterns Driven by a Gamma Intensity Distribution

As a final and most general case of speckled speckle, we consider a further generalization of the case analyzed above. In that case, we assumed that the conditional speckle statistics were negative exponential, and that the driving intensity was gamma distributed. We now consider the case in which the conditional intensity statistics are *gamma distributed* while the driving intensity statistics remain gamma distributed.

There are several physical situations in which such a case would arise in practice. To illustrate with one example, consider again the case of the small, rough reflective disk illuminated by N independent lasers through different diffusers, with the disk intercepting a single speckle from each of those diffusers. Thus the illumination intensity at the disk again obeys gamma statistics, with order N, as before. However, in this case, rather than detecting the reflected intensity at a single point in space, instead we use an integrating detector which integrates over a region considerably larger than one speckle. As will be shown in Chapter 4, the statistics of the detected speckle, conditioned by the intensity illuminating the small disk, are also gamma distributed, this time with order M that depends on the width of a speckle (to be defined in the next chapter) and the area of integration of the detector. Analogous to Eq. (3-96), we now have for the density function of the detected intensity

$$p_I(I) = \int_0^\infty p_{I|x}(I|x) p_x(x)\, dx = \int_0^\infty \frac{M^M I^{M-1}}{\Gamma(M) x^M} \exp\left(-M\frac{I}{x}\right) p_x(x)\, dx$$

$$= \int_0^\infty \left[\frac{M^M I^{M-1}}{\Gamma(M) x^M} \exp\left(-M\frac{I}{x}\right)\right] \left[\frac{N^N x^{N-1}}{\Gamma(N) \bar{I}^N} \exp\left(-N\frac{x}{\bar{I}}\right)\right] dx$$

$$= \frac{2(MN)^{\frac{M+N}{2}}}{\bar{I}\Gamma(M)\Gamma(N)} \left(\frac{I}{\bar{I}}\right)^{\frac{M+N-2}{2}} K_{|N-M|}\left(2\sqrt{NM\frac{I}{\bar{I}}}\right) \qquad (3\text{-}106)$$

valid for $I > 0$. This distribution was introduced by Al-Habash, Andrews and Phillips [2] as a model that fits the statistics of intensity fluctuations encountered in propagation through the atmosphere under moderately strong turbulence conditions. They called it the "gamma–gamma" distribution, but we prefer here to view it as a natural generalization of the K-distribution discussed in the previous subsection.

The qth moment of this distribution is found to be

$$\overline{I^q} = \left(\frac{\bar{I}}{MN}\right)^q \frac{\Gamma(M+q)\Gamma(N+q)}{\Gamma(N)\Gamma(M)} \qquad (3\text{-}107)$$

and the ratio of standard deviation to mean intensity is

$$\frac{\sigma_I}{\bar{I}} = \sqrt{\frac{\Gamma(M)\Gamma(2+M)\Gamma(N)\Gamma(2+N)}{\Gamma^2(1+M)\Gamma^2(1+N)} - 1} = \sqrt{\frac{N+M+1}{NM}} \qquad (3\text{-}108)$$

where the final simplification is possible when M and N are integers.

4 Higher-Order Statistical Properties of Speckle

Based on the material presented in Chapter 3, the statistical properties of optical speckle at a single point in space (or, for dynamically changing speckle, a single point in time) are understood. Now we turn to the joint properties of speckle at two points, rather than one. The two values of speckle can represent values at two points in space, at two points in time, or at a single point in two different speckle patterns. Such considerations will allow us to deduce information about the scale sizes (coarseness) of speckles in a speckle pattern, and the physical properties that influence those scale sizes, as well as to discuss other important properties of speckle patterns.

4.1 Multivariate Gaussian Statistics

The underlying statistical model for a fully developed speckle field is that of a circular complex Gaussian random process, with real and imaginary parts that are real-valued jointly Gaussian random processes. It is therefore necessary to begin with a brief discussion of multivariate Gaussian distributions.

The characteristic function of an M-dimensional set of Gaussian random variables represented by a column vector $\vec{u} = \{u_1, u_2, \ldots, u_M\}^t$ is given by

$$\mathbf{M}_u(\vec{\omega}) = \exp\left\{j\vec{\bar{u}}^t\vec{\omega} - \frac{1}{2}\vec{\omega}^t \underline{C} \vec{\omega}\right\}, \tag{4-1}$$

where a superscript t indicates a matrix transpose, $\vec{\bar{u}}$ is a column vector of the means of the u_m, and $\vec{\omega}$ is a column vector with components $\omega_1, \omega_2, \ldots, \omega_M$. The symbol $\underline{C}$ represents the covariance matrix, that is, a matrix with element $c_{n,m}$ at the intersection of the nth row and the mth column given by following expected value:

$$c_{n,m} = E[(u_n - \overline{u_n})(u_m - \overline{u_m})]. \tag{4-2}$$

An M-dimensional inverse Fourier transform of this characteristic function yields the M-dimensional Gaussian probability density function[1] (cf. [70], p. 42, and [151], p. 170)

$$p(\vec{u}) = p(u_1, u_2, \ldots, u_M) = \frac{1}{(2\pi)^{M/2}|\mathcal{C}|^{1/2}} \exp\left[-\frac{1}{2}(\vec{u} - \bar{\vec{u}})' \mathcal{C}^{-1} (\vec{u} - \bar{\vec{u}})\right]. \quad (4\text{-}3)$$

Here $|\mathcal{C}|$ is the determinant of the covariance matrix and $\mathcal{C}^{-1}$ is the inverse matrix.

For a multivariate distribution of any kind, a moment of any order can be found by differentiating the characteristic function,

$$\overline{u_1^p u_2^q \cdots u_m^k} = \frac{1}{j^{p+q+\cdots+k}} \frac{\partial^{p+q+\cdots+k}}{\partial \omega_1^p \partial \omega_2^q \cdots \partial \omega_n^k} \mathbf{M}_u(\vec{\omega})\Big|_{\vec{\omega}=\vec{0}}, \quad (4\text{-}4)$$

where p, q and k are integers. For the particular case of a multivariate zero-mean Gaussian distribution, application of this equation to (4-1) yields the so-called *moment theorem* for real Gaussian variables,

$$\begin{aligned}\overline{u_1 u_2 \cdots u_{2k+1}} &= 0 \\ \overline{u_1 u_2 \cdots u_{2k}} &= \sum_P (\overline{u_j u_m}\, \overline{u_l u_p} \cdots \overline{u_q u_s})_{j \neq m, l \neq p, q \neq s},\end{aligned} \quad (4\text{-}5)$$

where the symbol $\sum_P$ indicates the sum over all possible $(2k)!/2^k k!$ distinct groupings of the $2k$ variables in pairs. For the most common case of $k = 2$, this result becomes

$$\overline{u_1 u_2 u_3 u_4} = \overline{u_1 u_2}\, \overline{u_3 u_4} + \overline{u_1 u_3}\, \overline{u_2 u_4} + \overline{u_1 u_4}\, \overline{u_2 u_3}. \quad (4\text{-}6)$$

4.2 Application to Speckle Fields

Let $\mathbf{A}_1, \mathbf{A}_2, \ldots, \mathbf{A}_N$ represent the complex values of a fully developed speckle field at points $(x_1, y_1), (x_2, y_2), \ldots, (x_N, y_N)$ in space, or alternatively at points $t_1, t_2, \ldots, t_N$ in time. In addition, represent the real and imaginary parts of the complex field at the nth such point by

$$\mathcal{R}_n = \text{Re}\{\mathbf{A}_n\}$$
$$\mathcal{I}_n = \text{Im}\{\mathbf{A}_n\}.$$

In the case of interest here, there are $M = 2N$ components of the vector $\vec{u}$,

$$\vec{u} = \{\mathcal{R}_1, \mathcal{R}_2, \ldots, \mathcal{R}_N, \mathcal{I}_1, \mathcal{I}_2, \ldots, \mathcal{I}_N\}^t.$$

Because the fields of interest are *circular* complex random variables $\mathbf{A}_n$ and $\mathbf{A}_m$ we have

---

1. Here and throughout this chapter, we will abandon the custom of subscripting probability density functions with symbols for the associated random variables, except in the few cases where such labeling is needed for clarity.

4.2 Application to Speckle Fields

$$\overline{\mathcal{R}_n} = \overline{\mathcal{I}_n} = \overline{\mathcal{R}_m} = \overline{\mathcal{I}_m} = 0$$

$$\overline{\mathcal{R}_n^2} = \overline{\mathcal{I}_n^2} = \overline{\mathcal{R}_m^2} = \overline{\mathcal{I}_m^2} = \sigma^2$$

$$\overline{\mathcal{R}_n \mathcal{I}_n} = \overline{\mathcal{R}_m \mathcal{I}_m} = 0 \qquad (4\text{-}7)$$

$$\overline{\mathcal{R}_n \mathcal{I}_m} = -\overline{\mathcal{R}_m \mathcal{I}_n}$$

$$\overline{\mathcal{R}_n \mathcal{R}_m} = \overline{\mathcal{I}_n \mathcal{I}_m}.$$

These relations simplify the structure of the covariance matrix. First, note that the matrix has a block structure, represented by

$$\underline{\mathcal{C}} = \begin{bmatrix} \underline{\mathcal{C}}_{RR} & \underline{\mathcal{C}}_{RI} \\ \underline{\mathcal{C}}_{IR} & \underline{\mathcal{C}}_{II} \end{bmatrix}, \qquad (4\text{-}8)$$

where $\underline{\mathcal{C}}_{RR}$ is the auto-covariance matrix of the real parts of the fields, $\underline{\mathcal{C}}_{II}$ is the auto-covariance of the imaginary parts of the field, while $\underline{\mathcal{C}}_{RI}$ and $\underline{\mathcal{C}}_{IR}$ are the cross-covariance matrices of the real and imaginary parts of the field. The form of the probability density function in Eq. (4-3) becomes

$$p(\vec{u}) = \frac{1}{(2\pi)^N |\underline{\mathcal{C}}|^{1/2}} \exp\left[-\frac{1}{2} \vec{u}^t \underline{\mathcal{C}}^{-1} \vec{u}\right], \qquad (4\text{-}9)$$

where relations (4-7) imply

$$\underline{\mathcal{C}} = \begin{bmatrix} \underline{\mathcal{C}}_{RR} & \underline{\mathcal{C}}_{RI} \\ -\underline{\mathcal{C}}_{RI} & \underline{\mathcal{C}}_{RR} \end{bmatrix}. \qquad (4\text{-}10)$$

In addition, we know something about the structures of $\underline{\mathcal{C}}_{RR}$ and $\underline{\mathcal{C}}_{RI}$. All the diagonal elements of $\underline{\mathcal{C}}_{RR}$ equal σ^2 while the off-diagonal elements are symmetrical about the diagonal, and all the diagonal elements of $\underline{\mathcal{C}}_{RI}$ are zero and the off-diagonal elements have negative symmetry above and below the diagonal.

A special case of interest is for speckle at 2 points in space (or time), in which case $N = 2$ and we must use a 4-dimensional probability density function to describe the real and imaginary parts of the fields, $p(\mathcal{R}_1, \mathcal{R}_2, \mathcal{I}_1, \mathcal{I}_2)$. In this case the submatrices $\underline{\mathcal{C}}_{RR}$ and $\underline{\mathcal{C}}_{RI}$ have the forms

$$\underline{\mathcal{C}}_{RR} = \sigma^2 \begin{bmatrix} 1 & \rho_c \\ \rho_c & 1 \end{bmatrix} \qquad (4\text{-}11)$$

and

$$\underline{\mathcal{C}}_{RI} = \sigma^2 \begin{bmatrix} 0 & \rho_s \\ -\rho_s & 0 \end{bmatrix}. \qquad (4\text{-}12)$$

The determinant of $\underline{\mathcal{C}}$ is easily shown to be

$$|\underline{\mathcal{C}}| = \sigma^8 (1 - \rho_c^2 - \rho_s^2)^2 \qquad (4\text{-}13)$$

and the inverse of $\underline{\mathcal{C}}$ is

$$\underline{\underline{C}}^{-1} = \frac{\sigma^2}{\sqrt{|\underline{\underline{C}}|}} \begin{bmatrix} -1 & \rho_c & 0 & \rho_s \\ \rho_c & -1 & -\rho_s & 0 \\ 0 & -\rho_s & -1 & \rho_c \\ \rho_s & 0 & \rho_c & -1 \end{bmatrix}. \quad (4\text{-}14)$$

The relevant fourth-order Gaussian probability density function can then be shown to be

$$p(\vec{u}) = \frac{\exp\left[-\frac{\mathcal{R}_1^2 + \mathcal{R}_2^2 + \mathcal{I}_1^2 + \mathcal{I}_2^2 - 2\rho_s(\mathcal{R}_1 \mathcal{I}_2 - \mathcal{I}_1 \mathcal{R}_2) - 2\rho_c(\mathcal{R}_1 \mathcal{R}_2 + \mathcal{I}_1 \mathcal{I}_2)}{2\sigma^2(1 - \rho_c^2 - \rho_s^2)}\right]}{4\pi^2 \sigma^4 (1 - \rho_c^2 - \rho_s^2)}. \quad (4\text{-}15)$$

For the special case of uncorrelated speckle samples ($\rho_c = \rho_s = 0$), the result simplifies to

$$p(\vec{u}) = \frac{\exp\left[-\frac{\mathcal{R}_1^2 + \mathcal{R}_2^2 + \mathcal{I}_1^2 + \mathcal{I}_2^2}{2\sigma^2}\right]}{4\pi^2 \sigma^4} \quad (4\text{-}16)$$

which is simply the product of the marginal probability density functions of the four first order density functions of the random variables $\mathcal{R}_1, \mathcal{R}_2, \mathcal{I}_1, \mathcal{I}_2$.

A general result derived from Eq. (4-5) and the properties described above holds for joint moments of circular complex Gaussian random variables $\mathbf{A}_1, \mathbf{A}_2, \ldots, \mathbf{A}_{2k}$:

$$\overline{\mathbf{A}_1^* \mathbf{A}_2^* \cdots \mathbf{A}_k^* \mathbf{A}_{k+1} \mathbf{A}_{k+2} \cdots \mathbf{A}_{2k}} = \sum_\Pi \overline{\mathbf{A}_1^* \mathbf{A}_p} \, \overline{\mathbf{A}_2^* \mathbf{A}_q} \cdots \overline{\mathbf{A}_k^* \mathbf{A}_r}, \quad (4\text{-}17)$$

where $\sum_\Pi$ represents a summation over the $k!$ possible permutations $(p, \ldots, r)$ of $(1, 2, \ldots, k)$. This result will be called the *complex Gaussian moment theorem*. For the special case of four such variables ($k = 2$), we have

$$\overline{\mathbf{A}_1^* \mathbf{A}_2^* \mathbf{A}_3 \mathbf{A}_4} = \overline{\mathbf{A}_1^* \mathbf{A}_3} \, \overline{\mathbf{A}_2^* \mathbf{A}_4} + \overline{\mathbf{A}_1^* \mathbf{A}_4} \, \overline{\mathbf{A}_2^* \mathbf{A}_3}. \quad (4\text{-}18)$$

With the above results as background, we are now prepared to explore the multi-dimensional statistics of speckle amplitude and intensity.

4.3 Multidimensional Statistics of Speckle Amplitude, Phase and Intensity

As discussed in Chapter 3, the amplitude A of a speckle pattern at a point in space/time is defined as the length of the complex phasor $\mathbf{A}$, and the phase is defined as the phase of the phasor. We wish to calculate the multidimensional probability density function of N samples of the amplitude and phase of a speckle pattern, i.e. $p(A_1, A_2, \ldots, A_N)$ and $p(\theta_1, \theta_2, \ldots, \theta_N)$. These density functions depend on the $2N$th-order density function of the underlying real and imaginary parts of the field. Our approach is to find the joint density function of all $2N$ variables

4.3 Multidimensional Statistics of Speckle Amplitude, Phase and Intensity

$A_1, \theta_1, A_2, \theta_2, \ldots, A_N, \theta_N$, and then to integrate with respect to the appropriate variables to find the respective marginal density functions.

A formal approach to finding the density functions of interest rests on a change of variables as follows:

$$\begin{aligned} \mathcal{R}_1 &= A_1 \cos \theta_1 & \mathcal{I}_1 &= A_1 \sin \theta_1 \\ \mathcal{R}_2 &= A_2 \cos \theta_2 & \mathcal{I}_1 &= A_2 \sin \theta_2 \\ &\vdots & &\vdots \\ \mathcal{R}_N &= A_N \cos \theta_N & \mathcal{I}_N &= A_N \sin \theta_N \end{aligned} \quad (4\text{-}19)$$

These expressions must be substituted into the probability density function (4-9), and the result must be multiplied by the magnitude of the Jacobian determinant[2] of the transformation, which is given by

$$\|J\| = \begin{Vmatrix} \partial \mathcal{R}_1/\partial A_1 & \partial \mathcal{I}_1/\partial A_1 & \cdots & \partial \mathcal{R}_N/\partial A_1 & \partial \mathcal{I}_N/\partial A_1 \\ \partial \mathcal{R}_1/\partial \theta_1 & \partial \mathcal{I}_1/\partial \theta_1 & \cdots & \partial \mathcal{R}_N/\partial \theta_1 & \partial \mathcal{I}_N/\partial \theta_1 \\ \vdots & \vdots & \vdots & \vdots & \vdots \\ \partial \mathcal{R}_1/\partial A_N & \partial \mathcal{I}_1/\partial A_N & \cdots & \partial \mathcal{R}_N/\partial A_N & \partial \mathcal{I}_N/\partial A_N \\ \partial \mathcal{R}_1/\partial \theta_N & \partial \mathcal{I}_1/\partial \theta_N & \cdots & \partial \mathcal{R}_N/\partial \theta_N & \partial \mathcal{I}_N/\partial \theta_N \end{Vmatrix}. \quad (4\text{-}20)$$

In the above expression, the symbol $\|\cdot\|$ indicates the magnitude of the determinant. Many of the entries in this Jacobian matrix are zero; the matrix is in fact banded with only the diagonal and sub-diagonals above and below the diagonal nonzero. Finally we must integrate the expression for the density function $p(A_1, \theta_1, \ldots, A_N, \theta_N)$ with respect to all of the θ_n to find the joint density function of all of the A_n, and with respect to all the A_n to find the joint density function of all the θ_n. In the general Nth-order case, the calculation is daunting.

We satisfy ourselves here with the case of only two speckle samples, $\mathbf{A_1}$ and $\mathbf{A_2}$. The Jacobian matrix then is of size 4×4.

$$\begin{aligned} \|J\| &= \begin{Vmatrix} \partial \mathcal{R}_1/\partial A_1 & \partial \mathcal{I}_1/\partial A_1 & \partial \mathcal{R}_2/\partial A_1 & \partial \mathcal{I}_2/\partial A_1 \\ \partial \mathcal{R}_1/\partial \theta_1 & \partial \mathcal{I}_1/\partial \theta_1 & \partial \mathcal{R}_2/\partial \theta_1 & \partial \mathcal{I}_2/\partial \theta_1 \\ \partial \mathcal{R}_1/\partial A_2 & \partial \mathcal{I}_1/\partial A_2 & \partial \mathcal{R}_2/\partial A_2 & \partial \mathcal{I}_2/\partial A_2 \\ \partial \mathcal{R}_1/\partial \theta_2 & \partial \mathcal{I}_1/\partial \theta_2 & \partial \mathcal{R}_2/\partial \theta_2 & \partial \mathcal{I}_2/\partial \theta_2 \end{Vmatrix} \\ &= \begin{Vmatrix} \cos \theta_1 & \sin \theta_1 & 0 & 0 \\ -A_1 \sin \theta_1 & A_1 \cos \theta_1 & 0 & 0 \\ 0 & 0 & \cos \theta_2 & \sin \theta_2 \\ 0 & 0 & -A_2 \sin \theta_2 & A_2 \cos \theta_2 \end{Vmatrix}, \end{aligned} \quad (4\text{-}21)$$

---

2. We consistently use the symbol $|M|$ to represent the determinant of a matrix M and the symbol $\|M\|$ to represent the modulus of that determinant.

which can be simplified to

$$\|J\| = A_1 A_2. \tag{4-22}$$

Thus, using (4-15) and trigonometric simplifications, the joint density function of $A_1, \theta_1, A_2, \theta_2$ is

$$p(A_1, \theta_1, A_2, \theta_2) = \frac{A_1 A_2}{4\pi^2 \sigma^4 (1 - \rho_c^2 - \rho_s^2)}$$

$$\times \exp\left[-\frac{A_1^2 + A_2^2 + 2\rho_s A_1 A_2 \sin(\theta_1 - \theta_2) - 2\rho_c A_1 A_2 \cos(\theta_1 - \theta_2)}{2\sigma^2(1 - \rho_c^2 - \rho_s^2)}\right]. \tag{4-23}$$

One further simplification is possible. Define the complex correlation coefficient between the speckle fields $\mathbf{A}_1$ and $\mathbf{A}_2$ as

$$\boldsymbol{\mu} = \mu e^{j\phi} = \frac{\overline{\mathbf{A}_1 \mathbf{A}_2^*}}{\sqrt{\overline{|\mathbf{A}_1|^2} \, \overline{|\mathbf{A}_2|^2}}} = \frac{\overline{\mathcal{R}_1 \mathcal{R}_2} + \overline{\mathcal{I}_1 \mathcal{I}_2} + j\overline{\mathcal{R}_2 \mathcal{I}_1} - j\overline{\mathcal{R}_1 \mathcal{I}_2}}{\sqrt{(\overline{\mathcal{R}_1^2} + \overline{\mathcal{I}_1^2})(\overline{\mathcal{R}_2^2} + \overline{\mathcal{I}_2^2})}}$$

$$= \frac{2\sigma^2 \rho_c + j 2\sigma^2 \rho_s}{2\sigma^2} = \rho_c + j\rho_s. \tag{4-24}$$

Substituting $\rho_c = \mu \cos\phi$ and $\rho_s = \mu \sin\phi$, together with trigonometric simplifications, yields

$$p(A_1, \theta_1, A_2, \theta_2)$$

$$= \frac{A_1 A_2}{4\pi^2 \sigma^4 (1 - \mu^2)} \exp\left[-\frac{A_1^2 + A_2^2 - 2 A_1 A_2 \mu \cos(\phi + \theta_1 - \theta_2)}{2\sigma^2 (1 - \mu^2)}\right]. \tag{4-25}$$

This equation is valid for $A_1, A_2 \geq 0$ and $-\pi \leq \theta_1, \theta_2 < \pi$. The problem remains to integrate (4-25) to find the marginal density functions of interest.

4.3.1 Joint Density Function of the Amplitudes

To find the joint density function of the amplitudes A_1 and A_2, we integrate (4-25) with respect to θ_1 and θ_2. To simplify the integration, first hold θ_2 constant and consider the integration with respect to θ_1. Because we are integrating over one full period of the cos function, we can as well integrate a new variable $\alpha = \phi + \theta_1 - \theta_2$ over a full period of 2π radians. Thus the integral becomes

$$p(A_1, A_2)$$

$$= \int_{-\pi}^{\pi} d\theta_2 \int_{-\pi}^{\pi} \frac{A_1 A_2}{4\pi^2 \sigma^4 (1 - \mu^2)} \exp\left[-\frac{A_1^2 + A_2^2 - 2 A_1 A_2 \mu \cos(\alpha)}{2\sigma^2 (1 - \mu^2)}\right] d\alpha. \tag{4-26}$$

The integrals are readily performed, with the result

4.3 Multidimensional Statistics of Speckle Amplitude, Phase and Intensity

$$p(A_1, A_2) = \frac{A_1 A_2}{(1-\mu^2)\sigma^4} e^{-\frac{A_1^2+A_2^2}{2\sigma^2(1-\mu^2)}} I_0\left[\frac{\mu A_1 A_2}{(1-\mu^2)\sigma^2}\right], \qquad (4\text{-}27)$$

where I_0 is again a modified Bessel function of the first kind, order zero, and the result is valid for $0 \le A_1, A_2 \le \infty$.

As a check on this result, we find the marginal density function of a single amplitude, A_1,

$$p(A_1) = \int_0^\infty p(A_1, A_2)\, dA_2 = \frac{A_1}{\sigma^2} e^{-\frac{A_1^2}{2\sigma^2}} \qquad (4\text{-}28)$$

which is a *Rayleigh* density, in agreement with previous results.

Figure 4.1 illustrates the shape of the normalized joint density function $\sigma^2 p(A_1/\sigma, A_2/\sigma)$ for various values of μ. The figures on the right are contour plots of the figures on the left. It can be seen that as the correlation coefficient increases, the joint density function approaches a shaped delta-function sheet along the line $A_1 = A_2$.

Another quantity of interest is the *conditional* density of A_1 given that the value of A_2 is known. This density function is represented by $p(A_1|A_2)$, and can be found using Bayes' rule,

$$p(A_1|A_2) = \frac{p(A_1, A_2)}{p(A_2)} = \frac{A_1}{(1-\mu^2)\sigma^2} e^{-\frac{A_1^2+\mu^2 A_2^2}{2\sigma^2(1-\mu^2)}} I_0\left[\frac{\mu A_1 A_2}{(1-\mu^2)\sigma^2}\right]. \qquad (4\text{-}29)$$

When $\mu = 0$, the two amplitudes are independent and the density function for A_1 again reduces to a Rayleigh density.

4.3.2 Joint Density Function of the Phases

To find the joint density function $p(\theta_1, \theta_2)$, we must integrate the density $p(A_1, \theta_1, A_2, \theta_2)$ of Eq. (4-25) with respect to the variables A_1 and A_2 over the ranges $(0, \infty)$:

$$p(\theta_1, \theta_2) = \int_0^\infty \int_0^\infty \frac{A_1 A_2}{4\pi^2 \sigma^4 (1-\mu^2)}$$
$$\times \exp\left[-\frac{A_1^2 + A_2^2 - 2A_1 A_2 \mu \cos(\phi + \theta_1 - \theta_2)}{2\sigma^2(1-\mu^2)}\right] dA_1\, dA_2. \qquad (4\text{-}30)$$

To perform the required integrations, we change the variables of integration, substituting

$$\begin{aligned} A_1 &= \sigma\sqrt{1-\mu^2}\, z^{1/2} e^{\psi/2} \\ A_2 &= \sigma\sqrt{1-\mu^2}\, z^{1/2} e^{-\psi/2}, \end{aligned} \qquad (4\text{-}31)$$

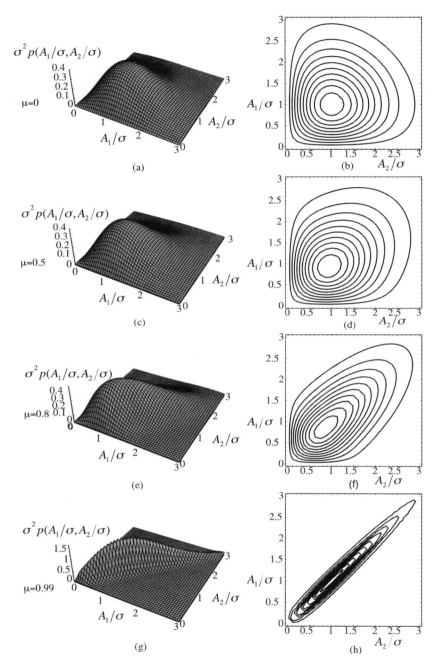

Figure 4.1 Normalized joint density function of the amplitudes A_1 and A_2 for various values of μ. Figures on the right are contour plots of figures on the left.

4.3 Multidimensional Statistics of Speckle Amplitude, Phase and Intensity

where $-\infty < \psi < \infty$ and $0 \leq z < \infty$. The Jacobian of the transformation is $\sigma^2(1-\mu^2)/2$, so the integral becomes

$$p(\theta_1, \theta_2) = \frac{1-\mu^2}{4\pi^2} \int_0^\infty z e^{\mu \cos(\phi+\theta_1-\theta_2)z} \frac{1}{2} \int_{-\infty}^\infty e^{-z \cosh \psi} \, d\psi \, dz. \quad (4\text{-}32)$$

The integral identity

$$\frac{1}{2} \int_{-\infty}^\infty e^{-z \cosh \psi} \, d\psi = K_0(z)$$

for $z > 0$ allows the density function to be rewritten

$$p(\theta_1, \theta_2) = \frac{1-\mu^2}{4\pi^2} \int_0^\infty z e^{\mu \cos(\phi+\theta_1-\theta_2)z} K_0(z) \, dz. \quad (4\text{-}33)$$

This remaining integral can be performed ([111], p. 404) with the result

$$p(\theta_1, \theta_2) = \left(\frac{1-\mu^2}{4\pi^2}\right) \frac{(1-\beta^2)^{1/2} + \beta\pi - \beta \cos^{-1}\beta}{(1-\beta^2)^{3/2}}, \quad (4\text{-}34)$$

where $\beta = \mu \cos(\phi + \theta_1 - \theta_2)$ and $-\pi \leq \theta_1, \theta_2 \leq \pi$, and $p(\theta_1, \theta_2)$ is zero otherwise.

Two observations about this result are important. First, for fixed ϕ, the joint density function of phases depends only on the phase difference $\theta_1 - \theta_2$. A second observation is that, while amplitude A and phase θ were seen to be independent at any one point, A_1 and A_2 are not jointly independent of θ_1 and θ_2, since

$$p(A_1, \theta_1, A_2, \theta_2) \neq p(A_1, A_2) p(\theta_1, \theta_2).$$

An exception occurs when $\mu = 0$, in which case the density function does factor.

The reader may be tempted to believe that the above result for $p(\theta_1, \theta_2)$ is also the density function for the phase difference $\Delta\theta = \theta_1 - \theta_2$. This is not quite true, as we shall now demonstrate. Let the following change of variables be made:

$$\begin{aligned} \Delta\theta &= \phi + \theta_1 - \theta_2 \\ \theta_2 &= \theta_2, \end{aligned} \quad (4\text{-}35)$$

or equivalently

$$\begin{aligned} \theta_1 &= \Delta\theta + \theta_2 - \phi \\ \theta_2 &= \theta_2. \end{aligned} \quad (4\text{-}36)$$

The magnitude of the Jacobian determinant of this transformation is unity, and hence we find that

$$p_{\Delta\theta}(\Delta\theta) = \int_{-\pi}^\pi p_{\theta_1, \theta_2}(\Delta\theta + \theta_2 - \phi, \theta_2) \, d\theta_2, \quad (4\text{-}37)$$

where the form of the density function on the right side is the same as the form of the joint density of θ_1 and θ_2. With this change of variables, the integrand proves to be

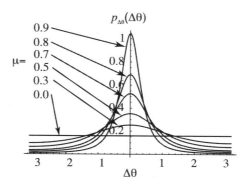

Figure 4.2 Probability density function $p_{\Delta\theta}(\Delta\theta)$ for various values of μ.

independent of θ_1, and the integration yields simply a factor of 2π, and a density function of phase difference $\Delta\theta$ given by

$$p_{\Delta\theta}(\Delta\theta) = \left(\frac{1-\mu^2}{2\pi}\right) \frac{(1-\beta^2)^{1/2} + \beta\pi - \beta\cos^{-1}\beta}{(1-\beta^2)^{3/2}}, \qquad (4\text{-}38)$$

where now $\beta = \mu\cos(\Delta\theta)$ and $-\pi \leq \Delta\theta \leq \pi$.

Figure 4.2 shows plots of the density function $p_{\Delta\theta}(\Delta\theta)$ for various values of μ. Note that as $\mu \to 0$, the density function becomes uniform with height $1/2\pi$, while as $\mu \to 1$, the density function approaches a delta function at $\Delta\theta = 0$.

A final quantity of interest is the standard deviation $\sigma_{\Delta\theta}$ of the phase difference $\Delta\theta$ as a function of μ, the magnitude of the complex correlation coefficient. An analytical solution to this problem is given by the expression[3]

$$\sigma_{\Delta\theta} = \sqrt{\frac{\pi^2}{3} - \pi\arcsin(\mu) + \arcsin^2(\mu) - \frac{1}{2}\text{Li}_2(\mu^2)}. \qquad (4\text{-}39)$$

where Li_2 is the dilogarithm function. This result is shown in Fig. 4.3, which plots the standard deviation of phase difference vs. the parameter μ. The standard deviation is seen to drop from a value of $\pi/\sqrt{3} = 1.81$ for $\mu = 0$ and approach zero as $\mu \to 1$.

We now turn attention to the joint density function of speckle intensity at two points in space.

4.3.3 Joint Density Function of the Intensities

The joint density function of speckle intensities can easily be found from the joint density function of amplitudes by a simple transformation of variables. Let $I_1 = A_1^2$ and $I_2 = A_2^2$, or equivalently

---

3. An analytic solution containing an infinite series was given by Middleton ([111], p. 405) and by Donati and Martini [34], but the infinite series sums to the dilogarithm function.

4.3 Multidimensional Statistics of Speckle Amplitude, Phase and Intensity 69

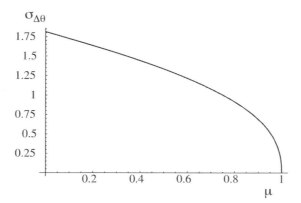

Figure 4.3 Standard deviation of the phase difference $\Delta\phi$ as a function of the magnitude μ of the complex correlation coefficient.

$$A_1 = \sqrt{I_1} \quad A_2 = \sqrt{I_2}$$
$$\theta_1 = \theta_1 \quad \theta_2 = \theta_2. \tag{4-40}$$

The magnitude of the Jacobian of this transformation is $1/(4\sqrt{I_1 I_2})$. Substitute for A_1 and A_2 in Eq. (4-27) and multiply by the magnitude of the Jacobian, giving the result

$$p(I_1, I_2) = \frac{1}{\bar{I}^2(1-\mu^2)} \exp\left[-\frac{I_1 + I_2}{\bar{I}(1-\mu^2)}\right] I_0\left[\frac{2\mu\sqrt{I_1 I_2}}{\bar{I}(1-\mu^2)}\right], \tag{4-41}$$

where we have noted from the first-order statistics of intensity that $2\sigma^2 = \bar{I}$. The above equation is valid for $0 \leq I_1, I_2 < \infty$ and is zero otherwise.

Figure 4.4 shows the joint density function for intensity for various values of the correlation coefficient μ. The figures on the right are contour plots of the figures on the left. Note that when $\mu = 0$, the joint density is the product of two exponential distributions. For μ approaching unity, the joint distribution approaches a shaped delta function sheet, as explained in more detail below.

The joint moments of I_1 and I_2 can be shown to be (see [111], p. 402)

$$\overline{I_1^n I_2^m} = \bar{I}^{n+m} n!\, m!\, _2F_1(-n, -m; 1; \mu^2) \tag{4-42}$$

where n and m are nonnegative integers, and $_2F_1$ is a Gaussian hypergeometric function. For the case $n = m = 1$, the moment is the first-order correlation of I_1 and I_2, and this expression reduces to

$$\Gamma_I = \overline{I_1 I_2} = \bar{I}^2(1 + \mu^2), \tag{4-43}$$

as can also be verified using the complex Gaussian moment theorem. Thus the intensity correlation can be found from the magnitude of the underlying complex field correlation.

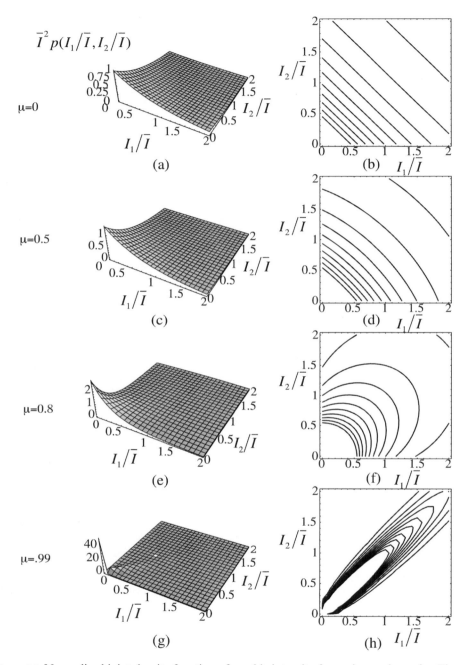

Figure 4.4 Normalized joint density function of speckle intensity for various values of μ. Figures on the right are contour plots of figures on the left.

4.3 Multidimensional Statistics of Speckle Amplitude, Phase and Intensity

We note that the *conditional* density function of I_1 given the value of I_2 is

$$p(I_1|I_2) = \frac{p(I_1, I_2)}{p(I_2)} = \frac{1}{\bar{I}(1-\mu^2)} \exp\left[-\frac{I_1 + \mu^2 I_2}{\bar{I}(1-\mu^2)}\right] I_0 \left[\frac{2\mu\sqrt{I_1 I_2}}{\bar{I}(1-\mu^2)}\right]. \quad (4\text{-}44)$$

For $\mu = 0$, this expression reduces to the usual negative exponential density for I_1. When $\mu \to 1$, the perfect correlation of the underlying fields assures that $I_1 = I_2$, and the conditional probability density function approaches a delta function,

$$p(I_1|I_2) \to \delta(I_1 - I_2).$$

Note in addition, this result implies that, as $\mu \to 1$, the joint density function for I_1 and I_2 of Eq. (4-41) approaches

$$p(I_1, I_2) = p(I_1|I_2)p(I_2) = p(I_2)\delta(I_1 - I_2) \to \frac{1}{\bar{I}} \exp\left(-\frac{I_2}{\bar{I}}\right)\delta(I_1 - I_2). \quad (4\text{-}45)$$

As a matter of curiosity, we also note that the form of Eq. (4-44) is identical with that of the modified Rician of Eq. (3-24), provided the following associations are made:

$$A_0^2 = I_0 \to \mu^2 I_2$$
$$2\sigma^2 = \bar{I}_n \to \bar{I}(1-\mu^2).$$

Finally we consider the density function of the *difference* of intensities, $\Delta I = I_1 - I_2$, as a function of the magnitude μ of the complex correlation coefficient. Following the procedure used to arrive at Eq. (4-37), we express the probability density function of ΔI in terms of the joint density function $p_{I_1, I_2}(I_1, I_2)$ as follows:

$$p_{\Delta I}(\Delta I) = \int_0^\infty p_{I_1, I_2}(\Delta I + I_2, I_2)\, dI_2. \quad (4\text{-}46)$$

Substitution of Eq. (4-41) into this integral should yield the results we are seeking. Again an analytic solution is elusive, but numerical integration yields the results shown in Fig. 4.5 for various values of μ. As expected, the difference of intensities can be positive or negative, with a density function that is symmetrical about the origin. When $\mu = 0$, it is a two-sided exponential (or *Laplace* density), while when $\mu \to 1$, the density function approaches a delta function at $\Delta I = 0$.

To close this section, we calculate the variance of the intensity difference ΔI,

$$\sigma_{\Delta I}^2 = \overline{\Delta I^2} = \overline{(I_1 - I_2)^2} = \overline{I_1^2} + \overline{I_2^2} - 2\overline{I_1 I_2}$$
$$= 4\bar{I}^2 - 2\bar{I}^2(1+\mu^2) = 2\bar{I}^2(1-\mu^2), \quad (4\text{-}47)$$

where we have assumed that $\overline{I_1^2} = 2\bar{I}^2$ and $\overline{I_2^2} = 2\bar{I}^2$, consistent with the known second-order moments of exponential variates. The ratio of standard deviation of intensity difference to mean intensity is thus

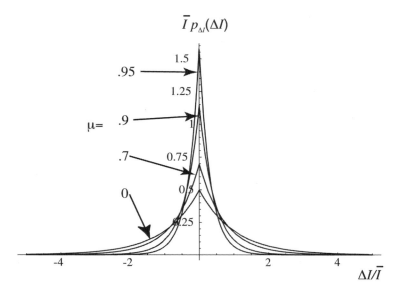

Figure 4.5 Probability density function of the intensity difference $\Delta I = I_1 - I_2$ for various values of the magnitude μ of the complex correlation coefficient.

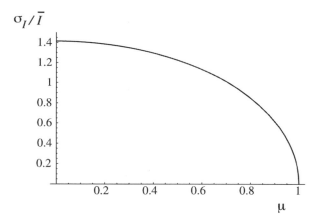

Figure 4.6 Normalized standard deviation of intensity difference as a function of the modulus of the complex correlation coefficient.

$$\frac{\sigma_{\Delta I}}{\bar{I}} = \sqrt{2(1-\mu^2)} \qquad (4\text{-}48)$$

which is shown plotted in Fig. 4.6 as a function of μ.

4.4 Autocorrelation Function and Power Spectrum of Speckle

Throughout our discussions in the previous section, we have frequently encountered the correlation coefficient μ, which describes the normalized correlation between the fields, say $\mathbf{A}_1$ and $\mathbf{A}_2$, at two points in a speckle pattern. We also saw that, for fully developed speckle, the correlation of the intensities, Γ_I, can be expressed in terms of μ, as seen in Eq. (4-43). In this section we explore the dependence of these correlations on the optical geometry in which the speckle pattern arises. We will consider two different cases, first a free-space propagation geometry, and second an imaging geometry.

The earliest publications concerning the autocorrelation function and power spectral density of speckle were those of Goldfischer [64] and Goodman [66].

4.4.1 Free-Space Propagation Geometry

We first consider the case when a coherent source illuminates a planar rough surface, and the scattered light is observed some distance z from that surface, as illustrated in Fig. 4.7. The results apply equally well to the case of transmission through a diffuser. In both cases it is assumed that the phase perturbations suffered by the incident wave are at least several times 2π radians. The wavelength of the incident radiation is

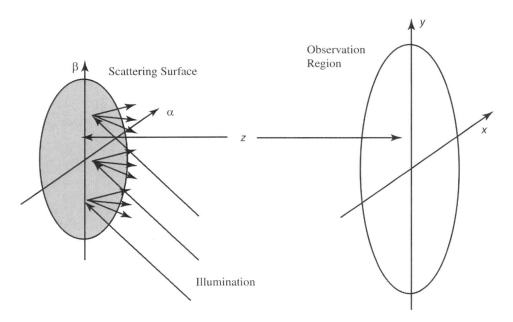

Figure 4.7 Free space scattering geometry.

assumed to be λ, the scattering surface is assumed to be stationary in time, and the shape of the scattering spot is for the moment unspecified. Assuming paraxial propagation (i.e. only small scattering angles are involved) the complex amplitude $\mathbf{A}(x, y)$ of the light at position (x, y) in the observation plane is related to the complex amplitude $\mathbf{a}(\alpha, \beta)$ of the scattered light at coordinates (α, β) in a plane just to the right of the scattering surface by the Fresnel diffraction integral (cf. [71], p. 67),

$$\mathbf{A}(x, y) = \frac{e^{jkz}}{j\lambda z} e^{j\frac{k}{2z}(x^2+y^2)} \int\!\!\!\int_{-\infty}^{\infty} \mathbf{a}(\alpha, \beta) e^{j\frac{k}{2z}(\alpha^2+\beta^2)} e^{-j\frac{2\pi}{\lambda z}(x\alpha+y\beta)} \, d\alpha \, d\beta. \quad (4\text{-}49)$$

Our initial goal is to calculate the autocorrelation function $\mathbf{\Gamma_A}$ of the speckle field $\mathbf{A}(x, y)$ at two points (x_1, y_1) and (x_2, y_2), i.e.

$$\mathbf{\Gamma_A}(x_1, y_1; x_2, y_2) = \overline{\mathbf{A}(x_1, y_1)\mathbf{A}^*(x_2, y_2)}.$$

If we substitute Eq. (4-49) for the $\mathbf{A}$'s, and interchange the orders of averaging and double integration, we obtain

$$\mathbf{\Gamma_A}(x_1, y_1; x_2, y_2) = \frac{1}{\lambda^2 z^2} e^{j\frac{k}{2z}(x_1^2+y_1^2-x_2^2-y_2^2)} \int\!\!\!\int_{-\infty}^{\infty} \int\!\!\!\int_{-\infty}^{\infty} \overline{\mathbf{a}(\alpha_1, \beta_1)\mathbf{a}^*(\alpha_2, \beta_2)}$$

$$\times \, e^{j\frac{k}{2z}(\alpha_1^2+\beta_1^2-\alpha_2^2-\beta_2^2)} e^{-j\frac{2\pi}{\lambda z}(x_1\alpha_1+y_1\beta_1-x_2\alpha_2-y_2\beta_2)} \, d\alpha_1 \, d\beta_1 \, d\alpha_2 \, d\beta_2, \quad (4\text{-}50)$$

or

$$\mathbf{\Gamma_A}(x_1, y_1; x_2, y_2) = \frac{1}{\lambda^2 z^2} e^{j\frac{k}{2z}(x_1^2+y_1^2-x_2^2-y_2^2)} \int\!\!\!\int_{-\infty}^{\infty} \int\!\!\!\int_{-\infty}^{\infty} \mathbf{\Gamma_a}(\alpha_1, \beta_1; \alpha_2, \beta_2)$$

$$\times \, e^{j\frac{k}{2z}(\alpha_1^2+\beta_1^2-\alpha_2^2-\beta_2^2)} e^{-j\frac{2\pi}{\lambda z}(x_1\alpha_1+y_1\beta_1-x_2\alpha_2-y_2\beta_2)} \, d\alpha_1 \, d\beta_1 \, d\alpha_2 \, d\beta_2, \quad (4\text{-}51)$$

where

$$\mathbf{\Gamma_a}(\alpha_1, \beta_1; \alpha_2, \beta_2) = \overline{\mathbf{a}(\alpha_1, \beta_1)\mathbf{a}^*(\alpha_2, \beta_2)} \quad (4\text{-}52)$$

is the correlation function of the fields immediately to the right of the scattering surface.

At this point we must adopt an assumption regarding the nature of $\mathbf{\Gamma_a}$. This correlation is partly determined by the surface-height correlation function and partly by the finite wavelength of light (see, for example, [70], Section 8.3.2). For the present we assume that, for all practical purposes, the correlation extent of the wavefield $\mathbf{a}(\alpha, \beta)$ is sufficiently small that it can be adequately represented by a delta function,[4]

$$\mathbf{\Gamma_a}(\alpha_1, \beta_1; \alpha_2, \beta_2) = \kappa I(\alpha_1, \beta_1)\delta(\alpha_1 - \alpha_2, \beta_1 - \beta_2), \quad (4\text{-}53)$$

---

4. We generalize this assumption in Section 4.5.

4.4 Autocorrelation Function and Power Spectrum of Speckle

where κ is a constant with dimensions length squared, $I(\alpha_1, \beta_1)$ is the intensity of the light just to the right of the scattering surface, and $\delta(\alpha_1 - \alpha_2, \beta_1 - \beta_2)$ is a two-dimensional delta function. Substituting this equation into Eq. (4-51) and using the sifting property of the delta function, we obtain

$$\Gamma_{\mathbf{A}}(x_1, y_1; x_2, y_2) = \frac{\kappa}{\lambda^2 z^2} e^{j\frac{k}{2z}(x_1^2 + y_1^2 - x_2^2 - y_2^2)} \iint_{-\infty}^{\infty} I(\alpha, \beta) e^{-j\frac{2\pi}{\lambda z}[\alpha \Delta x + \beta \Delta y]} \, d\alpha \, d\beta, \quad (4\text{-}54)$$

where $\Delta x = x_1 - x_2$, $\Delta y = y_1 - y_2$, and we have replaced the symbols (α_1, β_1) by simply (α, β).

In most applications, it is the *modulus* of the correlation function of the complex fields that is critical, so the quadratic-phase complex exponentials preceding the integral in Eq. (4-54) can be ignored. For any case where they are needed, they will be reintroduced. Thus we have

$$\Gamma_{\mathbf{A}}(\Delta x, \Delta y) = \frac{\kappa}{\lambda^2 z^2} \iint_{-\infty}^{\infty} I(\alpha, \beta) e^{-j\frac{2\pi}{\lambda z}[\alpha \Delta x + \beta \Delta y]} \, d\alpha \, d\beta, \quad (4\text{-}55)$$

which, aside from scaling factors, is a simple two-dimensional Fourier transform of the intensity distribution of the light leaving the scattering spot. In many cases this intensity distribution is proportional to the intensity distribution incident on the scattering spot, but we shall mention this assumption specifically when it is used. This result is entirely equivalent to the van Cittert–Zernike theorem of classical coherence theory (see Ref. [70], Section 5.6).

The complex correlation coefficient, which is defined by

$$\mu_{\mathbf{A}}(\Delta x, \Delta y) = \Gamma_{\mathbf{A}}(\Delta x, \Delta y) / \Gamma_{\mathbf{A}}(0, 0),$$

takes the form

$$\mu_{\mathbf{A}}(\Delta x, \Delta y) = \frac{\iint_{-\infty}^{\infty} I(\alpha, \beta) e^{-j\frac{2\pi}{\lambda z}[\alpha \Delta x + \beta \Delta y]} \, d\alpha \, d\beta}{\iint_{-\infty}^{\infty} I(\alpha, \beta) \, d\alpha \, d\beta} \quad (4\text{-}56)$$

and has the property that $\mu_{\mathbf{A}}(0, 0) = 1$.

Note that Eqs. (4-55) and (4-56) hold in both the near and far fields of the scattering spot, subject only to the assumption that the Fresnel diffraction equation (4-49) holds.

We are now prepared to find the autocorrelation function of the *intensity* distribution in a speckle pattern. With reference to Eq. (4-43), the correlation function of the intensity distribution is related to the normalized correlation function of the amplitude through

$$\Gamma_I(\Delta x, \Delta y) = \bar{I}^2[1 + |\boldsymbol{\mu_A}(\Delta x, \Delta y)|^2]$$

$$= \bar{I}^2\left[1 + \frac{\left|\iint_{-\infty}^{\infty} I(\alpha,\beta)e^{-j\frac{2\pi}{\lambda z}[\alpha\Delta x + \beta\Delta y]}\,d\alpha\,d\beta\right|^2}{\iint_{-\infty}^{\infty} I(\alpha,\beta)\,d\alpha\,d\beta}\right]. \quad (4\text{-}57)$$

The *power spectral density function* $\mathcal{G}_I(v_X, v_Y)$ of the speckle pattern represents the distribution of intensity fluctuation power over the two-dimensional frequency plane, and is given by the Fourier transform of the autocorrelation function of intensity

$$\mathcal{G}_I(v_X, v_Y) = \iint_{-\infty}^{\infty} \Gamma_I(\Delta x, \Delta y) e^{-j2\pi(v_X \Delta x + v_Y \Delta y)}\,d\Delta x\,d\Delta y. \quad (4\text{-}58)$$

With the help of the autocorrelation theorem ([71], p. 8) and proper attention to scaling constants, the power spectral density can be reduced to

$$\mathcal{G}_I(v_X, v_Y) = \bar{I}^2\left[\delta(v_X, v_Y) + (\lambda z)^2 \frac{\iint_{-\infty}^{\infty} I(\alpha,\beta)I(\alpha+\lambda z v_X, \beta+\lambda z v_Y)\,d\alpha\,d\beta}{\left[\iint_{-\infty}^{\infty} I(\alpha,\beta)\,d\alpha\,d\beta\right]^2}\right]. \quad (4\text{-}59)$$

The delta function in this expression corresponds to the zero-frequency discrete power contributed by the average intensity $\bar{I}$. The second term, consisting of the normalized and scaled autocorrelation function of the intensity distribution $I(\alpha,\beta)$ at the scattering spot, represents the distribution of fluctuation power over spatial frequency for the varying part of the speckle pattern intensity.

The second term of the power spectrum (for which we use the symbol $\tilde{\mathcal{G}}$) yields the continuous portion of the spectrum, and has a value at $v_X = v_Y = 0$ of

$$\tilde{\mathcal{G}}(0,0) = \bar{I}^2(\lambda z)^2 \frac{\iint_{-\infty}^{\infty} I^2(\alpha,\beta)\,d\alpha\,d\beta}{\left[\iint_{-\infty}^{\infty} I(\alpha,\beta)\,d\alpha\,d\beta\right]^2}. \quad (4\text{-}60)$$

The ratio of the integrals has the dimensions of inverse area. Indeed, when the intensity at the scattering spot, $I(\alpha,\beta)$, is uniform over a geometrical area, the ratio of integrals reduces to $1/A$, where A is the area of the scattering spot. Thus in this special case,

$$\tilde{\mathcal{G}}(0,0) = \bar{I}^2\frac{(\lambda z)^2}{A} = \bar{I}^2\frac{\lambda^2}{\Omega_s}, \quad (4\text{-}61)$$

where Ω_s is the solid angle subtended by the scattering spot when viewed from the observation region.

4.4 Autocorrelation Function and Power Spectrum of Speckle

We now illustrate these results with two examples, a square ($L \times L$), uniform distribution of intensity and a circular uniform distribution of intensity at the scattering spot. First, let the distribution of intensity be

$$I(\alpha, \beta) = I_0 \, \text{rect}\left(\frac{\alpha}{L}\right) \text{rect}\left(\frac{\beta}{L}\right), \tag{4-62}$$

where the rectangle function $\text{rect}(x)$ is unity for $|x| \leq \frac{1}{2}$, and zero otherwise. The Fourier transform of this function is

$$\iint_{-\infty}^{\infty} I_0 \, \text{rect}\left(\frac{\alpha}{L}\right) \text{rect}\left(\frac{\beta}{L}\right) e^{-j\frac{2\pi}{\lambda z}(\alpha \Delta x + \beta \Delta y)} \, d\alpha \, d\beta = L^2 I_0 \, \text{sinc}\left(\frac{L \Delta x}{\lambda z}\right) \text{sinc}\left(\frac{L \Delta y}{\lambda z}\right), \tag{4-63}$$

where $\text{sinc}(x) = \sin(\pi x)/(\pi x)$. It follows that the autocorrelation function of the speckle intensity in this case is given by

$$\Gamma_I(\Delta x, \Delta y) = \bar{I}^2 \left[1 + \text{sinc}^2\left(\frac{L \Delta x}{\lambda z}\right) \text{sinc}^2\left(\frac{L \Delta y}{\lambda z}\right)\right], \tag{4-64}$$

and the corresponding power spectral density of speckle intensity is

$$\mathcal{G}(\nu_X, \nu_Y) = \bar{I}^2 \left[\delta(\nu_X, \nu_Y) + \frac{(\lambda z)^2}{A} \Lambda\left(\frac{\lambda z}{L} \nu_X\right) \Lambda\left(\frac{\lambda z}{L} \nu_Y\right)\right], \tag{4-65}$$

where the triangle function $\Lambda(x) = 1 - |x|$ for $|x| \leq 1$, 0 otherwise, and $A = L^2$ is the area of the scattering spot. Figure 4.8 illustrates these two results.

Our second example is for a scattering spot with a uniform circular intensity distribution of diameter D,

$$I(\alpha, \beta) = I(\Omega) = \text{circ}\left(\frac{2\rho}{D}\right), \tag{4-66}$$

where $\rho = \sqrt{\alpha^2 + \beta^2}$ and

$$\text{circ}(\rho) = \begin{cases} 1 & \rho \leq 1 \\ 0 & \text{otherwise.} \end{cases}$$

To calculate the autocorrelation function of the speckle intensity, we first must Fourier transform the circular spot. The scaled Fourier transform of such a distribution is

$$\iint_{-\infty}^{\infty} \text{circ}\left(\frac{2\rho}{D}\right) e^{-j\frac{2\pi}{\lambda z}(\alpha \Delta x + \beta \Delta y)} \, d\alpha \, d\beta$$

$$= 2\pi \int_0^{D/2} \rho J_0\left(2\pi \frac{r}{\lambda z} \rho\right) d\rho = A \left[2 \frac{J_1\left(\frac{\pi D}{\lambda z} r\right)}{\frac{\pi D}{\lambda z} r}\right], \tag{4-67}$$

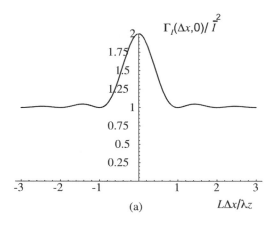

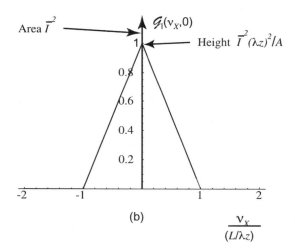

Figure 4.8 Cross sections of speckle intensity (a) autocorrelation and (b) power spectral density for free-space propagation and a uniform rectangular scattering spot.

where A is again the area of the scattering spot, this time given by $A = \pi(D/2)^2$, $J_1(\ )$ is a Bessel function of the first kind, order one, and $r = \sqrt{(\Delta x)^2 + (\Delta y)^2}$. With proper normalization, the autocorrelation function of the speckle intensity becomes

$$\Gamma_I(r) = \bar{I}^2 \left[1 + \left| 2 \frac{J_1\left(\frac{\pi D r}{\lambda z}\right)}{\frac{\pi D r}{\lambda z}} \right|^2 \right]. \tag{4-68}$$

The corresponding power spectral density of speckle intensity can then be shown to be

4.4 Autocorrelation Function and Power Spectrum of Speckle

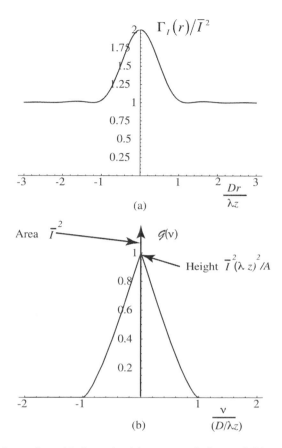

Figure 4.9 Cross sections of speckle intensity (a) autocorrelation and (b) power spectral density for free-space propagation and a uniform circular scattering spot.

$$\mathcal{G}(\nu) = \bar{I}^2 \left\{ \delta(\nu_X, \nu_Y) + \frac{(\lambda z)^2}{A} \frac{2}{\pi} \left[\arccos\left(\frac{\lambda z}{D}\nu\right) - \frac{\lambda z}{D}\nu \sqrt{1 - \left(\frac{\lambda z}{D}\nu\right)^2} \right] \right\}, \quad (4\text{-}69)$$

where $\nu = \sqrt{\nu_X^2 + \nu_Y^2}$. Figure 4.9 shows slices through the origin of the autocorrelation function of speckle intensity and the power spectral density of speckle intensity.

Having seen two examples of the calculation of the autocorrelation function and power spectral density of speckle intensity, we turn now to developing a measure of the average "size" of a speckle. The measure we adopt here is the *equivalent area* of the normalized covariance function of speckle intensity (cf. [20], chapter 8), which we call the "correlation area" or the "coherence area" and represent it by the symbol

$\mathcal{A}_c$. The normalized covariance function of speckle intensity is related to the autocorrelation function by

$$c_I(\Delta x, \Delta y) = \frac{\Gamma_I(\Delta x, \Delta y) - \bar{I}^2}{\bar{I}^2}.$$

The equivalent area of this quantity is given by

$$\mathcal{A}_c = \iint_{-\infty}^{\infty} c_I(\Delta x, \Delta y) \, d\Delta x \, d\Delta y = \iint_{-\infty}^{\infty} |\mu_A(\Delta x, \Delta y)|^2 \, d\Delta x \, d\Delta y. \tag{4-70}$$

For both of the examples presented above, this quantity is

$$\mathcal{A}_c = \frac{(\lambda z)^2}{A} = \frac{\lambda^2}{\Omega_s}, \tag{4-71}$$

where A is the area of the scattering spot and again Ω_s is the solid angle subtended by that spot, a result that is true for a uniformly bright spot of any shape. More generally, for a spot that is not uniformly bright,

$$\mathcal{A}_c = (\lambda z)^2 \frac{\iint_{-\infty}^{\infty} I^2(\alpha, \beta) \, d\alpha \, d\beta}{\left[\iint_{-\infty}^{\infty} I(\alpha, \beta) \, d\alpha \, d\beta\right]^2}. \tag{4-72}$$

Because the geometry is fundamentally two-dimensional, there is no unique way to define the one-dimensional *width* of a speckle, but a reasonable approximation is simply the square root of the equivalent area discussed above.

4.4.2 Imaging Geometry

We turn attention now to an *imaging* geometry shown in Fig. 4.10. A rough scattering object on the left is illuminated with coherent light.[5] A portion of the scattered light is collected by the simple positive lens, and that light is then brought to a focus in the image plane on the right. The object distance from the lens is z_o and the image distance from the lens is z_i. The lens has focal length f. For the following arguments, we assume that the object is a uniform rough surface, with no variations of intensity reflectance.

One possible approach to the problem of finding the autocorrelation function Γ_A of the fields in the (x, y) plane is to propagate the field autocorrelation function through the entire system, starting with Eq. (4-51), making the delta-function correlation approximation at the scattering surface (i.e. at the object), using a quadratic phase transmission function for the lens (see [71], Section 5.1) and thereby calculating

---

5. For a transmissive object illuminated through a diffuser, similar arguments to those that follow apply.

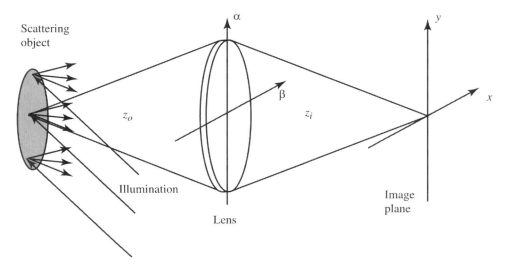

Figure 4.10 Imaging geometry leading to a speckled image.

the correlation function behind the lens, and then again applying Eq. (4-51) (without the delta-function approximation) to pass from the lens to the image plane. Such an approach is possible but tedious.

Instead we use an approximation first introduced by Zernike in a closely related problem [183]. The assumptions behind this approximation are (1) the illuminated area on the object is large compared with the correlation area of the wavefront reflected from the surface, and (2) the illuminated region on the object is very broad compared with the extent of a single lens resolution cell on the object. The first assumption assures that the wavefront incident on the lens aperture has phase variations that are approximately uniformly distributed over 2π radians due to interference of many randomly-phased contributions from a multitude of object correlation areas, and the second assumption guarantees that the speckle size incident on the lens aperture is very small compared with the lens aperture. As a byproduct of the analysis of a different problem in Appendix D.2, a proof is found at the end of that appendix that Zernike's approximation is correct under most conditions of interest.

With these assumptions, it is possible to treat the *lens pupil itself* as a δ-correlated, uniformly bright effective scattering source, and to consider only propagation from the lens to the image plane using Eq. (4-56). Thus the speckle amplitude correlation is determined in the image plane by a Fourier transform of the intensity distribution across the lens pupil. For the usual case of a circular lens pupil, with uniform average intensity across its face, the results of the second example at the end of the previous section apply, with the sole change that the distance z is replaced by the distance z_i. Thus the intensity autocorrelation function and the power spectral density of the image-plane speckle are as shown earlier in Fig. 4.9.

Two additional comments are needed here. First, since the lens pupil behaves essentially as a randomly phased, δ-correlated source, lens aberrations, which are usually represented as phase errors across the lens pupil, have no effect on the properties of the speckle correlation observed in the image plane. Second, while we have considered only a simple thin lens in this analysis, an identical result will be obtained for any imaging system, provided the *exit pupil*[6] of the imaging system is regarded as the equivalent δ-correlated source, with the distance z_i now being the distance from the exit pupil to the image plane. Implicit in this result is the conclusion that aberrations of the imaging lens have no effect on the speckle correlation properties, a fact that is true provided that (1) the speckle is fully developed and therefore has no significant specular component (see Ref. [116] for a discussion of the case where this is not true), and (2) the aberrations do not change phase appreciably over a distance comparable with the correlation area of the scattered wavefront.[7]

4.4.3 Speckle Size in Depth

Until now, we have considered only the transverse properties of speckle in a measurement plane parallel to the scattering surface. We turn attention now to the properties of speckle in depth, that is, along the direction normal to the scattering plane.[8] In particular, we wish to find the distance in depth over which a speckle pattern remains correlated, just as we found the transverse correlation width previously. While correlation depth will vary, depending on the particular (x, y) coordinates chosen, nonetheless the results at the origin $(x = 0, y = 0)$ are representative of the results elsewhere, particularly if the angles involved are small. Therefore we will calculate the cross-correlation of the observed fields at transverse coordinates $(x = 0, y = 0)$, but at different depths z and $z + \Delta z$, which we represent by

$$\Gamma_A(0, 0; \Delta z) = \overline{\mathbf{A}(0, 0; z)\mathbf{A}^*(0, 0; z + \Delta z)}.$$

The fields in question can be written (cf. Eq. (4-49))

$$\mathbf{A}(0, 0; z) = \frac{1}{j\lambda z} \iint_{-\infty}^{\infty} \mathbf{a}(\alpha, \beta) e^{j\frac{k}{2z}(\alpha^2 + \beta^2)} \, d\alpha \, d\beta$$

$$\mathbf{A}^*(0, 0; z + \Delta z) = \frac{1}{-j\lambda(z + \Delta z)} \iint_{-\infty}^{\infty} \mathbf{a}^*(\alpha, \beta) e^{-j\frac{k}{2(z + \Delta z)}(\alpha^2 + \beta^2)} \, d\alpha \, d\beta.$$

(4-73)

The assumption $\Delta z \ll z$ is made, allowing $z + \Delta z$ to be replaced by z in the multiplier in front of the second set of integrals, but not in the exponent, since λ is a small number. Substituting these equations into the expression for Γ_A yields

---

6. For a definition of the exit pupil of an imaging system, see Ref. [71], Appendix B, Section 5.
7. In the present case, the scattered wavefront is delta-correlated, but in a subsequent generalization, this will not be the case. The idealization of a delta-correlated wavefront implies that the image plane is uniformly illuminated, which of course is not true when an object is being imaged by the system. See Section 4.5.
8. For a more detailed treatment of this problem, see [100].

4.4 Autocorrelation Function and Power Spectrum of Speckle

$$\boldsymbol{\Gamma}_\mathbf{A}(0,0;\Delta z) = \frac{1}{\lambda^2 z^2} \iint\limits_{-\infty}^{\infty} \iint\limits_{-\infty}^{\infty} \overline{\mathbf{a}(\alpha_1,\beta_1)\mathbf{a}^*(\alpha_2,\beta_2)}$$

$$\times e^{j\frac{k}{2z}(\alpha_1^2+\beta_1^2)} e^{-j\frac{k}{2(z+\Delta z)}(\alpha_2^2+\beta_2^2)} \, d\alpha_1 \, d\beta_1 \, d\alpha_2 \, d\beta_2. \quad (4\text{-}74)$$

At this point we again make the δ-correlated assumption

$$\boldsymbol{\Gamma}_\mathbf{a}(\alpha_1,\beta_1;\alpha_2,\beta_2) = \overline{\mathbf{a}(\alpha_1,\beta_1)\mathbf{a}^*(\alpha_2,\beta_2)} = \kappa I(\alpha_1,\beta_1)\delta(\alpha_1-\alpha_2,\beta_1-\beta_2),$$

and also make the approximation

$$\frac{1}{z+\Delta z} \approx \frac{1}{z}\left(1 - \frac{\Delta z}{z}\right)$$

valid for $\Delta z \ll z$, with the result

$$\boldsymbol{\Gamma}_\mathbf{A}(0,0;\Delta z) = \frac{\kappa}{\lambda^2 z^2} \iint\limits_{-\infty}^{\infty} I(\alpha,\beta) e^{j\frac{k\Delta z}{2z^2}(\alpha^2+\beta^2)} \, d\alpha \, d\beta. \quad (4\text{-}75)$$

We can also define a normalized correlation coefficient by normalizing this quantity by $\boldsymbol{\Gamma}_\mathbf{A}(0,0;0)$, with the result

$$\boldsymbol{\mu}_\mathbf{A}(\Delta z) = \frac{\iint\limits_{-\infty}^{\infty} I(\alpha,\beta) e^{j\frac{k\Delta z}{2z^2}(\alpha^2+\beta^2)} \, d\alpha \, d\beta}{\iint\limits_{-\infty}^{\infty} I(\alpha,\beta) \, d\alpha \, d\beta}. \quad (4\text{-}76)$$

The normalized covariance of intensities is the magnitude squared of this quantity.

We illustrate with two specific cases, a uniform, square spot of dimensions $L \times L$ and a uniform circular spot of diameter D. The required manipulations are shown in the following equations:

$$\text{circular:} \quad |\boldsymbol{\mu}_\mathbf{A}(\Delta z)|^2 = \left| \frac{\int_0^{D/2} r e^{j\frac{k\Delta z}{2z^2}r^2} \, dr}{\int_0^{D/2} r \, dr} \right|^2$$

$$= \operatorname{sinc}^2\left(\frac{x}{8\pi}\right) \quad \text{with } x = \frac{\pi D^2}{\lambda z^2}\Delta z \quad (4\text{-}77)$$

$$\text{square:} \quad |\boldsymbol{\mu}_\mathbf{A}(\Delta z)|^2 = \frac{1}{L^2} \left| \iint\limits_{-L/2}^{L/2} e^{j\frac{k\Delta z}{2z^2}(\alpha^2+\beta^2)} \, d\alpha \, d\beta \right|^2$$

$$= \left| \sqrt{\frac{2\pi}{x}}\left[C\left(\sqrt{\frac{x}{2\pi}}\right) + jS\left(\sqrt{\frac{x}{2\pi}}\right) \right] \right|^4 \quad \text{with } x = \frac{\pi L^2}{\lambda z^2}\Delta z, \quad (4\text{-}78)$$

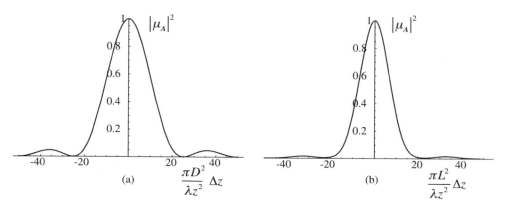

Figure 4.11 Plots of $|\mu_A(\Delta z)|^2$ for the cases of (a) a circular aperture, and (b) a square aperture.

where $C(z)$ and $S(z)$ are the Fresnel cosine and sine integrals, respectively. Figure 4.11 shows plots of the above results. The correlation widths of the intensity, measured by the width of the curves at their half-maxima, are given approximately by

$$\text{circular} \quad (\Delta z)_{1/2} = 7.1\lambda(z/D)^2$$
$$\text{square} \quad (\Delta z)_{1/2} = 5.0\lambda(z/L)^2. \quad (4\text{-}79)$$

For comparison purposes, the widths at half maxima for the *transverse* normalized correlation functions are

$$\text{circular} \quad (\Delta r)_{1/2} = 1.0\lambda(z/D)$$
$$\text{square} \quad (\Delta x)_{1/2} = (\Delta y)_{1/2} = 0.9\lambda(z/L). \quad (4\text{-}80)$$

If the spot sizes are small compared with the distance to the observation plane, then the extent of a speckle in the axial dimension is much greater than the extent in the transverse dimension.

While we have concentrated on the free-space geometry, the results for the imaging geometry are the same provided the exit pupil of the imaging system is regarded as the effective scattering spot.

4.5 Dependence of Speckle on Scatterer Microstructure

In this section, we explore how various properties of speckle depend on the microstructure of the rough surfaces or volume scatterers from which the light is scattered.

We will abandon the δ-correlation assumption for the scattered wave, and with a series of approximations, will find the effect of a finite spread of correlation of that wave on the properties of the observed speckle. In addition, we will explore various other properties of speckle that depend on scatterer microstructure.

4.5.1 Surface vs. Volume Scattering

The scattering phenomena that produce speckle can be broadly divided into two classes: (1) *surface scattering*, for which the dominant source of the scattered field is the random interface between a surface and air, and (2) *volume scattering*, for which the incident optical wave penetrates the surface and may undergo multiple scattering before emerging from the surface in the direction of a detector or collection optics. Scattering from rough metallic surfaces is to a good approximation surface scattering. Scattering from a glass plate, one side of which has been roughened on a scale that is large compared with a wavelength, is largely surface scattering. However, scattering from a plastic slab with imbedded small metallic or dielectric particles is usually volume scattering. In cases of predominantly volume scattering, there may also be a component of surface scattering present. In addition, even for scattering that is purely from a surface, multiple reflection off more than one surface facet can occur, in which case the situation is between the two extremes.

The important differences between these two types of scattering stem from the fact that for surface scattering, the delays imparted to an incident optical wave are primarily determined by the random surface height fluctuations encountered, while for volume scattering, the delays are determined by the random path lengths travelled by photons as they undergo multiple scattering. Surface height fluctuation in surface scattering, and the average distribution of path lengths in volume scattering, play similar roles.

The average distribution of path lengths in volume scattering can be determined by illuminating a small patch of the surface with an extremely short optical pulse, and measuring the average impulse response of the scattered light that results. This average impulse response, with argument time replaced by path length divided by the speed of light, yields the average distribution of path lengths for the scattered light. Such a distribution can be calculated theoretically under certain conditions using the diffusion equation [160].

4.5.2 Effect of a Finite Correlation Area of the Scattered Wave

For both surface and volume scattering, the light can be characterized by a spatial correlation function of the scattered light immediately leaving the scattering medium. In this section, we determine the effect of such a correlation function on the properties of speckle, considering two different geometries. The first geometry is for a scattering medium in front of a lens and the observation region in the rear focal plane of that lens. The second is a free-space geometry.

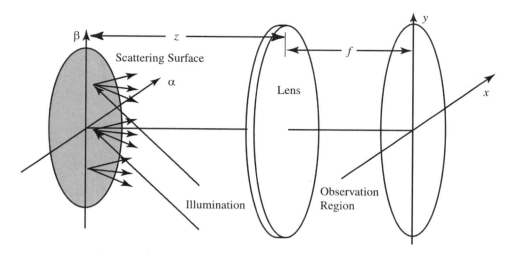

Figure 4.12 Observation region in the rear focal plane of a positive lens.

Observation Plane in the Rear Focal Plane of a Positive Lens

Suppose that the scattering surface lies in front of a positive lens and the observation region lies in the rear focal plane of that lens. The geometry is illustrated in Fig. 4.12. In this case the fields in the two planes are related by ([71], p. 106)

$$\mathbf{A}(x, y) = \frac{e^{j\frac{\pi}{\lambda f}\left(1-\frac{z}{f}\right)(x^2+y^2)}}{\lambda f} \iint_{-\infty}^{\infty} \mathbf{a}(\alpha, \beta) \exp\left[-j\frac{2\pi}{\lambda f}(\alpha x + \beta y)\right] d\alpha\, d\beta, \quad (4\text{-}81)$$

where, as usual, **a** is the field just to the right of the scattering surface, **A** is the field in the observation region, f is the focal length of the lens, and z is the distance from the scattering surface to the lens.

We can now calculate the correlation function of the observed fields in terms of the correlation function of the scattered fields,

$$\Gamma_\mathbf{A}(x_1, y_1; x_2, y_2) = \frac{e^{j\frac{\pi}{\lambda f}\left(1-\frac{z}{f}\right)(x_1^2+y_1^2-x_2^2-y_2^2)}}{\lambda^2 f^2}$$

$$\times \iint_{-\infty}^{\infty} \iint_{-\infty}^{\infty} \Gamma_\mathbf{a}(\alpha_1, \beta_1; \alpha_2, \beta_2) e^{-j\frac{2\pi}{\lambda f}(x_1\alpha_1+y_1\beta_1-x_2\alpha_2-y_2\beta_2)}\, d\alpha_1\, d\beta_1\, d\alpha_2\, d\beta_2. \quad (4\text{-}82)$$

To simplify this expression, it is helpful to make the following definitions:

$$\begin{aligned} \Delta x &= x_1 - x_2 & \Delta\alpha &= \alpha_1 - \alpha_2 \\ \Delta y &= y_1 - y_2 & \Delta\beta &= \beta_1 - \beta_2. \end{aligned} \quad (4\text{-}83)$$

It then follows that

4.5 Dependence of Speckle on Scatterer Microstructure

$$x_1\alpha_1 + y_1\beta_1 - x_2\alpha_2 - y_2\beta_2$$
$$= \alpha_2\Delta x + \beta_2\Delta y + x_2\Delta\alpha + y_2\Delta\beta + \Delta\alpha\Delta x + \Delta\beta\Delta y$$
$$= \alpha_2\Delta x + \beta_2\Delta y + \Delta\alpha x_1 + \Delta\beta y_1. \quad (4\text{-}84)$$

In addition, rather than using the approximation of a δ-correlated $\Gamma_\mathbf{a}$ as in Eq. (4-53), we assume instead that the correlation function takes the separable form

$$\Gamma_\mathbf{a}(\alpha_1,\beta_1;\alpha_2,\beta_2) = \sqrt{I(\alpha_1,\beta_1)I(\alpha_2,\beta_2)}\mu_\mathbf{a}(\Delta\alpha,\Delta\beta) \approx I(\alpha_2,\beta_2)\mu_\mathbf{a}(\Delta\alpha,\Delta\beta), \quad (4\text{-}85)$$

where two approximations are implicit in this form, one that the fluctuations of the scattered complex field at the surface are wide-sense stationary (i.e. $\mu_\mathbf{a}$ depends only on $\Delta\alpha$ and $\Delta\beta$), and second that the width of the scattering spot is much greater than the correlation width of the field $\mathbf{a}$, so that within that correlation width, $I(\alpha_1,\beta_1) \approx I(\alpha_2,\beta_2)$.

Thus we obtain

$$\Gamma_\mathbf{A}(x_1,y_1;x_2,y_2) = \frac{e^{j\frac{\pi}{\lambda f}\left(1-\frac{z}{f}\right)(x_1^2+y_1^2-x_2^2-y_2^2)}}{\lambda^2 f^2} \iint\limits_{-\infty}^{\infty} I(\alpha_2,\beta_2) e^{-j\frac{2\pi}{\lambda f}[\Delta x\alpha_2 + \Delta y\beta_2]}\, d\alpha_2\, d\beta_2$$

$$\times \iint\limits_{-\infty}^{\infty} \mu_\mathbf{a}(\Delta\alpha,\Delta\beta) e^{-j\frac{2\pi}{\lambda f}[x_1\Delta\alpha + y_1\Delta\beta]}\, d\Delta\alpha\, d\Delta\beta. \quad (4\text{-}86)$$

Thus the double integral has factored into a product of two integrals, one with respect to (α_2,β_2) and one with respect to $(\Delta\alpha,\Delta\beta)$. The first of these two integrals is a Fourier transform and in fact is identical to the one we obtained when we assumed a δ-correlated wave amplitude at the surface. Indeed the current result reduces to that same result if we substitute a δ-function normalized autocorrelation function for $\mu_\mathbf{a}$. The second integral is also a Fourier transform and is a direct result of the fact that the scattered wave is not δ-correlated. In fact, this second integral of the narrow normalized correlation function $\mu_\mathbf{a}$ yields a broad variation of average intensity across the observation region. We illustrate this result with a diagram in Fig. 4.13.

A useful interpretation of these results is that the amplitude correlation function of the scattered fields determines the spread of angles at which light propagates away from the scattering surface at each point, while the intensity distribution across the scattering spot determines how fine the speckle structure will be. Note that in the case of observation in the rear focal plane of the lens, angles in front of the lens are converted into spatial positions in the focal plane, and therefore for statistically stationary roughness on the surface, all correlation areas on the scattered spot contribute the same distribution of light in the focal plane, all centered on the optical axis.[9]

---

9. We have assumed that the scattering spot and the observation region are sufficiently small, and the lens sufficiently large, that vignetting is not a problem.

88   CHAPTER 4   Higher-Order Statistical Properties of Speckle

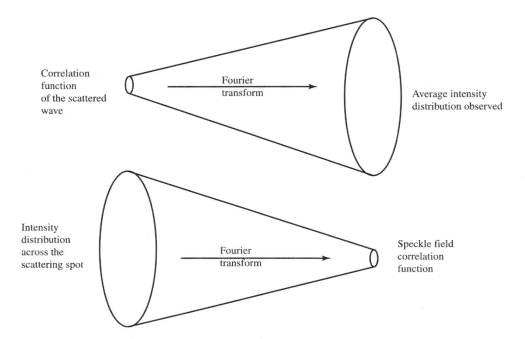

Figure 4.13 Fourier relations between correlation functions and intensity distributions.

This relationship between the normalized correlation function $\mu_\mathbf{a}$ of the scattered field and the average intensity across the observation region was first proved (in the case of a free-space geometry, as described in the following subsection) in Ref. [66] in the context of speckle. We shall refer to this relationship as the *generalized van Cittert–Zernike theorem*.

The generalized van Cittert–Zernike theorem provides a connection between the correlation function of the fields as they leave the scattering surface and the average intensity distribution across the observation region. Thus by measurements of the observed average intensity information, one can in principle deduce the form of the field correlation at the surface by means of a Fourier transform. But if we are interested in the character of the surface microstructure, we are missing information—how are the surface height fluctuations related to the fluctuations in the scattered field $\mathbf{a}(\alpha,\beta)$? After a brief detour in the next two subsections, we will address this question in a section that follows.

Free-Space Geometry
The geometry in the free-space propagation case was shown previously in Fig. 4.7. With reference to Eq. (4-51) for the autocorrelation function of the fields in the observation region in terms of the autocorrelation function of the fields immediately to the right of the scattering spot,

4.5 Dependence of Speckle on Scatterer Microstructure

$$\Gamma_A(x_1, y_1; x_2, y_2) = \frac{1}{\lambda^2 z^2} e^{j\frac{\pi}{\lambda z}(x_1^2+y_1^2-x_2^2-y_2^2)}$$

$$\times \iint_{-\infty}^{\infty} \iint_{-\infty}^{\infty} \Gamma_a(\alpha_1, \beta_1; \alpha_2, \beta_2) e^{j\frac{\pi}{\lambda z}(\alpha_1^2+\beta_1^2-\alpha_2^2-\beta_2^2)}$$

$$\times e^{-j\frac{2\pi}{\lambda z}(x_1\alpha_1+y_1\beta_1-x_2\alpha_2-y_2\beta_2)} \, d\alpha_1 \, d\beta_1 \, d\alpha_2 \, d\beta_2. \quad (4\text{-}87)$$

We again assume that the correlation function takes the separable form of Eq. (4-85). As before, if we substitute this expression for Γ_a into the equation for Γ_A, and change the variables of integration to (α_2, β_2) and $(\Delta\alpha, \Delta\beta)$, transforming all variables as before, the expression for the speckle field correlation becomes

$$\Gamma_A(x_1, y_1; x_2, y_2) = \frac{1}{\lambda^2 z^2} e^{j\frac{\pi}{\lambda z}(x_1^2+y_1^2-x_2^2-y_2^2)}$$

$$\times \iint_{-\infty}^{\infty} d\alpha_2 \, d\beta_2 \iint_{-\infty}^{\infty} d\Delta\alpha \, d\Delta\beta I(\alpha_2, \beta_2) \mu_a(\Delta\alpha, \Delta\beta)$$

$$\times e^{j\frac{\pi}{\lambda z}[2\alpha_2\Delta\alpha+\Delta\alpha^2+2\beta_2\Delta\beta+\Delta\beta^2]} e^{-j\frac{2\pi}{\lambda z}[x_1\Delta\alpha+y_1\Delta\beta]} e^{-j\frac{2\pi}{\lambda z}[\Delta x\alpha_2+\Delta y\beta_2]}. \quad (4\text{-}88)$$

At this point two approximations are required. First we assume that

$$z \gg \frac{\pi}{\lambda}(\Delta\alpha^2 + \Delta\beta^2)_{max}, \quad (4\text{-}89)$$

where $(\Delta\alpha^2 + \Delta\beta^2)_{max}$ is the maximum value of this quantity for which the correlation function μ_a has significant value. Equivalently we are assuming that the observation region is in the far field of the very small correlation area on the complex field **a** at the scattering surface. This allows the exponential terms in $\Delta\alpha^2$ and $\Delta\beta^2$ to be dropped. An equivalent expression of this condition is

$$z > 2\frac{L_c^2}{\lambda}, \quad (4\text{-}90)$$

where L_c is the linear dimension of the correlation area of the scattered wave.

Secondly, and more stringently, we assume that the distance z satisfies

$$z \gg \frac{2\pi}{\lambda}(\alpha_2\Delta\alpha + \beta_2\Delta\beta)_{max}. \quad (4\text{-}91)$$

An interpretation of this requirement is that the observation region lies in the far field of an aperture with an area that is the geometric mean of the correlation area of the scattered wave at the surface and the area of the scattering spot itself. Equivalently, if L_s is the linear dimension of the scattering spot, we require

$$z > 2\frac{L_s L_c}{\lambda}. \quad (4\text{-}92)$$

More on this approximation in a moment.

With these two approximations, the expression for the speckle field correlation becomes

$$\Gamma_A(x_1, y_1; x_2, y_2) = \frac{1}{\lambda^2 z^2} e^{j\frac{\pi}{\lambda z}(x_1^2 + y_1^2 - x_2^2 - y_2^2)}$$

$$\times \iint_{-\infty}^{\infty} I(\alpha_2, \beta_2) e^{-j\frac{2\pi}{\lambda z}[\Delta x \alpha_2 + \Delta y \beta_2]} d\alpha_2 \, d\beta_2$$

$$\times \iint_{-\infty}^{\infty} \mu_a(\Delta\alpha, \Delta\beta) e^{-j\frac{2\pi}{\lambda z}[x_1 \Delta\alpha + y_1 \Delta\beta]} d\Delta\alpha \, d\Delta\beta, \qquad (4\text{-}93)$$

which, aside from the details of the unimportant phase factor preceding the integrals, is the same result obtained for observation in the rear focal plane of a lens.

4.5.3 A Regime where Speckle Size Is Independent of Scattering Spot Size

There exists a regime within which the average size of a speckle is independent of the size of the intensity distribution across the scattering spot. In this regime, the speckle size is also independent of the distance between the plane of the scattering spot and the measurement plane, and independent of the wavelength of the light used. This is true only under a restricted set of conditions, as we shall explain. This phenomenon is closely related to the assumption of Eq. (4-91), and in fact occurs when it is violated. The regime was first recognized by M. Giglio and his co-workers (see [61] and [21]).

Unlike the case of measurement in the focal plane, the free-space geometry does not map angles in the scattering plane into spatial positions in the observation region. As a consequence, as illustrated in Fig. 4.14, the angular cone of light contributed by a particular correlation area on the scattering surface ends up in the observation plane centered at the geometrical projection of its original location, the projection being parallel to the optical axis. Thus the broad intensity distributions contributed by different correlation areas are not centered on the same observation point. This phenomenon, when it is significant, destroys the Fourier transform relation between the field correlation function in the scattering plane and the average intensity distribution in the observation plane. An exception occurs when Eq. (4-91) holds, for in this case the angular scatter from a correlation area is so broad, and the scattering spot size is sufficiently small, that the difference of the center locations of the contributed intensity distributions in the observation region is a small fraction of the width of the average intensity distribution. In such a case, the Fourier transform relation between μ_a and the average observed intensity distribution is valid.

For the subject at hand in this subsection, we want to consider the case when Eq. (4-91) is violated, namely when

$$z < 2\frac{L_s L_c}{\lambda}. \qquad (4\text{-}94)$$

4.5 Dependence of Speckle on Scatterer Microstructure

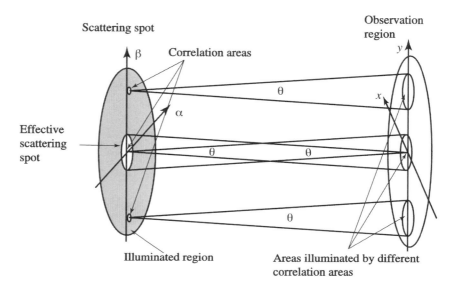

Figure 4.14 Illustration of situation under which the effective spot size is smaller than the actual spot size. θ is the angular spread of the light from a single correlation cell, and also the angle subtended by the effective spot size.

Suppose that the correlation area of the scattered field, that is, the equivalent area of the function μ_a, is sufficiently large that the angular width of the Fourier transform of μ_a is relatively narrow (see top Fourier transform relation in Fig. 4.13). In such a case, when the scattering spot is observed from a point in the observation plane, outer portions of that spot may not be visible, because the light from those portions has been scattered too narrowly to reach the observation point. Thus under these conditions there exists an *effective* spot size that is smaller than the actual spot size, and the speckle correlation in the observation plane will be broader than would otherwise be predicted. If the distance between the scattering surface and the observation plane is increased, the effective size of the scattering spot will be increased, exactly canceling what would otherwise be a growth of the speckle size, leaving the speckle size independent of distance. Similar cancellation effects occur when the wavelength is changed. Under the conditions hypothesized here, there is, in effect, a minimum size of the speckle, regardless of how large the scattering spot is made or how small the observation distance might be.

Considering the Fourier transform relations involved, some thought shows that *the minimum size of the speckle field correlation area is exactly the same as the size of the correlation area of the scattered fields at the rough surface*. This condition will hold only when the area represented on the right of the top part of Fig. 4.13 is smaller than the area represented on the left of the bottom part of that figure. In words, this regime requires that the area illuminated in the observation plane by the scattered wave from a single correlation cell be smaller than the area illuminated in the scattering plane by the incident light. In such a case, the effective spot size will be smaller than the actual spot size. In a one-dimensional approximation, Eq. (4-94) implies that this condition will hold only if

$$\frac{z}{L_s} < 2\frac{L_c}{\lambda}. \tag{4-95}$$

It is clear immediately that if the correlation width of the scattered wave is of the order of λ (as it often is in practice), the effects described here will be observed only when the distance z is smaller than about twice the actual spot size, $2L_s$.

This regime is interesting because it is unlike the usual regime, in which speckle size decreases with spot size and increases with distance. While this effect can be observed in practice, it is not the usual case encountered.

4.5.4 Relation between the Correlation Areas of the Scattered Wave and the Surface Height Fluctuations—Surface Scattering

We again assume a free-space reflection geometry, as shown in Fig. 4.15. The surface of the scattering spot is assumed to be rough and wide-sense stationary (i.e. its correlation function depends only on the differences of measurement coordinates). A typ-

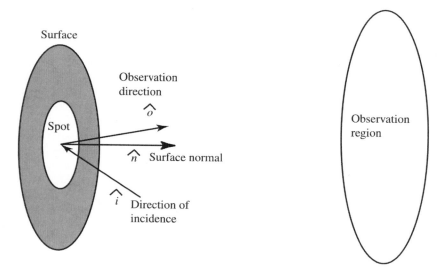

Figure 4.15 Geometry for consideration of surface effects. $\hat{n}$ represents a unit vector pointing in the direction of the average surface normal, $\hat{\imath}$ represents a unit vector pointing in the direction of incidence of the illumination, and $\hat{o}$ represents a unit vector pointing in the direction of the observation point in the plane on the right.

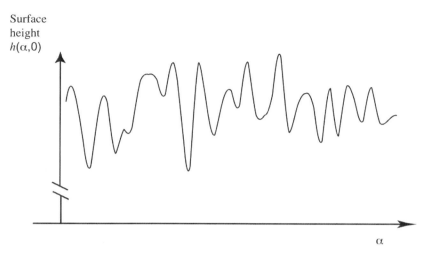

Figure 4.16 Example of a cross-section of a typical random surface height fluctuation.

ical cross section of the surface height $h(\alpha,\beta)$ at the microscale is shown in Fig. 4.16, with an arbitrary standard deviation of surface height fluctuations for the moment. The relationship between these height variations and the amplitude variations of the scattered wave is in general an extremely complex one, influenced by variations of surface slope on reflection, multiple scattering, shadowing, and the effects of propagation from the actual surface to the (α,β) plane just above the surface. For purposes of analysis here, we adopt an oversimplified model, which nonetheless gives some physical insight into the relationship between the surface height fluctuations and the fluctuations of the scattered wave. Let the scattered complex amplitude just above the surface, $\mathbf{a}(\alpha,\beta)$, be related to the surface-height by a purely geometrical approximation that assigns a phase to $\mathbf{a}$ that is the phase delay associated with propagating to the surface and scattering from the surface. Thus

$$\mathbf{a}(\alpha,\beta) = \mathbf{r}\,\mathbf{S}(\alpha,\beta)e^{j\phi(\alpha,\beta)} \tag{4-96}$$

with

$$\phi(\alpha,\beta) = \frac{2\pi}{\lambda}(-\hat{i}\cdot\hat{n} + \hat{o}\cdot\hat{n})h(\alpha,\beta), \tag{4-97}$$

where $\mathbf{r}$ is the average amplitude reflectivity of the surface, $\mathbf{S}(\alpha,\beta)$ represents the complex amplitude across the spot illumination,[10] and the dot product $\hat{p}\cdot\hat{q}$

---

10. Note that, in addition to defining the area and amplitude weighting across the scattering spot, $\mathbf{S}$ contains a term $e^{j\frac{2\pi}{\lambda}[(-\hat{i}\cdot\hat{\alpha})\alpha + (-\hat{i}\cdot\hat{\beta})\beta]}$ representing the complex field associated with the illumination wave.

represents the cosine of the angle between the unit vectors $\hat{p}$ and $\hat{q}$. The dot products in the expression for ϕ take account of the fact that the surface height variations are foreshortened when illuminated from or observed from an angle that departs from the surface normal.[11]

The variance σ_ϕ^2 of the phase shifts in Eq. (4-97) is related to the variance σ_h^2 of the surface-height fluctuations through

$$\sigma_\phi^2 = \left[\frac{2\pi}{\lambda}(-\hat{i}\cdot\hat{n} + \hat{o}\cdot\hat{n})\right]^2 \sigma_h^2, \tag{4-98}$$

and the correlation function $\Gamma_\phi(\Delta\alpha, \Delta\beta)$ of the phase shifts is related to the normalized correlation function $\mu_h(\Delta\alpha, \Delta\beta)$ of the surface-height fluctuations through

$$\Gamma_\phi(\Delta\alpha, \Delta\beta) = \sigma_\phi^2 \mu_h(\Delta\alpha, \Delta\beta). \tag{4-99}$$

The autocorrelation function of the complex field $\mathbf{a}(\alpha, \beta)$ is given by

$$\boldsymbol{\Gamma}_\mathbf{a}(\alpha_1,\beta_1;\alpha_2,\beta_2) = \overline{\mathbf{a}(\alpha_1,\beta_1)\mathbf{a}^*(\alpha_2,\beta_2)} = |\mathbf{r}|^2 \mathbf{S}(\alpha_1,\beta_1)\mathbf{S}^*(\alpha_2,\beta_2)\,\overline{\exp[j(\phi_1-\phi_2)]}.$$

Provided the variations of $\mathbf{S}$ are much coarser than the correlation width of the scattered field, this expression takes the form

$$\boldsymbol{\Gamma}_\mathbf{a}(\alpha_1,\beta_1;\alpha_2,\beta_2) = |\mathbf{r}|^2 I_{\text{inc}}(\alpha_2,\beta_2)\boldsymbol{\mu}_\mathbf{a}(\Delta\alpha,\Delta\beta) \tag{4-100}$$

where I_{inc} is the intensity incident on the scattering area, and the normalized autocorrelation function of the field becomes

$$\boldsymbol{\mu}_\mathbf{a}(\Delta\alpha, \Delta\beta) = \overline{\exp[j(\phi_1-\phi_2)]}. \tag{4-101}$$

We recognize the above equation as being closely related to the characteristic function of the random variable ϕ. In fact we can write

$$\boldsymbol{\mu}_\mathbf{a}(\Delta\alpha, \Delta\beta) = \mathbf{M}_{\phi_1,\phi_2}(1,-1) = \mathbf{M}_{\Delta\phi}(1), \tag{4-102}$$

where $\mathbf{M}_{\phi_1,\phi_2}(\omega_1,\omega_2)$ is the joint characteristic function of ϕ_1 and ϕ_2, while $\mathbf{M}_{\Delta\phi}(\omega)$ is the characteristic function of the random variable $\Delta\phi = \phi_1 - \phi_2$.

We can make little further progress unless an assumption about the statistics of the random phase $\phi(\alpha,\beta)$ (or equivalently the random surface height $h(\alpha,\beta)$) is made. The most common assumption is that ϕ is a Gaussian random variable, in which case the phase difference $\Delta\phi$ is also Gaussian, regardless of the correlation between ϕ_1 and ϕ_2. The characteristic function of this phase difference is

$$\mathbf{M}_{\Delta\phi}(\omega) = \exp\left(-\frac{\sigma_{\Delta\phi}^2 \omega^2}{2}\right)$$

and when evaluated at $\omega = 1$, it becomes

---

11. We have assumed that the incident field has a near-planar wavefront, and that the scattering spot is small enough and far enough away from the observation region that a single direction of observation is a reasonable approximation.

4.5 Dependence of Speckle on Scatterer Microstructure

$$\mathbf{M}_{\Delta\phi}(1) = \exp\left(-\frac{\sigma_{\Delta\phi}^2}{2}\right) = \exp\left(-\frac{\overline{\Delta\phi^2}}{2}\right) = \exp[-\sigma_\phi^2(1 - \mu_\phi(\Delta\alpha, \Delta\beta)].$$

Equation (4-99) shows that $\mu_\phi(\Delta\alpha, \Delta\beta) = \mu_h(\Delta\alpha, \Delta\beta)$, so we obtain the result

$$\boldsymbol{\mu}_{\mathbf{a}}(\Delta\alpha, \Delta\beta) = \exp[-\sigma_\phi^2(1 - \mu_h(\Delta\alpha, \Delta\beta))]. \tag{4-103}$$

Recall that the phase variance σ_ϕ^2 depends on the surface height variance σ_h^2 through Eq. (4-98).

Now we make a convenient mathematical assumption about the form of the surface height correlation function μ_h, namely let

$$\mu_h(\Delta\alpha, \Delta\beta) = \exp\left[-\left(\frac{r}{r_c}\right)^2\right] \tag{4-104}$$

where $r = \sqrt{\Delta\alpha^2 + \Delta\beta^2}$ and r_c is the radius at which the normalized surface correlation falls to $1/e$. We then have

$$\boldsymbol{\mu}_{\mathbf{a}}(r) = \exp[-\sigma_\phi^2(1 - e^{-(r/r_c)^2})]. \tag{4-105}$$

Figure 4.17 illustrates the normalized autocorrelation function of the surface height fluctuations and the normalized autocorrelation function of the scattered field when σ_ϕ takes on several different values. Note that for small values of σ_ϕ, the amplitude autocorrelation approaches an asymptote as r/r_c grows large. In fact, the value of that asymptote is easily seen to be $\exp(-\sigma_\phi^2)$, a value that rapidly approaches zero as σ_ϕ^2 grows. The asymptote, when significant, is an indication that there exists a non-negligible specular reflection of the incident wave, as can be seen by noting the flat correlation function when $\sigma_\phi = 0$, in which only a specular reflection exists. The fraction of intensity in the specular reflection is given by $\exp(-2\sigma_\phi^2)$.

To study the nonspecular component of the reflected wave, it is helpful to subtract out the asymptote of the autocorrelation function, and renormalize to value unity at the origin, yielding

$$\boldsymbol{\mu}_{\mathbf{a}}'(r) = \frac{\exp[-\sigma_\phi^2(1 - e^{-(r/r_c)^2})] - \exp(-\sigma_\phi^2)}{1 - \exp(-\sigma_\phi^2)}. \tag{4-106}$$

The coherence area of this nonspecular component can be found by evaluating

$$\mathcal{A}_{\mathbf{a}} = 2\pi \int_0^\infty r\boldsymbol{\mu}_{\mathbf{a}}'(r)\, dr, \tag{4-107}$$

which, for the case at hand, takes the form

$$\mathcal{A}_{\mathbf{a}} = \frac{2\pi e^{-\sigma_\phi^2}}{1 - e^{-\sigma_\phi^2}} \int_0^\infty r[\exp(\sigma_\phi^2 e^{-(r/r_c)^2}) - 1]\, dr.$$

To evaluate this integral, we expand the exp term in a power series and then interchange orders of integration and summation,

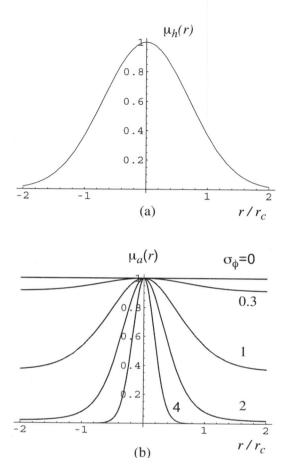

Figure 4.17 Normalized autocorrelation functions for (a) the surface height fluctuations, and (b) the field just above the rough surface.

$$\mathcal{A}_\mathbf{a} = \frac{2\pi e^{-\sigma_\phi^2}}{1 - e^{-\sigma_\phi^2}} \int_0^\infty r \left[\sum_{k=0}^\infty \frac{\sigma_\phi^{2k} e^{-k(r/r_c)^2}}{k!} - 1 \right] dr$$

$$= \frac{2\pi e^{-\sigma_\phi^2}}{1 - e^{-\sigma_\phi^2}} \sum_{k=1}^\infty \frac{\sigma_\phi^{2k}}{k!} \int_0^\infty r e^{-k(r/r_c)^2} dr. \qquad (4\text{-}108)$$

The last integral above equals $r_c^2/2k$ and therefore

$$\mathcal{A}_\mathbf{a} = \frac{\pi r_c^2 e^{-\sigma_\phi^2}}{1 - e^{-\sigma_\phi^2}} \sum_{k=1}^\infty \frac{\sigma_\phi^{2k}}{k \cdot k!}.$$

4.5 Dependence of Speckle on Scatterer Microstructure

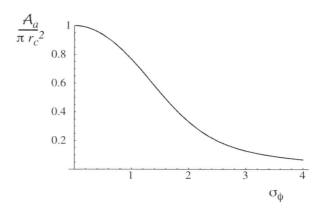

Figure 4.18 Normalized wave coherence area vs. standard deviation of the phase.

The series can be summed [1] with the result

$$\mathcal{A}_\mathbf{a} = \frac{\pi r_c^2 e^{-\sigma_\phi^2}}{1 - e^{-\sigma_\phi^2}} [\text{Ei}(\sigma_\phi^2) - \mathcal{E} - \ln(\sigma_\phi^2)], \tag{4-109}$$

where $\text{Ei}(x)$ represents the exponential integral and $\mathcal{E}$ is Euler's constant. Figure 4.18 shows a plot of the wave correlation area $\mathcal{A}_\mathbf{a}$ normalized by the surface-height correlation area πr_c^2 vs. the standard deviation σ_ϕ of the phase fluctuations. Note that when the standard deviation of the phase reaches the value π, the wave correlation area is about 1/10 of the surface-height correlation area. For normal illumination and normal observation direction, this corresponds to a surface-height standard deviation of about $\lambda/4$. The reason for the shrinking of the wave correlation area as the standard deviation of the phase increases lies in the fact that when the phase fluctuations begin to exceed 2π radians, they wrap back into the interval $(0, 2\pi)$. The correlation area of wrapped phase function, i.e. of $\phi(\alpha, \beta)$ modulo 2π, is smaller than the correlation area of the unwrapped phase. Figure 4.19 shows a 1-D unwrapped phase distribution that varies over several times 2π, and a version of the same phase distribution that is wrapped on $(0, 2\pi)$. It is easily seen that the correlation structure is finer in the wrapped version than in the unwrapped version.

In closing this section, we note that our goal was to explore the relationship between the surface-height correlation function and the wave-amplitude correlation function at the scattering surface. We have accomplished this goal, subject to some simplifying assumptions about the statistical properties of the surface height fluctuations. We have found that the wave-amplitude correlation function depends in a complex way on *both* the surface-height correlation function and the surface-height standard deviation. We will return to this subject in Section 8.5.4.

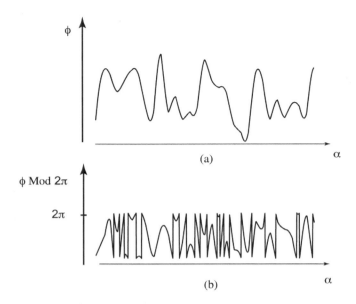

Figure 4.19 One-dimensional slices of (a) a phase function that fluctuates over several times 2π, and (b) the same phase function wrapped into the interval $(0, 2\pi)$.

4.5.5 Dependence of Speckle Contrast on Surface Roughness—Surface Scattering

We consider now the dependence of speckle contrast on surface roughness for the particular imaging geometry shown in Fig. 4.20. Surface scattering is again assumed. This geometry is chosen because it coincides with the experimental geometry used by Fuji and Asakura [54] in measurements of the quantities to be analyzed here.

Because of the normal incidence of the light on the surface, the phase and surface height are related by (cf. Eq. (4-97))

$$\phi(\alpha,\beta) = \frac{2\pi}{\lambda}[1 + (\hat{o} \cdot \hat{n})]h(\alpha,\beta), \qquad (4\text{-}110)$$

where again $\hat{n}$ is the outward surface normal and $\hat{o}$ is a unit vector pointing in the direction of the observation point, which, for small angles subtended by the first lens, is determined by the position of the image point of interest in the image plane. For an image point near the optical axis, the right-hand side of this equation reduces to $(4\pi/\lambda)h(\alpha,\beta)$. Under these same conditions, the variance of the phase fluctuations is related to the variance of the surface height fluctuations by

$$\sigma_\phi^2 = \left(\frac{4\pi}{\lambda}\right)^2 \sigma_h^2. \qquad (4\text{-}111)$$

4.5 Dependence of Speckle on Scatterer Microstructure

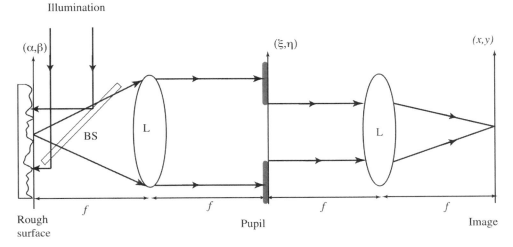

Figure 4.20 Side view of imaging geometry for study of speckle contrast. The lenses L have focal lengths f, and BS represents a beam splitter.

Since speckle contrast is the quantity of concern, we can, without loss of generality, assume the average intensity reflectivity and the incident intensity to be unity. The amplitude of the field immediately to the right of the rough surface is therefore given by

$$\mathbf{a}(\alpha, \beta) = \exp[j\phi(\alpha, \beta)] = \exp\left[j\frac{4\pi}{\lambda}h(\alpha, \beta)\right]. \tag{4-112}$$

Since the focal lengths of the two lenses are identical, the magnification of the imaging system is unity, and (neglecting image inversion) the field at image coordinates (x, y) can be written as a convolution,

$$\mathbf{A}(x, y) = \iint_{-\infty}^{\infty} \mathbf{k}(x - \alpha, y - \beta)\mathbf{a}(\alpha, \beta)\, d\alpha\, d\beta, \tag{4-113}$$

where the point-spread function $\mathbf{k}(\alpha, \beta)$ is related to the pupil function $\mathbf{P}(\xi, \eta)$ of the imaging system through (see [71], p. 114)

$$\mathbf{k}(\alpha, \beta) = \frac{1}{(\lambda f)^2} \iint_{-\infty}^{\infty} \mathbf{P}(\xi, \eta) \exp\left[j\frac{2\pi}{\lambda f}(\alpha\xi + \beta\eta)\right] d\xi\, d\eta. \tag{4-114}$$

The system is assumed to be free of aberrations, and therefore $\mathbf{P}(\xi, \eta) = P(\xi, \eta)$ is real-valued; we also assume that P is unity at the origin. Finally, for pupils with either circular symmetry or separable symmetry in (ξ, η) about the optical axis, the point-spread function **k** is real-valued, an assumption we adopt at this point.

If a particular image point (x, y) is chosen, then the real-valued point-spread function $k(x - \alpha, y - \beta)$ should be thought of as a weighting function on the (α, β) plane that weights the scattered contributions from the surface to that particular image point. If the pupil represented by P is stopped down, then the weighting function broadens and more of the surface contributes scattered light to the image point in question. Likewise if the pupil is widened, then the weighting function becomes narrower, and less of the scattering surface influences the intensity at the chosen image point. In effect, by varying the pupil size, we can control how many correlation areas of the surface contribute to the observed light intensity.

At the given image point, the field consists of independent contributions from several or many different correlation areas of the wavefront leaving the scattering surface. The number N of such contributions can be approximated as the ratio of the equivalent area of an imaging resolution spot $(\mathcal{A}_k)$ on the object to the correlation area $(\mathcal{A}_a)$ of the wave scattered at the surface,

$$N \approx \begin{cases} \mathcal{A}_k/\mathcal{A}_\mathbf{a} & \mathcal{A}_k \geq \mathcal{A}_\mathbf{a} \\ 1 & \mathcal{A}_k < \mathcal{A}_\mathbf{a}. \end{cases} \tag{4-115}$$

Here $\mathcal{A}_k$ is the equivalent area of the point-spread function $k(\alpha, \beta)$ on the object, and $\mathcal{A}_\mathbf{a}$ is the equivalent area of the amplitude correlation function (after the specular reflection has been removed, cf. Eq. (4-106)), that is,

$$\mathcal{A}_k = \frac{\iint\limits_{-\infty}^{\infty} k(\alpha, \beta)\, d\alpha\, d\beta}{k(0,0)} = \frac{P(0,0)}{\frac{\mathcal{A}_p}{\lambda^2 f^2}} = \frac{\lambda^2 f^2}{\mathcal{A}_p}$$

$$\mathcal{A}_\mathbf{a} = \iint\limits_{-\infty}^{\infty} \mu'_\mathbf{a}(\Delta\alpha, \Delta\beta)\, d\Delta\alpha\, d\Delta\beta, \tag{4-116}$$

where we have used the fact that $\mu'_\mathbf{a}(0,0) = 1$, and for simplicity we have taken the pupil function to be a clear opening $(P(0,0) = 1)$ of area $\mathcal{A}_p$. The quantity $\mathcal{A}_\mathbf{a}$ has already been calculated in the previous section, Eq. (4-109), for Gaussian surface-height fluctuations, and, in the case of a circularly symmetric Gaussian-shaped surface-height correlation function, was found to be given by

$$\mathcal{A}_\mathbf{a} = \frac{\pi r_c^2 e^{-\sigma_\phi^2}}{1 - e^{-\sigma_\phi^2}} [\mathrm{Ei}(\sigma_\phi^2) - \mathcal{E} - \ln(\sigma_\phi^2)],$$

where, again, $\mathrm{Ei}(x)$ represents the exponential integral function, and $\mathcal{E}$ is Euler's constant. Taking the ratio of $\mathcal{A}_k$ to $\mathcal{A}_\mathbf{a}$, we find that

$$N = \frac{N_0(e^{\sigma_\phi^2} - 1)}{\mathrm{Ei}(\sigma_\phi^2) - \mathcal{E} - \ln(\sigma_\phi^2)} \tag{4-117}$$

where N_0 represents the number of *surface-height* correlation areas lying within the equivalent area of the point-spread function,

$$N_0 = \frac{\mathcal{A}_k}{\pi r_c^2}.$$

With these definitions, we can now apply the results of Section 3.5 to the problem at hand. From that section, in particular Eq. (3-94), the contrast is known to be given by

$$C = \sqrt{\frac{8(N-1)\left[N-1+\cosh\left(\left(4\pi\frac{\sigma_h}{\lambda}\right)^2\right)\right]\sinh^2\left(\left(4\pi\frac{\sigma_h}{\lambda}\right)^2/2\right)}{N(N-1+e^{\left(4\pi\frac{\sigma_h}{\lambda}\right)^2})^2}}, \quad (4\text{-}118)$$

where we have substituted $\sigma_\phi = (4\pi/\lambda)\sigma_h$. Now substituting Eq. (4-117) into the equation for C, we find an expression for C that depends only on N_0 and σ_h/λ.

These results are illustrated in Fig. 4.21, which plots contrast vs. σ_h/λ for various values of N_0, and contrast vs. N_0 for various values of σ_h/λ. From part (a) of this figure, we note two expected results: first, for a perfectly flat surface ($\sigma_h/\lambda = 0$), speckle contrast vanishes, and second, as the surface height fluctuations rise (σ_h/λ grows), eventually the contrast saturates at unity, as expected for fully developed speckle. We also note that the amount of surface roughness required to cause saturation of the contrast at unity depends on the value of N_0, with more fluctuations required the larger N_0. This phenomenon is associated with the different rates of growth of the specular and scattered (or "diffuse") components of light, as explained with respect to part (b) of the figure in what follows.

In part (b) we see the curious behavior that, for small values of surface roughness, the contrast first grows with the parameter N_0, then falls towards zero. The

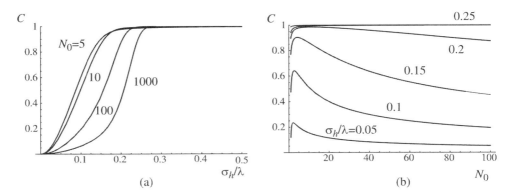

Figure 4.21 Contrast C vs. (a) σ_h/λ for various values of N_0, and (b) contrast vs. N_0 for various values of σ_h/λ.

explanation for this phenomenon is related to that presented following Fig. 3.15, and lies in the different dependencies of the specular and diffuse components of light on the value of N_0. In particular, for the geometry analyzed here, the specular component of reflected light always appears as a small spot on the optical axis in the focal plane of the first lens, while the diffusely scattered light fills the pupil. To change N_0, we change the size of the pupil stop, making it smaller to increase N_0. Since the specular component is always confined to the region of the optical axis, it is unaffected as we reduce the pupil diameter, at least until the pupil is extremely small. On the other hand, as we decrease the pupil, the average intensity of the transmitted scattered light is reduced in proportion to $\mathcal{A}_p$, the area of the pupil. Thus for small values of σ_h/λ, an increase of N_0 from 1 (for which there is no speckle) and beyond causes a rise in speckle contrast from zero, while eventually as N_0 increases further, the value of σ_I drops while the value of $\bar{I}$ stays constant, causing the contrast to drop. This explains the behavior seen in part (b) for contrast vs. N_0 for small σ_h/λ. Eventually, when the pupil size is small enough to affect the specularly-reflected light as well as the scattered light, this drop in C with N_0 will not continue.

Although it is perhaps not obvious in the previous analysis, the statistics of the scattered light are decidedly *noncircular* for the geometry we have analyzed; i.e. the variance of the real part of the observed field $A(x, y)$ is not equal to the variance of the imaginary part of that field. More discussion of this fact can be found in [68]. This noncircular behavior is unique to the image plane, where we are forming a low-pass filtered image of a pure phase distribution. In planes that are not near the image plane, circularity of the statistics is restored, due primarily to the fact that the point-spread function $k(\alpha, \beta)$ becomes complex-valued, rather than real-valued as it is in the image plane for the particular imaging geometry analyzed here.

4.5.6 Properties of Speckle Resulting from Volume Scattering

The theory of speckle that is generated by volume scattering is less developed and in many ways more difficult than the corresponding theory for surface scattering. Figure 4.22 shows an oversimplified and exaggerated view of what happens when light scatters within the volume of a dielectric slab containing a dense random set of tiny dielectric inhomogeneities. Three light rays entering the medium are shown. If the particle density is sufficiently high, each incident ray travels a different path due to multiple scattering. The path shown entering the farthest to the left in this illustration travels a particularly long path due to total internal reflection at the top slab–air interface. At each scattering location, there is generally also a finite probability of absorption. Some rays will escape from the scattering medium at angles not entering the collecting lens, and therefore will not contribute to the measurement. Rays that undergo many scattering events will have a large probability of being absorbed during one of those events.

Several general observations are possible. First, multiple scattering lengthens the spread of different path-lengths (and therefore time delays) involved in propagation

4.5 Dependence of Speckle on Scatterer Microstructure

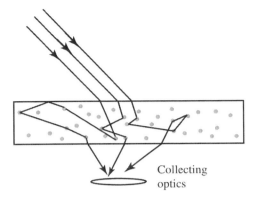

Figure 4.22 Three paths through a volume scattering medium showing multiple scattering.

from the source to the detector. This spread can be far greater than is the case for surface scattering from a moderately rough surface. As discussed in the following chapter, the result can be a medium for which the speckle is far more sensitive to frequency or wavelength change than would be the case for surface scattering. Of course if the delay times are comparable with or exceed the coherence time of the source, a reduction in speckle contrast can be expected.

Second, the size of a speckle observed in the focal plane of the collecting lens is influenced primarily by the angular spread of the light striking that plane, and the angular spread depends on many different factors. Not only can the geometrical properties of the measurement system and the dimension of the spot incident on the volume scatterer influence speckle size, but so too can the spread of the light beyond those boundaries due to internal multiple scattering. Which of these factors plays a dominant role depends on whether the measurement is in a free-space geometry or an imaging geometry. In a free-space geometry, any phenomenon internal to the medium that causes the exiting spot of light to be larger than the entering spot will diminish the dimensions of the speckle in the observation plane. The size of the index inhomogeneities compared with a wavelength can influence the angular spread of the scattered light (smaller particles scatter more widely[12]), and therefore can influence the spread of both the time delay and the angular extent of the exiting light. In an imaging geometry, the angular spread of the arriving light depends primarily on the numerical aperture of the imaging system, and is generally not influenced by the size of the scattering spot.

---

12. In Ref. [62]), the authors show that if speckle size is measured in the near field of the scattering spot under the conditions described in Section 4.5.3, then the average size of the scatterers can be determined from the speckle size.

Several papers have been published on the characteristics of speckle generated by volume scattering under various scattering conditions (see, for example, [127], [62], [160] and [161]). The mathematical formalism that should be used to analyze the scattering depends on the density of the scatterers. The latter reference assumes that the particle density is large enough to allow a diffusion approximation for the spread of light in the scattering medium, and uses an image method to satisfy the boundary conditions. Using these methods, the average time-spread of a δ-function light pulse is calculated numerically.

Let (x, y) be the transverse coordinates parallel to the front and back surfaces of the scattering medium, and let z be the depth coordinate, with an origin taken at the surface where light enters the medium. The light exits the medium at the surface with $z = L$, L being the thickness of the medium. Now imagine a point (δ-function) source at coordinates $(x', 0, 0)$ being scanned across the surface in the x direction, and a point detector at $(x', 0, L)$ being scanned in synchronism with the source, again in the x direction. On a grid of points x', the time responses of the detector to unit impulse optical pulses are measured. If the medium is homogeneous, and the source and detector are always far from the edge of the sample, these experimental responses can be averaged to produce an average impulse response $\bar{I}(t)$ (this is the impulse response predicted by the diffusion analysis mentioned above). Then the probability density function governing the optical path lengths travelled by the light from input to output of the scattering medium can be written

$$p_l(l) = \frac{\bar{I}(l/c)}{\int_0^\infty \bar{I}(l/c)\, dl}. \tag{4-119}$$

The probability density function $p_l(l)$ plays the same role as the surface-height probability density function in the case of surface scattering, although its shape may be very different from that observed in the surface scattering case.

It has been assumed above that the time-resolution of the measurement system is sufficiently great to measure the very short impulse responses of the medium. If this is not the case, methods for measuring this quantity using correlations of speckle patterns obtained at three different optical frequencies [171] [172] using CW light have been developed.

Figure 4.23 shows three different calculated path-length probability density functions for a slab of thickness 3.6 cm, from Ref. [161]. Note that the path-length density functions peak at distances that are many times the physical thickness of the medium. In this figure, μ_a represents the absorption coefficient, and μ_s' represents a reduced scattering coefficient, with

$$\mu_s' = (1 - g)\mu_s,$$

where μ_s is the linear scattering coefficient and g is the mean cosine of the scattering angle. The longest tails of the probability density function result when the absorption coefficient is small and the reduced scattering coefficient is large.

We return to the subject of volume scattering when we consider the frequency (or wavelength) dependence of speckle patterns in Chapter 6.

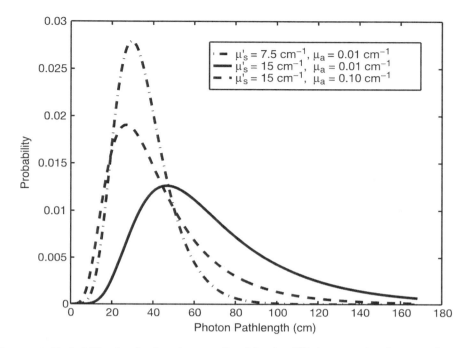

Figure 4.23 Probability density functions predicted by the diffusion equation for several sets of medium parameters (from [161]). The medium thickness is 3.6 cm, but the density functions extend to path lengths that are many times that thickness, due to multiple scattering. Reprinted with the permission of K.J. Webb and the Optical Society of America.

4.6 Statistics of Integrated and Blurred Speckle

We have studied the statistical properties of speckle at one or more points in space/time. However, in an experiment, speckle can not be measured at an ideal point, but rather is integrated over some finite area of a detector element. In addition, even with an extremely tiny detector, we may be measuring a moving speckle pattern in which case the measured quantity is again an integral over some portion of a speckle pattern. For these reasons, we turn attention now to the statistical properties of integrated speckle.

In the case of a uniform detector of finite size, let the measured intensity, represented here by W, be defined by

$$W = \frac{1}{A_D} \iint_{-\infty}^{\infty} D(x, y) I(x, y) \, dx \, dy, \qquad (4\text{-}120)$$

where $D(x, y)$ is a real and positive weighting function representing the distribution of the detector's photosensitivity over space, $\mathcal{A}_D = \iint\limits_{-\infty}^{\infty} D(x, y)\, dx\, dy$, and $I(x, y)$ is the intensity distribution of the speckle pattern that is being detected. For a uniformly sensitive photodetector, the weighting function has the form

$$D(x, y) = \begin{cases} 1 & \text{in the sensitive area} \\ 0 & \text{outside the sensitive area,} \end{cases} \quad (4\text{-}121)$$

and $\mathcal{A}_D$ is the area of the detector.

In the case of a moving speckle pattern that is detected by a very small detector, the measured intensity can again be represented in the form of the above equation, but the weighting function must be chosen appropriately for the situation. For example, if the speckle pattern is moving along the x-axis with uniform speed v, the weighting function takes the form

$$D(x, y) = \text{rect}\left(\frac{x}{vT}\right)\delta(y) \quad (4\text{-}122)$$

where T is the integration time and $\delta(y)$ is a delta function.

When a temporally changing speckle pattern is measured with a detector size that is small compared with a speckle and with an integration time T, then the integrated intensity of interest can be written

$$W = \frac{1}{T}\int_0^T I(t)\, dt = \frac{1}{T}\int_{-\infty}^{\infty} \text{rect}\left(\frac{t - T/2}{T}\right) I(t)\, dt, \quad (4\text{-}123)$$

where $I(t)$ represents the temporally changing intensity of the speckle pattern.

4.6.1 Mean and Variance of Integrated Speckle

In this section we consider first the case of spatial integration of speckle, and then present corresponding results for temporal integration.

Spatial Integration

Our first goal is to find exact expressions for the mean and variance of the measured intensity W. To this end, we write mean intensity as

$$\overline{W} = \frac{1}{\mathcal{A}_D} \iint\limits_{-\infty}^{\infty} D(x, y)\bar{I}\, dx\, dy = \bar{I}, \quad (4\text{-}124)$$

where the orders of averaging and integration have been interchanged, and $\bar{I}$ is the mean intensity of the incident speckle pattern (assumed independent of (x, y)). Thus the mean of the detected intensity is identical with the true mean of the speckle pattern.

4.6 Statistics of Integrated and Blurred Speckle

To find the variance of W, we first find its second moment,

$$\overline{W^2} = \frac{1}{\mathcal{A}_D^2} \iint_{-\infty}^{\infty} \iint_{-\infty}^{\infty} D(x_1, y_1) D(x_2, y_2) \overline{I(x_1, y_1) I(x_2, y_2)} \, dx_1 \, dy_1 \, dx_2 \, dy_2, \quad (4\text{-}125)$$

where, again, the orders of averaging and integration have been interchanged. For a wide-sense stationary speckle pattern, the average of the product of intensities depends only on the differences of coordinates $\Delta x = x_1 - x_2$ and $\Delta y = y_1 - y_2$, and the equation for the second moment can be directly reduced to

$$\overline{W^2} = \frac{1}{\mathcal{A}_D^2} \iint_{-\infty}^{\infty} K_D(\Delta x, \Delta y) \Gamma_I(\Delta x, \Delta y) \, d\Delta x \, d\Delta y, \quad (4\text{-}126)$$

where

$$K_D(\Delta x \Delta y) = \iint_{-\infty}^{\infty} D(x_1, y_1) D(x_1 - \Delta x, y_1 - \Delta y) \, dx_1 \, dy_1. \quad (4\text{-}127)$$

The function K_D is the deterministic autocorrelation function of the function $D(x, y)$, and the function Γ_I is the statistical autocorrelation function of the intensity $I(x, y)$.

We now invoke the circular complex Gaussian statistics of the fields of a fully developed speckle pattern to write

$$\Gamma_I(\Delta x, \Delta y) = \bar{I}^2 [1 + |\mu_\mathbf{A}(\Delta x, \Delta y)|^2]. \quad (4\text{-}128)$$

Substituting into Eq. (4-126), we obtain

$$\overline{W^2} = \frac{\bar{I}^2}{\mathcal{A}_D^2} \iint_{-\infty}^{\infty} K_D(\Delta x, \Delta y) \, d\Delta x \, d\Delta y$$

$$+ \frac{\bar{I}^2}{\mathcal{A}_D^2} \iint_{-\infty}^{\infty} K_D(\Delta x, \Delta y) |\mu_\mathbf{A}(\Delta x, \Delta y)|^2 \, d\Delta x \, d\Delta y. \quad (4\text{-}129)$$

The first term above reduces to $\bar{I}^2$, with the result that the variance of W can be written

$$\sigma_W^2 = \frac{\bar{I}^2}{\mathcal{A}_D^2} \iint_{-\infty}^{\infty} K_D(\Delta x, \Delta y) |\mu_\mathbf{A}(\Delta x, \Delta y)|^2 \, d\Delta x \, d\Delta y. \quad (4\text{-}130)$$

Quantities of considerable interest are the contrast and the rms signal-to-noise ratio of the detected speckle, defined by

$$C = \sigma_W / \overline{W}, \quad (S/N)_{\text{rms}} = \overline{W} / \sigma_W. \quad (4\text{-}131)$$

Defining

$$M = \left[\frac{1}{A_D^2} \iint_{-\infty}^{\infty} K_D(\Delta x, \Delta y)|\mu_A(\Delta x, \Delta y)|^2 \, d\Delta x \, d\Delta y\right]^{-1}, \quad (4\text{-}132)$$

we find

$$C = 1/\sqrt{M}, \quad (S/N)_{\text{rms}} = \sqrt{M}. \quad (4\text{-}133)$$

The parameter M is of fundamental importance in determining the statistics of the detected intensity, and for this reason we devote some considerable discussion to it. To gain some physical insight into its meaning, we first consider two limiting cases, one a detection area that is very large compared with the average size of a speckle, and the other the opposite—a detection area that is small compared with the average size of a speckle. In the first case, the function $K_D(\Delta x, \Delta y)$ is much wider than the function $|\mu_A(\Delta x, \Delta y)|^2$, and therefore we can factor $K_D(0,0)$ outside the integral, with the result

$$M \approx \left[\frac{K_D(0,0)}{A_D^2} \iint_{-\infty}^{\infty} |\mu_A(\Delta x, \Delta y)|^2 \, d\Delta x \, d\Delta y\right]^{-1}. \quad (4\text{-}134)$$

The double integral,

$$K_D(0,0) = \iint_{-\infty}^{\infty} D^2(x_1, y_1) \, dx_1 \, dx_2,$$

has the dimensions of area. It follows that $A_D^2/K_D(0,0)$ has the dimensions of area, and we accordingly name this the effective measurement area, A_m:

$$A_m = \frac{A_D^2}{\iint_{-\infty}^{\infty} D^2(x_1, y_1) \, dx_1 \, dx_2}.$$

Note that, in the usual case of a uniformly sensitive detector, D is unity inside the detector aperture and zero outside that aperture, in which case $A_m = A_D$. In addition, recognizing that $|\mu_A(\Delta x, \Delta y)|^2$ is the autocovariance function of intensity, we define

$$A_c = \iint_{-\infty}^{\infty} |\mu_A(\Delta x, \Delta y)|^2 \, d\Delta x \, d\Delta y \quad (4\text{-}135)$$

as the correlation area of the speckle intensity. With this definition, we see that the parameter M is given by

$$M \approx \mathcal{A}_m/\mathcal{A}_c \quad (\mathcal{A}_m \gg \mathcal{A}_c). \tag{4-136}$$

Thus in the limit of a large measurement area compared with the speckle size, the value of M increases in direct proportion to the ratio of $\mathcal{A}_m$ to $\mathcal{A}_c$, which can be interpreted as the average number of speckles influencing the measurement.

For the opposite case of the average size of a speckle much wider than the measurement area, we replace $|\mu_A|^2$ with unity, yielding

$$M \approx \left[\frac{1}{\mathcal{A}_D^2} \iint_{-\infty}^{\infty} K_D(\Delta x, \Delta y)\, d\Delta x\, d\Delta y \right]^{-1} = 1 \quad (\mathcal{A}_m \ll \mathcal{A}_c). \tag{4-137}$$

In this case, M can never fall below unity. This is consistent with the interpretation of M as being the average number of speckles influencing the measurement, for no matter how small the measurement aperture, a minimum of one speckle will influence the result.

Exact expressions for M can be found once the detector aperture function $D(x, y)$ and the intensity covariance function $|\mu_A|^2$ are specified. In the case of a square, uniform detector, $L \times L$ meters in size, and a Gaussian-shaped intensity pattern on the scattering spot, we have

$$\mu_A(\Delta x, \Delta y) = \exp\left[-\frac{\pi}{2\mathcal{A}_c}(\Delta x^2 + \Delta y^2) \right], \tag{4-138}$$

and we can perform the required integration to find

$$M = \left[\sqrt{\frac{\mathcal{A}_c}{\mathcal{A}_m}}\, \text{erf}\left(\sqrt{\frac{\pi \mathcal{A}_m}{\mathcal{A}_c}} \right) - \left(\frac{\mathcal{A}_c}{\pi \mathcal{A}_m} \right)\left(1 - \exp\left(-\frac{\pi \mathcal{A}_m}{\mathcal{A}_c} \right) \right) \right]^{-2}, \tag{4-139}$$

where $\text{erf}(x)$ is a standard error function. A closed form expression can also be found in the case of a Gaussian spot and a uniform circular detector,

$$M = (\mathcal{A}_m/\mathcal{A}_c)[1 - e^{-2(\mathcal{A}_m/\mathcal{A}_c)}[I_0(2\mathcal{A}_m/\mathcal{A}_c) + I_1(2\mathcal{A}_m/\mathcal{A}_c)]]^{-1}, \tag{4-140}$$

where $I_0(x)$ and $I_1(x)$ are modified Bessel functions of the first kind, orders zero and one. In other cases, M can be found by numerical integration. Figure 4.24 shows plots of M vs. the ratio $\mathcal{A}_m/\mathcal{A}_c$ for three different labeled cases. The curves on the right are simply blown-up versions of the curves on the left. It can be seen that, as predicted, for all cases, M approaches unity as $\mathcal{A}_m/\mathcal{A}_c$ becomes small, while M increases as $\mathcal{A}_m/\mathcal{A}_c$ when that ratio is large.

Temporal Integration
When spatially constant but time-varying speckle is integrated over time, results very similar to those derived above result (for a reference, see [70], Section 6.1). In particular, we can again use a similar analysis, this time with respect to the temporal fluctuations of intensity. Again we find a parameter M representing the number of degrees of freedom. By analogy with the results obtained for spatial integration, the number of degrees of freedom within the integrated speckle is

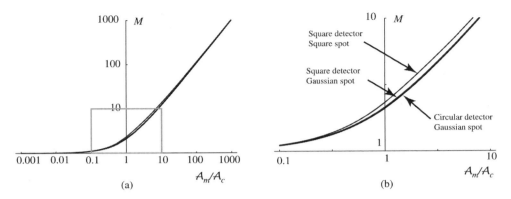

Figure 4.24 Plots of M vs. $\mathcal{A}_m/\mathcal{A}_c$ for various combinations of spot intensity distribution and detector shape. Part (b) of the figure is a magnified version of the portion of the curve lying in the box in part (a).

$$M = \left[\frac{\left(\int_{-\infty}^{\infty} P_T(\tau) \, d\tau \right)^2}{\int_{-\infty}^{\infty} K_T(\tau) |\boldsymbol{\mu}_A(\tau)|^2 \, d\tau} \right], \tag{4-141}$$

a choice that will result in the mean and variance of the integrated speckle being exactly correct. Here $P_T(t)$ is the integration window weighting function and $K_T(\tau)$ is the autocorrelation function of $P_T(t)$; that is, the integrated intensity is

$$W = \int_{-\infty}^{\infty} P_T(t) I(t) \, dt, \tag{4-142}$$

and $K_T(\tau)$ is

$$K_T(\tau) = \int_{-\infty}^{\infty} P_T(t) P_T(t - \tau) \, dt. \tag{4-143}$$

The contrast C of the integrated speckle is given by

$$C = \sqrt{\frac{1}{M}}. \tag{4-144}$$

For a uniform integration window $P_T(t) = \text{rect}\left(\frac{t - T/2}{T}\right)$, the result becomes

$$C = \left[\frac{2}{T} \int_0^T \left(1 - \frac{\tau}{T} \right) |\boldsymbol{\mu}_A(\tau)|^2 \, d\tau \right]^{1/2}. \tag{4-145}$$

Now we have completed our exploration of the exact moments and contrast of integrated and blurred speckle patterns. We turn now to a formalism that provides excellent approximations to the actual probability density functions of such measurements.

4.6.2 Approximate Result for the Probability Density Function of Integrated Intensity

While we have successfully found exact expressions for the moments and contrast of integrated intensity, in many problems a more complete description of the statistics of this quantity is needed. Accordingly in this section we find some excellent approximations to the first-order probability density function of integrated intensity. This approximation originated with S.O. Rice in his analysis of the properties of Gaussian noise [132], was used by L. Mandel in his analysis of the time statistics of the intensity of thermal light [105], and was first applied to two-dimensional speckle by Goodman [66].

The key approximation made is to replace the continuous speckle intensity pattern falling on the detector by a two-dimensional "boxcar" function, consisting of m contiguous rectangular pillboxes with different heights. A single pillbox can be thought of as a single correlation area of the speckle pattern, and within that correlation area we assume that the intensity takes on a constant value. Different pillboxes have different values of intensity. Indeed, we assume that the intensities present in the various pillboxes are all statistically independent and, for fully developed polarized speckle, we assume that the values of the intensities of the pillboxes all obey negative-exponential densities with identical mean intensities. Figure 4.25 shows a slice through the 2D intensity pattern and the approximate intensity pattern it is replaced with.

The integrated intensity may be approximated as the normalized volume under the boxcar function,

$$W = \frac{1}{\mathcal{A}_D} \iint_{-\infty}^{\infty} D(x,y) I(x,y) \, dx \, dy \approx \frac{1}{\mathcal{A}_D} \sum_{k=1}^{m} \mathcal{A}_0 I_k, \qquad (4\text{-}146)$$

where $\mathcal{A}_0$ is the area of the base of all of the pillboxes (the same for all k) and I_k is the height of the kth pillbox. The sum is assumed to run over the entire 2D array of m pillboxes.

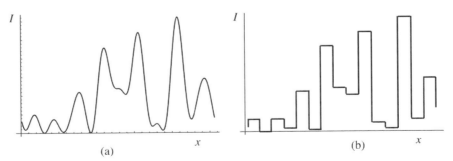

Figure 4.25 One dimensional slice through the intensity distribution falling on the detector: (a) actual, and (b) boxcar approximation.

Given that the intensities I_k for each pillbox obey independent negative-exponential statistics, with identical means $\bar{I}$, the characteristic function of the approximate value of W must be

$$M_W(\omega) = \prod_{k=1}^{m} \frac{1}{1 - j\omega \frac{A_0}{A_D} \overline{W}} = \left[\frac{1}{1 - j\omega \frac{\overline{W}}{m}} \right]^m, \qquad (4\text{-}147)$$

where we have used the fact that $A_0/A_D = 1/m$. The probability density function corresponding to this characteristic function is the gamma density,

$$p(W) = \begin{cases} \dfrac{(m/\bar{I})^m W^{m-1} \exp(-mW/\bar{I})}{\Gamma(m)} & W \geq 0 \\ 0 & \text{otherwise,} \end{cases} \qquad (4\text{-}148)$$

where $\Gamma(m)$ is a gamma function with argument m.

The gamma density function has two parameters, m and $\overline{W}$. A reasonable strategy is to choose those two parameters to make the approximate density fit as closely as possible to the exact density, whatever it may be. Towards that end, we choose m and $\overline{W}$ such that the mean and variance of the approximate distribution are exactly equal to the true mean and variance calculated in the previous section. This is accomplished if we set $m = M$ (where M is the same parameter M calculated in the previous section) and let $\overline{W}$ represent the true mean of the exact distribution, again as calculated in the previous section. Thus the form of the gamma density approximation is changed slightly to read

$$p(W) = \begin{cases} \dfrac{(M/\overline{W})^M W^{M-1} \exp(-MW/\overline{W})}{\Gamma(M)} & W \geq 0 \\ 0 & \text{otherwise.} \end{cases} \qquad (4\text{-}149)$$

Note that the parameter M calculated in the previous section need not be an integer, so the quasi-physical idea of an integer number of pillboxes in the boxcar approximation should be abandoned. Whatever value of M is calculated by the methods of the previous section, integer or non-integer, it should be used in the gamma density function above. There is, of course, no guarantee that choosing the two parameters to make the mean and variance exactly correct is the best way to choose the parameters, but it is the simplest way and we shall use it here.

Plots of the gamma density function have already been presented in Fig. 3.12, and we will not repeat that figure here. It suffices to mention that for $M = 1$ the density function is the familiar exponential density, while as $M \to \infty$, the density function approaches a delta function centered on the mean value $W = \overline{W}$. For large but finite values of M, the Central Limit Theorem of statistics implies that the density function of integrated intensity approaches a Gaussian density with mean $\overline{W}$ and standard deviation $\overline{W}/\sqrt{M}$. It is worth noting that the nth moment of the gamma density with parameter M is easily shown to be

$$\overline{W^n} = \frac{\Gamma(M+n)}{\Gamma(M)} \left(\frac{\overline{W}}{M} \right)^n. \qquad (4\text{-}150)$$

Keeping in mind the picture of the detector area occupied by M different independent pillboxes, we will often refer to M as the "number of degrees of freedom" of the portion of the speckle pattern captured by the detector.

4.6.3 "Exact" Result for the Probability Density Function of Integrated Intensity

A mathematical formalism exists that allows the calculation of more accurate probability density functions for integrated speckle intensity in some cases. Often these solutions are called "exact" solutions, but as we shall see, there is always some form of approximation involved in the calculation. The method is known as the Karhunen–Loève expansion (see [111], p. 380 and [32], p. 96), and it was first applied in a speckle context by Condie [27] who calculated the cumulants of the exact density function and compared them with the cumulants of the gamma density. Dainty [29] and Barakat [12] applied the same method to find general expressions for the probability density function of W. Barakat plotted exact density functions for the case of a slit aperture and a Gaussian intensity distribution across the scattering spot. Scribot [143] did the same for a slit aperture and a rectangular strip intensity distribution.

The Exact Part of the Solution

We begin with the portion of the "exact" solution that is truly exact. To find a more accurate form of the probability density function of W, begin by expanding the speckle field across the detector aperture Σ in an orthonormal series,

$$\mathbf{A}(x, y) = \begin{cases} \dfrac{1}{\sqrt{\mathcal{A}_D}} \sum_{n=0}^{\infty} \mathbf{b}_n \psi_n(x, y) & (x, y) \text{ in detector aperture } \Sigma \\ 0 & \text{otherwise,} \end{cases} \quad (4\text{-}151)$$

where again $\mathcal{A}_D$ is the detector aperture area, and

$$\iint_{\Sigma} \psi_m^*(x, y) \psi_n(x, y) \, dx \, dy = \begin{cases} 1 & n = m \\ 0 & n \neq m. \end{cases}$$

The coefficients of the expansion are given by

$$\mathbf{b}_n = \sqrt{\mathcal{A}_D} \iint_{\Sigma} \mathbf{A}(x, y) \psi^*(x, y) \, dx \, dy. \quad (4\text{-}152)$$

There are many possible orthogonal expansions that might satisfy the above conditions, but we wish to choose one that will have *uncorrelated* expansion coefficients. Thus we require

$$E[\mathbf{b}_n \mathbf{b}_m^*] = \begin{cases} \lambda_n & m = n \\ 0 & m \neq n. \end{cases} \quad (4\text{-}153)$$

Note that, since the coefficients are defined by a weighted integral of a circular complex Gaussian process, they are circular complex Gaussian variables themselves, and thus, in addition to being uncorrelated, they are independent.

Substituting Eq. (4-152) into Eq. (4-153) yields

$$E[\mathbf{b}_n \mathbf{b}_m^*] = \mathcal{A}_D \iint_\Sigma \iint_\Sigma \overline{\mathbf{A}(x_1, y_1)\mathbf{A}^*(x_2, y_2)} \psi_n(x_2, y_2) \psi_m^*(x_1, y_1) \, dx_1 \, dy_1 \, dx_2 \, dy_2$$

$$= \iint_\Sigma \left[\mathcal{A}_D \iint_\Sigma \Gamma_\mathbf{A}(x_1, y_1; x_2, y_2) \psi_n(x_2, y_2) \, dx_2 \, dy_2 \right] \psi_m^*(x_1, y_1) \, dx_1 \, dy_1$$

$$= \bar{I} \mathcal{A}_D \iint_\Sigma \left[\iint_\Sigma \mu_\mathbf{A}(x_1, y_1; x_2, y_2) \psi_n(x_2, y_2) \, dx_2 \, dy_2 \right] \psi_m^*(x_1, y_1) \, dx_1 \, dy_1. \quad (4\text{-}154)$$

The constraint of Eq. (4-153) will be satisfied provided

$$\bar{I} \mathcal{A}_D \iint_\Sigma \mu_\mathbf{A}(x_1, y_1; x_2, y_2) \psi_n(x_2, y_2) \, dx_2 \, dy_2 = \lambda_n \psi_n(x_1, y_1). \quad (4\text{-}155)$$

Thus to achieve uncorrelated coefficients, it is necessary to choose the orthonormal functions to be eigenfunctions of the integral equation (4-155), and the constants λ_n are the corresponding eigenvalues. Neglecting possible quadratic phase factors, which are unimportant because we are interested only in intensity, the speckle fields will be wide-sense stationary, and the integral equation reduces to

$$\bar{I} \mathcal{A}_D \iint_\Sigma \mu_\mathbf{A}(x_1 - x_2, y_1 - y_2) \psi_n(x_2, y_2) \, dx_2 \, dy_2 = \lambda_n \psi_n(x_1, y_1), \quad (4\text{-}156)$$

the left side of which is recognized as a convolution of $\mu_\mathbf{A}$ and ψ_n. Because the normalized amplitude correlation function $\mu_\mathbf{A}$ is Hermitian and positive definite, the eigenfunctions ψ can be shown to be real-valued and nonnegative.

The detected intensity is given by

$$W = \frac{1}{\mathcal{A}_D} \iint_\Sigma \mathbf{A}(x, y) \mathbf{A}^*(x, y) \, dx \, dy.$$

Substituting Eqs. (4-151) and (4-153) into this equation yields

$$W = \sum_{n=0}^{\infty} |\mathbf{b}_n|^2. \quad (4\text{-}157)$$

But since the $\mathbf{b}_n$ are independent circular Gaussian random variables, this is the sum of an infinite number of independent random variables with negative-exponential distributions, and the characteristic function of W must be of the form

$$M_W(\omega) = \prod_{n=0}^{\infty}(1 - j\omega\lambda_n)^{-1}. \tag{4-158}$$

If the λ_n are distinct, then the corresponding probability density function is

$$p(W) = \sum_{n=0}^{\infty} \frac{d_n}{\lambda_n} \exp(-W/\lambda_n) \tag{4-159}$$

for $W \geq 0$, where

$$d_n = \prod_{\substack{m=0 \\ m \neq n}}^{\infty} \left(1 - \frac{\lambda_m}{\lambda_n}\right)^{-1}. \tag{4-160}$$

Since $p(W)$ is a probability density function, its area must be unity, from which it follows that $\sum_{n=0}^{\infty} d_n = 1$. The kth moment of detected intensity is

$$\overline{W^k} = \sum_{n=0}^{\infty} d_n \lambda_n^k k!, \tag{4-161}$$

and the mean and variance of the exact density are seen to be

$$\overline{W} = \sum_{n=0}^{\infty} d_n \lambda_n$$

$$\sigma_W^2 = \sum_{n=0}^{\infty} 2 d_n \lambda_n^2 - \left(\sum_{n=0}^{\infty} d_n \lambda_n\right)^2. \tag{4-162}$$

Further progress requires solution of the integral equation for a specific spot intensity distribution and detector aperture of interest, so that the λ_n can be specified.

Approximations to the Exact Solution
While the derivations to this point have been exact, finding the eigenvalues always involves some level of approximation. Nonetheless, the density functions derived by this method will usually be more accurate than the approximate (gamma) density functions discussed earlier. While the gamma density function has been chosen to have the true mean and variance, higher-order moments will in general not be correct, whereas the solutions we are about to present should be more accurate for high-order moments.

While solutions to some two-dimensional problems are possible, we focus here on one-dimensional problems because the computation of accurate eigenvalues is generally more easily accomplished. We present the solution for a very specific 1-D spot shape and 1-D detector shape. Let the illuminated spot on the rough surface be a narrow uniform strip of light of length L_s, modeled by

$$I(\alpha, \beta) = \text{rect}(\alpha/L_s)\delta(\beta).$$

Then, from the van Cittert–Zernike theorem, the normalized correlation function of the fields at the detector will be

$$\boldsymbol{\mu}_\mathbf{A}(\Delta x) = \text{sinc}(L_s \Delta x). \tag{4-163}$$

After some changes of variables, the one-dimensional eigen-equation can be written

$$\int_{-1}^{1} \frac{\sin[c(x_1 - x_2)]}{\pi(x_1 - x_2)} \psi_n(x_2) \, dx_2 = \lambda_n \psi_n(x_1), \tag{4-164}$$

where $c = \frac{\pi}{2} L_D / L_c$, with L_D being the length of the detector and L_c being the correlation length of the fields at the detector (i.e. the equivalent width of $\boldsymbol{\mu}_\mathbf{A}$).

This particular eigen-equation has been studied extensively in the literature, particularly by D. Slepian and his colleagues (see, for example, [148] and [149]). The eigenvalues have been tabulated for various values of c, and have been shown in graphical form as functions of c. Thus for this particular equation, solutions for the eigenvalues are available. This is not, in general, true for other forms of the normalized correlation function, although Barakat [12] did study the case of a Gaussian correlation function. It is possible to numerically calculate the eigenvalues with a desk-top PC, provided the kernel is discretized and matrix eigenvalues are found [90]. We illustrate this procedure with the eigen-equation given above as follows:

1. Sample the function $\frac{\sin[c(x_1-x_2)]}{\pi(x_1-x_2)}$ on the interval $(-1, 1)$ for both x_1 and x_2. The function should be sampled in both variables with spacings of $1/N$ between samples. We have thus created the $2N \times 2N$ matrix

$$\underline{K} = \frac{\sin[c(k-n)/N]}{\pi(k-n)/N} \quad \text{for} \quad k = (-N, -N+1, \ldots N)$$

$$\text{and} \quad n = (-N, -N+1, \ldots N).$$

2. Calculate the eigenvalues of this matrix using a matrix eigenvalue routine.
3. Multiply the eigenvalues so obtained by $1/N$ to obtain the λ_n.

The approximations associated with this approach result from the finite numerical accuracy of the eigenvalue calculations, and from decisions regarding how many eigenvalues to retain when calculating the probability density functions (we can not use an infinite number of them).

Figure 4.26 shows a plot of the values of the first five eigenvalues for the above kernel obtained with a 4000×4000 matrix. The values agree well with those obtained by Slepian.

Once the eigenvalues are in hand, it is then possible to plot the probability density functions of integrated intensity using Eq. (4-159) and to compare those with the

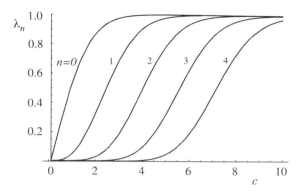

Figure 4.26 Plots of the first five eigenvalues of Eq. (4-164) as a function of c.

corresponding gamma distributions obtained in the earlier approximate analysis. Thus we plot together the density functions defined by

$$p(W) = \sum_{n=0}^{\infty} \frac{d_n}{\lambda_n} \exp(-W/\lambda_n)$$

$$p(W) = \frac{(M/\overline{W})^M W^{M-1} \exp(-MW/\overline{W})}{\Gamma(M)},$$
(4-165)

where the parameter M corresponding to a given c is found by numerical integration of

$$M = 2\left[\int_0^1 (1-y) \operatorname{sinc}^2(2cy/\pi)\, dy\right]^{-1},$$
(4-166)

taking into account that $c = \frac{\pi}{2}\frac{L_D}{L_c}$. The results are shown in Fig. 4.27, which are similar to those found by Scribot [143].

Comparison of the "Exact" and Approximate Solutions

From Fig. 4.27 we can see that when c is small, both probability density functions approach a negative exponential distribution, while when c is large, both approach a common limit. In fact, for large c, both distributions approach a Gaussian distribution, centered on $\overline{W}$, a consequence of the Central Limit Theorem. Thus it is only in the ranges of c near unity for which significant differences exist between the approximate and "exact" distributions. Remembering that both distributions have the same mean and variance, use of the approximate distribution is justified in most problems, especially since the contrast of speckle predicted by the approximate model is exact.

Finally, some insight into the relationship between the approximate and the "exact" solutions was provided by Condie [27], Scribot [143] and Barakat [12], who all pointed out that the approximate solution in essence assumes that for integer M, all

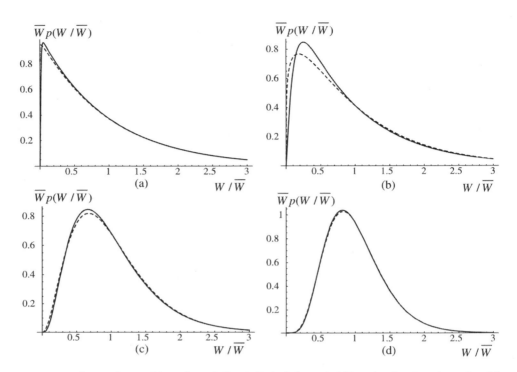

Figure 4.27 Approximate (dotted) and "exact" (solid) probability density functions for (a) $c = 0.25$, $M = 1.01$, $L_D/L_c = 0.16$ (b) $c = 1.0$, $M = 1.22$, $L_D/L_c = 0.64$ (c) $c = 4.0$, $M = 3.08$, $L_D/L_c = 2.54$ (d) $c = 8.0$, $M = 5.66$, $L_D/L_c = 5.10$.

eigenvalues with index M or smaller are unity, and all with larger indices are zero, which assumptions yield the gamma distribution with parameter M. On the other hand, the "exact" solution uses calculated distinct eigenvalues, arriving at a solution that is a sum of negative exponential distributions. For example, as Fig. 4.28 illustrates for the case of $M = 3$ or $c = 3.88$, the approximate solution replaces the true values of the eigenvalues ($\lambda_0, \lambda_1, \lambda_2$) by unity, and all higher indexed eigenvalues by zero, whereas their true values are shown in the figure.

4.6.4 Integration of Partially Polarized Speckle Patterns

When the speckle patterns of interest are partially polarized, a generalization of the preceding analysis is needed. The first step in the analysis is to find a linear transformation $\mathcal{L}_0$ that diagonalizes the coherency matrix. With this transformation, the total intensity $I(x, y)$ of the speckle can be expressed as the sum of two statistically independent intensities,

$$I(x, y) = I_\parallel(x, y) + I_\perp(x, y), \tag{4-167}$$

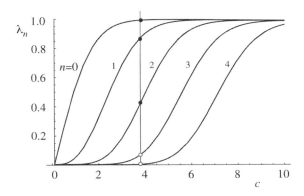

Figure 4.28 True eigenvalues for $M = 3$ and $c = 3.88$. The eigenvalues represented by the solid circles are all replaced by unity in the approximate solution, and the eigenvalues represented by open circles are replaced by zero.

where both $I_\|$ and $I_\perp$ obey negative exponential statistics. From Eq. (3-86), we know that the averages of the two intensity components are

$$\overline{I}_\| = \frac{1}{2}\overline{I}(1 + \mathcal{P})$$
$$\overline{I}_\perp = \frac{1}{2}\overline{I}(1 - \mathcal{P}). \tag{4-168}$$

The integrated intensity now takes the form

$$W = \frac{1}{\mathcal{A}_D} \iint\limits_{-\infty}^{\infty} D(x,y)[I_\|(x,y) + I_\perp(x,y)]\,dx\,dy$$

where the meaning of the symbols should be clear from Eq. (4-120).

The mean of W is easily shown to be $\overline{W}$. Following arguments similar to those used to obtain Eq. (4-130), the variance of the total intensity is given by

$$\sigma_W^2 = \frac{1}{4}\frac{\overline{W}^2(1+\mathcal{P})^2}{\mathcal{A}_D^2} \iint\limits_{-\infty}^{\infty} K_D(\Delta x, \Delta y)|\boldsymbol{\mu}_{\mathbf{A}_\|}(\Delta x, \Delta y)|^2\,d\Delta x\,d\Delta y$$

$$+ \frac{1}{4}\frac{\overline{W}^2(1-\mathcal{P})^2}{\mathcal{A}_D^2} \iint\limits_{-\infty}^{\infty} K_D(\Delta x, \Delta y)|\boldsymbol{\mu}_{\mathbf{A}_\perp}(\Delta x, \Delta y)|^2\,d\Delta x\,d\Delta y \tag{4-169}$$

where $\boldsymbol{\mu}_{\mathbf{A}_\|}$ and $\boldsymbol{\mu}_{\mathbf{A}_\perp}$ are the normalized correlation functions of the fields that underlie $I_\|$ and $I_\perp$. In general, $\boldsymbol{\mu}_{\mathbf{A}_\|}$ and $\boldsymbol{\mu}_{\mathbf{A}_\perp}$ need not be identical (often the brightness distributions of a 3D diffuse object are different when viewed through a polarization analyzer oriented in two orthogonal directions). Thus it is necessary to define two different parameters,

$$M_\| = \left[\frac{1}{\mathcal{A}_D^2} \iint_{-\infty}^{\infty} K_D(\Delta x, \Delta y) |\boldsymbol{\mu}_{\mathbf{A}_\|}(\Delta x, \Delta y)|^2 \, d\Delta x \, d\Delta y \right]^{-1}$$

$$M_\perp = \left[\frac{1}{\mathcal{A}_D^2} \iint_{-\infty}^{\infty} K_D(\Delta x, \Delta y) |\boldsymbol{\mu}_{\mathbf{A}_\perp}(\Delta x, \Delta y)|^2 \, d\Delta x \, d\Delta y \right]^{-1}$$

(4-170)

which represent the "degrees of freedom" associated with each of the independent polarization components. In terms of these parameters, the contrast of the speckle is given by

$$C = \frac{1}{2} \left[\frac{(1+\mathcal{P})^2}{M_\|} + \frac{(1-\mathcal{P})^2}{M_\perp} \right]^{1/2}, \qquad (4\text{-}171)$$

which for completely polarized speckle ($\mathcal{P} = 1$) reduces to $1/\sqrt{M_\|}$, and for unpolarized speckle ($\mathcal{P} = 0$) becomes $C = \frac{1}{2}\sqrt{1/M_\| + 1/M_\perp}$. In this latter case, if $M_\| = M_\perp = M$, then $C = 1/\sqrt{2M}$.

With respect to probability density functions of W, both the approximate and the exact approach can be taken. The simplest method to find the probability density functions in both cases is to multiply together two characteristic functions, one for $I_\|$ and one for $I_\perp$, and then to Fourier invert those products. Results from previous sections allow one to specify the two characteristic functions in both the approximate and the exact cases. We will not pursue detailed expressions here, since the results depend on many parameters, including detector shape, the form of the two normalized correlation functions, and the degree of polarization.

This concludes our discussion of integrated and blurred speckle patterns, and we now turn attention to other properties of speckle.

4.7 Statistics of Derivatives of Speckle Intensity and Phase

In a number of problems, the derivatives of speckle intensity and phase play an important role. For example, in some problems the properties of local maxima of speckle intensity are of interest, and for such problems we need to explore the zeros of the derivative of speckle intensity. In another case of interest, we are interested in the statistical properties of the geometrical ray directions in a speckle pattern. Since ray directions are determined by the local gradient of phase, such derivatives must be understood. We begin with a general development that will let us specialize to the cases of intensity and phase. The general development is followed by a discussion of the statistical properties of the derivatives of intensity, and finally the derivatives of phase are explored.

4.7.1 Background

In order to find the statistical properties of the derivatives of intensity and phase in a speckle pattern, we first must study the statistical properties of the derivatives of the real and imaginary parts of the speckle field. In this regard, we digress for a discussion of some notation.

The following equations indicate an abbreviated notation for derivatives (the symbols $\mathcal{R}$ and $\mathcal{I}$ stand for the real and imaginary parts of the complex speckle field):

$$\mathcal{R}_x = \frac{\partial}{\partial x}\mathcal{R} \qquad \mathcal{R}_y = \frac{\partial}{\partial y}\mathcal{R}$$

$$\mathcal{I}_x = \frac{\partial}{\partial x}\mathcal{I} \qquad \mathcal{I}_y = \frac{\partial}{\partial y}\mathcal{I}$$

$$\theta_x = \frac{\partial}{\partial x}\theta \qquad \theta_y = \frac{\partial}{\partial y}\theta$$

Our initial goal will be to find the joint probability density function $p(\mathcal{R}, \mathcal{I}, \mathcal{R}_x, \mathcal{I}_x, \mathcal{R}_y, \mathcal{I}_y)$.

Since for fully developed speckle, $\mathcal{R}$ and $\mathcal{I}$ are Gaussian with zero means and equal variances σ^2, derivatives of $\mathcal{R}$ and $\mathcal{I}$ are likewise Gaussian, since any linear transformation of a Gaussian retains Gaussian statistics. They also have zero mean. As a consequence, the six random variables of interest obey a multi-dimensional Gaussian distribution (cf. Eq. (4-3)),

$$p(\mathcal{R}, \mathcal{I}, \mathcal{R}_x, \mathcal{I}_x, \mathcal{R}_y, \mathcal{I}_y) = \frac{1}{8\pi^3\sqrt{\det \mathcal{C}}} \exp\left[-\frac{1}{2}(\vec{u}^t \mathcal{C}^{-1} \vec{u})\right], \tag{4-172}$$

where $\vec{u}^t$ is a row vector with entries $(\mathcal{R}, \mathcal{I}, \mathcal{R}_x, \mathcal{I}_x, \mathcal{R}_y, \mathcal{I}_y)$, and $\mathcal{C}$ is the covariance matrix,

$$\mathcal{C} = \begin{bmatrix}
\overline{\mathcal{R}\mathcal{R}} & \overline{\mathcal{R}\mathcal{I}} & \overline{\mathcal{R}\mathcal{R}_x} & \overline{\mathcal{R}\mathcal{I}_x} & \overline{\mathcal{R}\mathcal{R}_y} & \overline{\mathcal{R}\mathcal{I}_y} \\
\overline{\mathcal{I}\mathcal{R}} & \overline{\mathcal{I}\mathcal{I}} & \overline{\mathcal{I}\mathcal{R}_x} & \overline{\mathcal{I}\mathcal{I}_x} & \overline{\mathcal{I}\mathcal{R}_y} & \overline{\mathcal{I}\mathcal{I}_y} \\
\overline{\mathcal{R}_x\mathcal{R}} & \overline{\mathcal{R}_x\mathcal{I}} & \overline{\mathcal{R}_x\mathcal{R}_x} & \overline{\mathcal{R}_x\mathcal{I}_x} & \overline{\mathcal{R}_x\mathcal{R}_y} & \overline{\mathcal{R}_x\mathcal{I}_y} \\
\overline{\mathcal{I}_x\mathcal{R}} & \overline{\mathcal{I}_x\mathcal{I}} & \overline{\mathcal{I}_x\mathcal{R}_x} & \overline{\mathcal{I}_x\mathcal{I}_x} & \overline{\mathcal{I}_x\mathcal{R}_y} & \overline{\mathcal{I}_x\mathcal{I}_y} \\
\overline{\mathcal{R}_y\mathcal{R}} & \overline{\mathcal{R}_y\mathcal{I}} & \overline{\mathcal{R}_y\mathcal{R}_x} & \overline{\mathcal{R}_y\mathcal{I}_x} & \overline{\mathcal{R}_y\mathcal{R}_y} & \overline{\mathcal{R}_y\mathcal{I}_y} \\
\overline{\mathcal{I}_y\mathcal{R}} & \overline{\mathcal{I}_y\mathcal{I}} & \overline{\mathcal{I}_y\mathcal{R}_x} & \overline{\mathcal{I}_y\mathcal{I}_x} & \overline{\mathcal{I}_y\mathcal{R}_y} & \overline{\mathcal{I}_y\mathcal{I}_y}
\end{bmatrix}. \tag{4-173}$$

or equivalently

$$\mathcal{C} = \begin{bmatrix}
\Gamma_{\mathcal{R}\mathcal{R}} & \Gamma_{\mathcal{R}\mathcal{I}} & \Gamma_{\mathcal{R}\mathcal{R}_x} & \Gamma_{\mathcal{R}\mathcal{I}_x} & \Gamma_{\mathcal{R}\mathcal{R}_y} & \Gamma_{\mathcal{R}\mathcal{I}_y} \\
\Gamma_{\mathcal{I}\mathcal{R}} & \Gamma_{\mathcal{I}\mathcal{I}} & \Gamma_{\mathcal{I}\mathcal{R}_x} & \Gamma_{\mathcal{I}\mathcal{I}_x} & \Gamma_{\mathcal{I}\mathcal{R}_y} & \Gamma_{\mathcal{I}\mathcal{I}_y} \\
\Gamma_{\mathcal{R}_x\mathcal{R}} & \Gamma_{\mathcal{R}_x\mathcal{I}} & \Gamma_{\mathcal{R}_x\mathcal{R}_x} & \Gamma_{\mathcal{R}_x\mathcal{I}_x} & \Gamma_{\mathcal{R}_x\mathcal{R}_y} & \Gamma_{\mathcal{R}_x\mathcal{I}_y} \\
\Gamma_{\mathcal{I}_x\mathcal{R}} & \Gamma_{\mathcal{I}_x\mathcal{I}} & \Gamma_{\mathcal{I}_x\mathcal{R}_x} & \Gamma_{\mathcal{I}_x\mathcal{I}_x} & \Gamma_{\mathcal{I}_x\mathcal{R}_y} & \Gamma_{\mathcal{I}_x\mathcal{I}_y} \\
\Gamma_{\mathcal{R}_y\mathcal{R}} & \Gamma_{\mathcal{R}_y\mathcal{I}} & \Gamma_{\mathcal{R}_y\mathcal{R}_x} & \Gamma_{\mathcal{R}_y\mathcal{I}_x} & \Gamma_{\mathcal{R}_y\mathcal{R}_y} & \Gamma_{\mathcal{R}_y\mathcal{I}_y} \\
\Gamma_{\mathcal{I}_y\mathcal{R}} & \Gamma_{\mathcal{I}_y\mathcal{I}} & \Gamma_{\mathcal{I}_y\mathcal{R}_x} & \Gamma_{\mathcal{I}_y\mathcal{I}_x} & \Gamma_{\mathcal{I}_y\mathcal{R}_y} & \Gamma_{\mathcal{I}_y\mathcal{I}_y}
\end{bmatrix}. \tag{4-174}$$

Evaluation of the terms in this 6×6 matrix is a formidable task, and is carried out in Appendix C, Section C.1. The result is

$$C = \begin{bmatrix} \sigma^2 & 0 & 0 & 0 & 0 & 0 \\ 0 & \sigma^2 & 0 & 0 & 0 & 0 \\ 0 & 0 & b_x & 0 & 0 & 0 \\ 0 & 0 & 0 & b_x & 0 & 0 \\ 0 & 0 & 0 & 0 & b_y & 0 \\ 0 & 0 & 0 & 0 & 0 & b_y \end{bmatrix}, \quad (4\text{-}175)$$

where

$$\sigma^2 = \frac{\kappa}{2\lambda^2 z^2} \iint_{-\infty}^{\infty} I(\alpha, \beta)\, d\alpha\, d\beta$$

$$b_x = \frac{2\kappa\pi^2}{\lambda^4 z^4} \iint_{-\infty}^{\infty} \alpha^2 I(\alpha, \beta)\, d\alpha\, d\beta \qquad (4\text{-}176)$$

$$b_y = \frac{2\kappa\pi^2}{\lambda^4 z^4} \iint_{-\infty}^{\infty} \beta^2 I(\alpha, \beta)\, d\alpha\, d\beta$$

and, with reference to Appendix C, Section C.1, it has been assumed that the intensity distribution of the scattering spot has sufficient symmetry to assure that

$$\iint_{-\infty}^{\infty} \alpha\beta I(\alpha, \beta)\, d\alpha\, d\beta = 0.$$

The above correlation matrix, having no nonzero off-diagonal terms, implies that all six random variables of interest are uncorrelated, and by virtue of the fact that they are all Gaussian variates, are also statistically independent. Their variances are given by the diagonal terms. With this information, we can immediately write the six-dimensional probability density function as the product of six marginal densities, one for each variable. When these are multiplied together, the joint probability density function for $\vec{u}^t = (\mathcal{R}, \mathcal{I}, \mathcal{R}_x, \mathcal{I}_x, \mathcal{R}_y, \mathcal{I}_y)$ can then be seen to be

$$p(\mathcal{R}, \mathcal{I}, \mathcal{R}_x, \mathcal{I}_x, \mathcal{R}_y, \mathcal{I}_y) = \left(\frac{\exp\left[-\frac{\mathcal{R}^2 + \mathcal{I}^2}{2\sigma^2}\right]}{2\pi\sigma^2} \right) \left(\frac{\exp\left[-\frac{\mathcal{R}_x^2 + \mathcal{I}_x^2}{2b_x}\right]}{2\pi b_x} \right) \left(\frac{\exp\left[-\frac{\mathcal{R}_y^2 + \mathcal{I}_y^2}{2b_y}\right]}{2\pi b_y} \right)$$

$$= \frac{\exp\left[-\frac{\sigma^2 b_y(\mathcal{R}_x^2 + \mathcal{I}_x^2) + \sigma^2 b_x(\mathcal{R}_y^2 + \mathcal{I}_y^2) + b_x b_y(\mathcal{R}^2 + \mathcal{I}^2)}{2\sigma^2 b_x b_y}\right]}{8\pi^3 \sigma^2 b_x b_y}. \quad (4\text{-}177)$$

At this point we make a change of variables to find the density function $p(I, \theta, I_x, \theta_x, I_y, \theta_y)$. The required transformation is

$$\mathcal{R} = \sqrt{I} \cos \theta$$

$$\mathcal{I} = \sqrt{I} \sin \theta$$

$$\mathcal{R}_x = \frac{I_x}{2\sqrt{I}} \cos \theta - \sqrt{I} \theta_x \sin \theta$$

$$\mathcal{I}_x = \frac{I_x}{2\sqrt{I}} \sin \theta + \sqrt{I} \theta_x \cos \theta \qquad (4\text{-}178)$$

$$\mathcal{R}_y = \frac{I_y}{2\sqrt{I}} \cos \theta - \sqrt{I} \theta_y \sin \theta$$

$$\mathcal{I}_y = \frac{I_y}{2\sqrt{I}} \sin \theta + \sqrt{I} \theta_y \cos \theta.$$

The magnitude of the Jacobian determinant of this transformation is $1/8$, with the result that the joint density function of the variables $(I, \theta, I_x, I_y, \theta_x, \theta_y)$ is given by

$$p(I, \theta, I_x, I_y, \theta_x, \theta_y) = \frac{\exp\left[-\frac{4b_x b_y I^2 + \sigma^2(b_y I_x^2 + b_x I_y^2) + 4\sigma^2 I^2(b_y \theta_x^2 + b_x \theta_y^2)}{8I\sigma^2 b_x b_y}\right]}{64\pi^3 \sigma^2 b_x b_y}. \qquad (4\text{-}179)$$

The ranges of variables in this equation are

$$0 \leq I < \infty \qquad -\infty < I_x < \infty \qquad -\infty < I_y < \infty$$
$$-\pi \leq \theta < \pi \qquad -\infty < \theta_x < \infty \qquad -\infty < \theta_y < \infty. \qquad (4\text{-}180)$$

The task remains to integrate this general expression to find the marginal densities of interest in the two cases discussed below.

4.7.2 Parameters for Various Scattering Spot Shapes

Throughout the previous subsection, the parameters σ^2, b_x and b_y have appeared. In this subsection, we calculate the values of these parameters for some important intensity distributions at the scattering spot. The definitions of Eq. (4-176) are used in these calculations. Let I_0 be the intensity of the scattering spot at its center. The integrations are straightforward and we present the results in Figure 4.29.[13] In the right-hand column and the second from the right, we have assumed (in the rectangular case) that $b_x = b_y = b$ (i.e. that the rectangular spot is actually square). The column second from the right shows the correlation area for the particular spot shape illustrated,

---

13. When considering these results, remember that κ has the dimensions of (length)^2. See Eq. (4-53).

Spot shape	σ^2	b_x	b_y	A_c	b/σ^2
Rectangular, $L_x \times L_y$	$I_0 \dfrac{\kappa L_x L_y}{2\lambda^2 z^2}$	$I_0 \dfrac{\kappa \pi^2 L_x^3 L_y}{6\lambda^4 z^4}$	$I_0 \dfrac{\kappa \pi^2 L_x L_y^3}{6\lambda^4 z^4}$	$\dfrac{\lambda^2 z^2}{L^2}$	$\dfrac{\pi^2}{3 A_c}$
Circular, diameter D	$I_0 \dfrac{\kappa \pi D^2}{8\lambda^2 z^2}$	$I_0 \dfrac{\kappa \pi^3 D^4}{32\lambda^4 z^4}$	$I_0 \dfrac{\kappa \pi^3 D^4}{32\lambda^4 z^4}$	$\dfrac{4}{\pi}\dfrac{\lambda^2 z^2}{D^2}$	$\dfrac{\pi^3}{16 A_c}$
Gaussian, $1/e$ intensity at r_0	$I_0 \dfrac{\kappa \pi r_0^2}{2\lambda^2 z^2}$	$I_0 \dfrac{\kappa \pi^3 r_0^4}{\lambda^4 z^4}$	$I_0 \dfrac{\kappa \pi^3 r_0^4}{\lambda^4 z^4}$	$\dfrac{\lambda^2 z^2}{2\pi r_0^2}$	$\dfrac{\pi}{A_c}$

Figure 4.29 Parameter values for various spot shapes.

as calculated from Eq. (4-71). The last column shows the quantity b/σ^2, which is rather fundamental in most results. It has dimensions $1/\text{length}^2$ and can be seen to be inversely proportional to the intensity correlation area A_c in each case. These results can be substituted into any of the results to find the quantities of interest for a particular spot shape.

4.7.3 Derivatives of Speckle Phase: Ray Directions in a Speckle Pattern

Our first goal is to find the joint probability density function of the partial derivatives of the phase, namely, $p(\theta_x, \theta_y)$, as well as the marginal densities of θ_x and θ_y and the density function of the phase gradient $|\nabla \theta|$. We follow the analysis of Ochoa and Goodman [119] in this task. Before beginning this task, we should note that knowledge of the statistics of the phase derivatives is equivalent to knowledge of the statistics of the geometrical-optics ray directions in a speckle pattern, since the partial derivatives of phase are related to the direction cosines of the wavefront through

$$\theta_x(x, y) = \frac{2\pi}{\lambda} \chi$$
$$\theta_y(x, y) = \frac{2\pi}{\lambda} \psi, \qquad (4\text{-}181)$$

χ and ψ representing the direction cosines of the rays with respect to the x and y axes, respectively.

To find the density $p(\theta_x, \theta_y)$, we must perform the following integrations:

$$p(\theta_x, \theta_y) = \int_0^\infty dI \int_{-\infty}^\infty dI_x \int_{-\infty}^\infty dI_y \int_{-\pi}^\pi d\theta$$
$$\times \frac{\exp\left[-\dfrac{4 b_x b_y I^2 + \sigma^2 (b_y I_x^2 + b_x I_y^2) + 4\sigma^2 I^2 (b_y \theta_x^2 + b_x \theta_y^2)}{8 I \sigma^2 b_x b_y}\right]}{64 \pi^3 \sigma^2 b_x b_y}. \qquad (4\text{-}182)$$

In principle, the order in which we perform these integrations is immaterial, but in practice, the order may influence whether the integrals can be found in integration tables or whether the integrals are recognized by a symbolic mathematics program such as *Mathematica*. The order in which the integrals are listed above is recommended for these purposes. We refer the reader to Appendix C, Section C.2 for details about the results of these integrations. The resulting expression for the joint density function we are seeking is

$$p(\theta_x, \theta_y) = \frac{\sigma^2/\pi\sqrt{b_x b_y}}{\left(1 + \frac{\sigma^2}{b_x}\theta_x^2 + \frac{\sigma^2}{b_y}\theta_y^2\right)^2}. \tag{4-183}$$

The marginal densities of θ_x and θ_y are found by integrating the joint density with respect to θ_y and θ_x, respectively, with the result

$$p(\theta_x) = \frac{\sigma/(2\sqrt{b_x})}{\left(1 + \frac{\sigma^2}{b_x}\theta_x^2\right)^{3/2}} \qquad p(\theta_y) = \frac{\sigma/(2\sqrt{b_y})}{\left(1 + \frac{\sigma^2}{b_y}\theta_y^2\right)^{3/2}}. \tag{4-184}$$

The joint density and marginal densities above have parameters $\sigma/\sqrt{b_x}$ and $\sigma/\sqrt{b_y}$, which depend on the shape of the scattering spot. We have evaluated these parameters for various shapes in the previous subsection, from which it can be seen that they depend on the square root of the intensity correlation area, or in other words, the intensity correlation lengths in each of the (x, y) directions. Here we present plots of normalized versions of the joint and marginal densities in Fig. 4.30.

The *gradient* of the phase is also a quantity of interest. The gradient of a function $f(x, y)$ is defined by

$$\nabla f(x, y) = f_x(x, y)\hat{x} + f_y(x, y)\hat{y}, \tag{4-185}$$

where $\hat{x}$ and $\hat{y}$ are unit vectors in the x and y directions, respectively. The gradient has a magnitude $|\nabla f|$ and an angle φ given by

$$|\nabla f| = \sqrt{f_x^2 + f_y^2}$$

$$\varphi = \arctan \frac{f_y}{f_x}. \tag{4-186}$$

To find the probability density of $|\nabla \theta|$, we first find the joint density of $|\nabla \theta|$ and φ, which requires a transformation of variables,

$$\theta_x = |\nabla \theta| \cos \varphi$$
$$\theta_y = |\nabla \theta| \sin \varphi, \tag{4-187}$$

with the magnitude of the Jacobian determinant of the transformation being $|\nabla \theta|$. From Eq. (4-183), the joint density function of $|\nabla \theta|$ and φ is therefore

$$p(|\nabla \theta|, \varphi) = \frac{\sigma^2 |\nabla \theta|/\pi \sqrt{b_x b_y}}{\left(1 + \frac{\sigma^2}{b_x}(|\nabla \theta|^2 \cos^2 \varphi)^2 + \frac{\sigma^2}{b_y}(|\nabla \theta|^2 \sin^2 \varphi)^2\right)^2}. \tag{4-188}$$

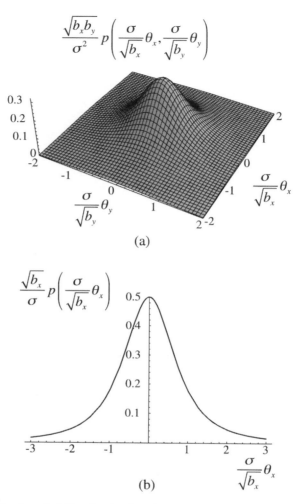

Figure 4.30 Normalized probability density functions of derivatives of speckle phase: (a) joint density of (θ_x, θ_y) and (b) marginal density function of θ_x.

To make further progress, we must assume sufficient symmetry[14] to the scattering spot to assure that $b_x = b_y = b$, in which case the joint density function becomes

---

14. This assumption is satisfied for a square scattering spot, for a circular scattering spot, and for a circularly symmetric Gaussian scattering spot, as well as many other spot shapes of interest.

4.7 Statistics of Derivatives of Speckle Intensity and Phase

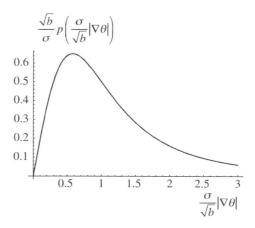

Figure 4.31 Probability density function of the gradient of phase.

$$p(|\nabla\theta|, \varphi) = \frac{\sigma^2}{\pi b} \frac{|\nabla\theta|}{\left(1 + \frac{\sigma^2}{b}|\nabla\theta|^2\right)^2}. \qquad (4\text{-}189)$$

Since the expression no longer depends on φ, integration over that variable results in multiplication by a factor of 2π, with the final result

$$p(|\nabla\theta|) = 2\frac{\sigma^2}{b} \frac{|\nabla\theta|}{\left(1 + \frac{\sigma^2}{b}|\nabla\theta|^2\right)^2}. \qquad (4\text{-}190)$$

The angle of the gradient is uniformly distributed on $(-\pi, \pi)$. The mean value of $|\nabla\theta|$ is found to be

$$\overline{|\nabla\theta|} = \frac{\pi\sqrt{b}}{2\sigma}, \qquad (4\text{-}191)$$

which again is seen to be scaled by $\sqrt{b}/\sigma$, a quantity proportional to the reciprocal of the intensity correlation length of the speckle. Figure 4.31 shows a plot of the probability density function of the gradient of phase.

4.7.4 Derivatives of Speckle Intensity

We turn now to finding the joint density function of the derivatives of the intensity, $p(I_x, I_y)$, the marginal densities of I_x and I_y, and the density function of the gradient of intensity, $|\nabla I|$. We again begin with the joint density of Eq. (4-179). The integrations required are

$$p(I_x, I_y) = \int_0^\infty dI \int_{-\infty}^\infty d\theta_y \int_{-\infty}^\infty d\theta_x \int_{-\pi}^\pi d\theta$$

$$\times \frac{\exp\left[-\frac{4b_x b_y I^2 + \sigma^2(b_x I_x^2 + b_y I_y^2) + 4\sigma^2 I^2(b_y \theta_x^2 + b_x \theta_y^2)}{8I\sigma^2 b_x b_y}\right]}{64\pi^3 \sigma^2 b_x b_y}. \qquad (4\text{-}192)$$

The required calculations are carried out in Appendix C, Section C.3, with the result

$$p(I_x, I_y) = \frac{K_0\left(\frac{1}{2\sigma}\sqrt{\frac{I_x^2}{b_x} + \frac{I_y^2}{b_y}}\right)}{8\pi\sigma^2\sqrt{b_x b_y}}, \qquad (4\text{-}193)$$

where $K_0(x)$ is a modified Bessel function of the second kind, order zero.

The marginal densities of I_x and I_y are shown in Appendix C, Section C.3, to be

$$p(I_x) = \frac{1}{4\sigma\sqrt{b_x}} \exp\left[-\frac{|I_x|}{2\sigma\sqrt{b_x}}\right]$$

$$p(I_y) = \frac{1}{4\sigma\sqrt{b_y}} \exp\left[-\frac{|I_y|}{2\sigma\sqrt{b_y}}\right], \qquad (4\text{-}194)$$

in agreement with the results of Ebeling [36]. Figure 4.32 shows plots of the joint density $p(I_x, I_y)$ and one marginal density $p(I_x)$.

Finally we find the probability density function of the magnitude of the gradient of intensity in the case when the symmetry of the spot assures that $b_x = b_y = b$. Following the procedure used in the case of phase derivatives, we make a transformation of variables

$$I_x = |\nabla I| \cos \varphi$$

$$I_y = |\nabla I| \sin \varphi, \qquad (4\text{-}195)$$

with a Jacobian of the transformation being $|\nabla I|$. Using the above joint density function of I_x and I_y, we obtain a joint density function for $|\nabla I|$ and φ,

$$p(|\nabla I|, \varphi) = \frac{|\nabla I| K_0\left(\frac{|\nabla I|}{2\sigma\sqrt{b}}\right)}{8\pi\sigma^2 b}. \qquad (4\text{-}196)$$

The result is independent of φ, which is uniformly distributed, so the marginal density function of $|\nabla I|$ is found to be

$$p(|\nabla I|) = \frac{|\nabla I| K_0\left(\frac{|\nabla I|}{2\sigma\sqrt{b}}\right)}{4\sigma^2 b}, \qquad (4\text{-}197)$$

a result first found by Kowalczyk [94]. A plot of this density function is shown in Fig. 4.33. The mean value of the magnitude of the gradient is

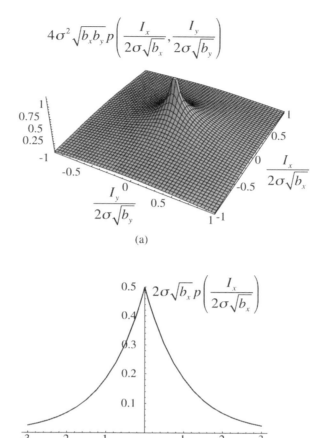

Figure 4.32 Probability density functions: (a) joint density of I_x and I_y, and (b) marginal density function of I_x.

$$\overline{|\nabla I|} = \pi\sigma\sqrt{b}. \qquad (4\text{-}198)$$

Normalizing out the effects of overall brightness, an alternative quantity of interest is

$$\frac{\overline{|\nabla I|}}{\bar{I}} = \frac{\pi\sigma\sqrt{b}}{2\sigma^2} = \frac{\pi}{2}\frac{\sqrt{b}}{\sigma}, \qquad (4\text{-}199)$$

which is identical with the mean value of the phase gradient presented earlier.

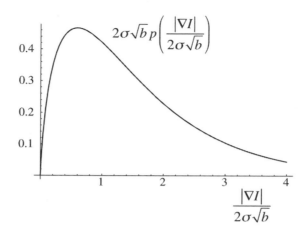

Figure 4.33 Probability density function of the magnitude of the gradient of intensity.

4.7.5 Level Crossings of Speckle Patterns

Let $I(x, y)$ represent the intensity of a fully developed, completely polarized speckle pattern with stationary statistics. It is of some interest to know how often this intensity can be expected to cross a given intensity level. The so-called *level-crossing problem* has been extensively studied in the literature, mostly based on the pioneering work of Rice [132]. For relevant literature, see [89], [9], [36], [13], and [181].

Consider a one-dimensional slice through the intensity distribution in a speckle pattern. For the moment, we assume that the slice is along the x-direction. We represent that slice by the function $I(x)$, temporarily ignoring the y-direction.

Following Middleton ([111], Section 9.4-1), we construct a *counting functional* that generates a unit step function each time the intensity I crosses through the particular level I_0. If $U(x)$ represents a unit step function ($U(x) = 1$ for $x > 0$, and $= 0$ for $x < 0$), then the counting functional is defined by $U(I(x) - I_0)$. Each time the intensity crosses the level I_0, the functional jumps by value unity. Differentiating this expression yields

$$\frac{dU}{dx} = \frac{dU}{dI}\frac{dI}{dx} = I_x(x)\delta(I - I_0),$$

where $\delta(x)$ is a unit-area delta function and $I_x(x)$ can be either positive or negative, depending on whether the crossing is upward or downward, respectively. To count upwards and downwards crossings equally, we must replace $I_x(x)$ by $|I_x(x)|$, yielding

$$n_0(I_0; x) = |I_x(x)|\delta(I(x) - I_0)$$

This quantity can be interpreted as the number of crossings (upwards or downwards) per unit distance on the x-axis, as measured at position x. The number of crossings between x_1 and x_2 is given by the integral of this quantity

$$N(I_0; \Delta x) = \int_{x_1}^{x_2} n_0(I_0, \xi)\, d\xi = \int_{x_1}^{x_2} |I_x(\xi)| \delta(I(\xi) - I_0)\, d\xi, \qquad (4\text{-}200)$$

where ξ is a dummy variable of integration and $\Delta x = x_2 - x_1$. If, on the other hand, we wish to count only those crossings with positive slope, the above equation must be modified to read

$$N(I_0; \Delta x) = \int_{x_1}^{x_2} I_x(\xi)\delta(I(\xi) - I_0)\, d\xi \quad \text{for } I_x(x) > 0. \qquad (4\text{-}201)$$

We shall focus on counting only the crossings with positive slope.

Given the above result, the average number of positive-slope crossings per unit interval in x is

$$\frac{\overline{N(I_0; \Delta x)}}{\Delta x} = \int_0^\infty \int_0^\infty I_x \delta(I - I_0) p(I, I_x)\, dI\, dI_x = \int_0^\infty I_x p(I_0, I_x)\, dI_x. \qquad (4\text{-}202)$$

Ebeling [36] started with this equation and calculated the average indicated above. We do as well, using the first line of Eq. (C-21) as our starting point, or more precisely,

$$p(I, I_x) = \frac{\exp\left(-\frac{I_x^2}{8I b_x} - \frac{I}{2\sigma^2}\right)}{4\sqrt{2\pi}\sigma^2 \sqrt{I b_x}}. \qquad (4\text{-}203)$$

The integral with respect to I_x can be performed and yields

$$\eta(I_0) = \frac{\overline{N(I_0; \Delta x)}}{\Delta x} = \frac{\sqrt{b_x I_0}\exp\left(-\frac{I_0}{2\sigma^2}\right)}{\sqrt{2\pi}\sigma^2} = \sqrt{\frac{b_x}{\pi\sigma^2}}\sqrt{\frac{I_0}{\bar{I}}}\exp\left(-\frac{I_0}{\bar{I}}\right), \qquad (4\text{-}204)$$

which is a factor of 2 larger than Ebeling's result, but which we believe to be correct. Figure 4.34 shows a plot of normalized level crossing rate vs $I_0/\bar{I}$. The maximum value of $\eta(I_0)$, $\eta_{\max}$, occurs at $I_0 = \bar{I}/2$ and is given by

$$\eta_{\max} = \sqrt{\frac{b_x}{4\pi\sigma^2 e}}. \qquad (4\text{-}205)$$

As a slight generalization, we can calculate the positive-slope crossing rates for both the x-axis and y-axis directions at the same time, and then generalize to an arbitrary direction. We start with the counting functional $U(I(x, y) - I_0)$, where again $U(\ )$ represents the unit step function. We then find the gradient of the counting functional

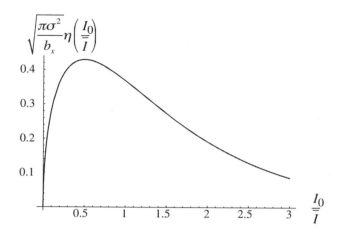

Figure 4.34 Normalized level crossing rate as a function of normalized level.

$$\nabla U(I(x,y) - I_0) = \frac{\partial}{\partial x} U(I(x,y) - I_0)\hat{x} + \frac{\partial}{\partial y} U(I(x,y) - I_0)\hat{y}$$
$$= I_x(x,y)\delta(I - I_0)\hat{x} + I_y(x,y)\delta(I - I_0)\hat{y}, \quad (4\text{-}206)$$

where $\hat{x}$ and $\hat{y}$ are unit vectors along the x and y axes, respectively. By analogy with the previous case, the *vector* crossing rate can then be written

$$\vec{\eta}(I_0) = \eta_x(I_0)\hat{x} + \eta_y(I_0)\hat{y}$$
$$= \int_0^\infty \int_0^\infty I_x \delta(I - I_0) p(I, I_x) \, dI \, dI_x \hat{x}$$
$$+ \int_0^\infty \int_0^\infty I_y \delta(I - I_0) p(I, I_y) \, dI \, dI_y \hat{y}$$
$$= \int_0^\infty I_x p(I_0, I_x) \, dI_x \hat{x} + \int_0^\infty I_y p(I_0, I_y) \, dI_y \hat{y}$$
$$= \sqrt{\frac{b_x}{\pi\sigma^2}} \sqrt{\frac{I_0}{\bar{I}}} \exp\left(-\frac{I_0}{\bar{I}}\right)\hat{x} + \sqrt{\frac{b_y}{\pi\sigma^2}} \sqrt{\frac{I_0}{\bar{I}}} \exp\left(-\frac{I_0}{\bar{I}}\right)\hat{y}. \quad (4\text{-}207)$$

The crossing rate with positive slope along the x-axis is the $\hat{x}$-component of this vector and the crossing rate with positive slope along the y-axis is the $\hat{y}$ component of this vector. The average crossing rate with positive slope along a line oriented at angle χ to the x-axis is given by [89]

$$\eta_\chi(I_0) = \sqrt{(\eta_x(I_0))^2 \cos^2 \chi + (\eta_y(I_0))^2 \sin^2 \chi}. \quad (4\text{-}208)$$

This concludes our discussion of the level-crossing problem for speckle.

4.8 Zeros of Speckle Patterns: Optical Vortices

The negative-exponential probability density function for the intensity of a fully polarized, fully developed speckle pattern has its maximum value at zero intensity. However, the probability that a specific value of speckle intensity occurs is zero; only the probability that the intensity falls within a given interval is nonzero. Nonetheless, we have seen that crossings of a given level of intensity occur at certain rates, even though the probabilities of those precise levels occurring are zero. The message to keep in mind is that, even though an event has zero probability, it can still occur.

The occurrence of precisely zero intensity at a point in a speckle pattern is an event that can and does occur. Such an occurrence is an example of a more general phenomenon known as an *optical singularity* or a *wavefront dislocation*, or an *optical vortex*, and a considerable literature exists on the properties of such singularities (see, for example, [118], [17], and [14]).

In what follows, we first explore the condition under which zeros of the intensity occur. We then explore the properties of the phase in the vicinity of such a zero. Finally we consider the density of such zeros in a speckle pattern.[15]

4.8.1 Conditions Required for a Zero of Intensity to Occur

The speckle intensity is, of course, related to the real ($\mathcal{R}$) and imaginary ($\mathcal{I}$) parts of the underlying field through

$$I = \mathcal{R}^2 + \mathcal{I}^2.$$

For fully-developed speckle, the real and imaginary parts of the field at a given point are identically distributed, zero-mean Gaussian random variables, and are statistically independent. For a zero of intensity to occur, *both* the real part and the imaginary parts of the field must be zero at the same point in space.

To illustrate the process by which zeros of intensity occur, Fig. 4.35(a) shows a simulated speckle pattern (uniform square scattering spot assumed), and part (b) shows contours of zero values for the real (solid lines) and imaginary (dotted lines) parts of the underlying field. Each intersection of a contour of real-part zero with the contour of imaginary-part zero is circled. At the center of these circles are the locations of the zeros of intensity. As can be seen, such zeros are by no means rare. In fact, if one counts the number of visible maxima of intensity in part (a) of the figure and the number of zeros evident in part (b), these two numbers are not very different.

4.8.2 Properties of Speckle Phase in the Vicinity of a Zero of Intensity

At the precise location of a zero of intensity, the phase is undefined. Where there is no light, there can be no phase. However, in the close vicinity of such a zero, the phase

---

15. I am indebted to Prof. Isaac Freund and Dr. Mark Dennis for educating me on the properties of the zeros of a speckle pattern.

134   CHAPTER 4   Higher-Order Statistical Properties of Speckle

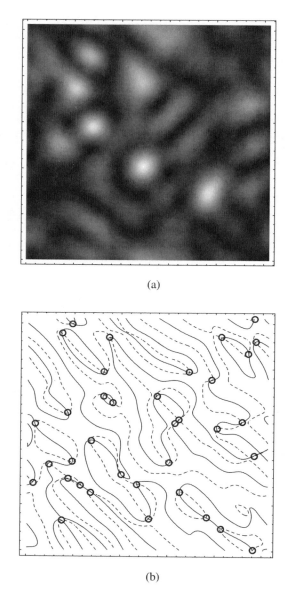

Figure 4.35 (a) A simulated speckle pattern and (b) contours of zero real and imaginary parts of the underlying field. The locations of zeros of the intensity are indicated by circles.

has interesting properties. Here we focus attention on a single zero and explore the various phase distributions that can exist very close to that zero. Figure 4.36 shows four examples of a zero. The solid lines represent part of a contour of zero of the real

4.8 Zeros of Speckle Patterns: Optical Vortices   135

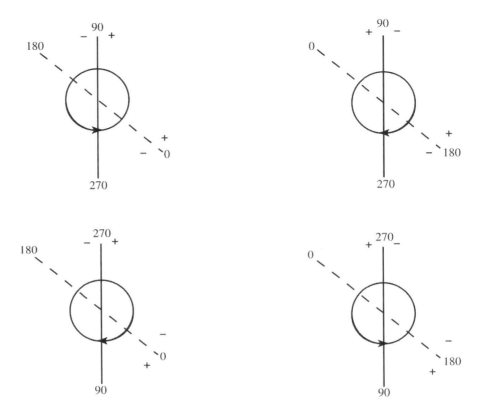

Figure 4.36 Four types of simultaneous real and imaginary part transitions through zero are possible. In each of these four cases, the solid line represents a zero of the real part and a dotted line represents a zero of the imaginary part. The "+" and "−" signs represent the positive and negative sides of the transition, respectively. The value of the phase at each transition is labeled (in degrees) in the figures. The transition of the real part has been shown to be along a vertical line, but the line can be nonvertical.

part of the field, and the dotted lines represent a part of the contour of zero for the imaginary part. For any particular set of two intersecting lines, there are four different possible ways the zero-contours of the real and imaginary parts could have intersected, depending on the directions in which the real and imaginary parts are transitioning through zero. The result is that there are two different ways the phase can circulate around the zero, clockwise and counterclockwise. The circulation of the phase around the zero is what has led to the name *vortex* for a phase singularity such as this. Just as there are two different types of electric charge, positive and negative, so too are the two types of vortex circulation. A vortex for which the phase circulates counterclockwise is called a "positive" vortex, while one for which the phase circulates clockwise is called a "negative" vortex. Figure 4.37 shows a contour

Figure 4.37 Contour plot of phase values in the speckle pattern of Fig. 4.35. There are 10 contours of phase separated by $2\pi/10$. One can clearly see the rotation of the phase about the points of zero intensity.

plot of the phase for the same speckle pattern shown in Fig. 4.35. The rotation of the phase about points of zero intensity can be seen clearly in this figure.

Another interesting property of the vortices is called the "sign principle" [51], which states that on a contour of either $\mathcal{R} = 0$ or $\mathcal{I} = 0$, adjacent vortices have opposite sign. This is best understood by considering a specific example. Suppose we are at a vortex on a contour of $\mathcal{R} = 0$, and that this vortex occurred as $\mathcal{I}$ transitioned from positive to negative through zero. Then any directly adjacent vortex on the contour $\mathcal{R} = 0$ must have occurred as $\mathcal{I}$ transitioned in the opposite sense through zero, that is, from negative to positive. The result will be opposite senses of phase rotation at the adjacent vortices, or equivalently, opposite signs.

The similarity of Fig. 4.37 to an electric field distribution is striking, with lines of field emanating from positive charge (a positive vortex) and ending on negative charge (a negative vortex).

4.8.3 The Density of Vortices in Fully Developed Speckle

A subject of particular interest is the spatial density of vortices: how many vortices are expected, on average, per unit area. In what follows, we follow closely the analyses found in [17] and also in [33].

Consider the speckle complex amplitude $\mathbf{A}(\mathbf{x}, \mathbf{y})$ in the vicinity of the center of a vortex. Let the center of the (x, y) coordinate system be placed at the center of the vortex, in which case the field in the vicinity of the singularity can be written

4.8 Zeros of Speckle Patterns: Optical Vortices

$$\mathbf{A}(x, y) = (\mathcal{R}_x x + \mathcal{R}_y y) + j(\mathcal{I}_x x + \mathcal{I}_y y) + \cdots$$
$$= \nabla \mathcal{R} \cdot \vec{r} + j \nabla \mathcal{I} \cdot \vec{r} + \cdots, \quad (4\text{-}209)$$

where terms higher order than linear have been neglected, and

$$\nabla \mathcal{R} = \frac{\partial}{\partial x} \mathcal{R} \hat{x} + \frac{\partial}{\partial y} \mathcal{R} \hat{y}$$

$$\nabla \mathcal{I} = \frac{\partial}{\partial x} \mathcal{I} \hat{x} + \frac{\partial}{\partial y} \mathcal{I} \hat{y} \quad (4\text{-}210)$$

$$\vec{r} = x\hat{x} + y\hat{y}.$$

Now we wish to convert the expression for **A** from a coordinate system expressed in terms of $(\mathcal{R}, \mathcal{I})$ to the (x, y) coordinate system. This transformation has a Jacobian

$$J = \left| \det \begin{bmatrix} \mathcal{R}_x & \mathcal{R}_y \\ \mathcal{I}_x & \mathcal{I}_y \end{bmatrix} \right| = |\mathcal{R}_x \mathcal{I}_y - \mathcal{R}_y \mathcal{I}_x|.$$

It is now possible to write the number of singularities in an area $\mathcal{A}$ as

$$N_{\mathcal{A}} = \iint_{\mathcal{A}} dx\, dy |\mathcal{R}_x \mathcal{I}_y - \mathcal{R}_y \mathcal{I}_x| \delta(\mathcal{R}) \delta(\mathcal{I}) \quad (4\text{-}211)$$

where the delta functions pick out the contours of $\mathcal{R} = 0$ and $\mathcal{I} = 0$. It follows that the (average) density of vortices, n_v, is

$$n_v = \lim_{\mathcal{A} \to \infty} \frac{N_{\mathcal{A}}}{\mathcal{A}} = \overline{\delta(\mathcal{R})\delta(\mathcal{I})|\mathcal{R}_x \mathcal{I}_y - \mathcal{R}_y \mathcal{I}_x|}, \quad (4\text{-}212)$$

where we have assumed ergodicity and replaced an area average by a statistical average. The appropriate probability density function for this average is that of Eq. (4-177). The integrations over $\mathcal{R}$ and $\mathcal{I}$ are easily performed, since we can use the sifting property of the delta functions, with the result

$$n_v = \frac{1}{2\pi\sigma^2} \int_{-\infty}^{\infty} \int_{-\infty}^{\infty} \int_{-\infty}^{\infty} \int_{-\infty}^{\infty} |\mathcal{R}_x \mathcal{I}_y - \mathcal{R}_y \mathcal{I}_x|$$
$$\times \frac{\exp\left[-\frac{\mathcal{R}_x^2 + \mathcal{I}_x^2}{2b_x}\right]}{2\pi b_x} \frac{\exp\left[-\frac{\mathcal{R}_y^2 + \mathcal{I}_y^2}{2b_y}\right]}{2\pi b_y} d\mathcal{R}_x\, d\mathcal{I}_x\, d\mathcal{R}_y\, d\mathcal{I}_y. \quad (4\text{-}213)$$

Integration is facilitated by making a change of variables. Representing $\nabla \mathcal{R}$ by a magnitude and phase $(|\nabla \mathcal{R}|, \varphi_0)$ and likewise $\nabla \mathcal{I}$ by $(|\nabla \mathcal{I}|, \varphi_0 + \varphi)$, we find that

$$|\mathcal{R}_x \mathcal{I}_y - \mathcal{R}_y \mathcal{I}_x| = |\nabla \mathcal{R}| |\nabla \mathcal{I}| |\cos \varphi_0 \sin(\varphi_0 + \varphi) - \cos(\varphi_0 + \varphi) \sin \varphi_0|$$
$$= |\nabla \mathcal{R}| |\nabla \mathcal{I}| |\sin \varphi|,$$

and the integral of interest becomes

$$n_v = \frac{1}{2\pi\sigma^2} \left(2\pi \int_0^\infty |\nabla\mathcal{R}|^2 \frac{e^{-\frac{|\nabla\mathcal{R}|^2}{2b_x}}}{2\pi b_x} d|\nabla\mathcal{R}| \right) \left(\int_0^\infty |\nabla\mathcal{I}|^2 \frac{e^{-\frac{|\nabla\mathcal{I}|^2}{2b_y}}}{2\pi b_y} d|\nabla\mathcal{I}| \right)$$

$$\times \left(\int_{-\pi}^{\pi} |\sin\varphi|\, d\varphi \right). \tag{4-214}$$

The first parenthesis evaluates to $\sqrt{\pi b_x/2}$, the second to $(1/(2\pi))\sqrt{\pi b_y/2}$ and the third to 4, which after combination yields

$$n_v = \frac{\sqrt{b_x b_y}}{2\pi\sigma^2}. \tag{4-215}$$

In the event that the brightness of the scattering spot has sufficient symmetry that $b_x = b_y = b$, the result becomes

$$n_v = \frac{b}{2\pi\sigma^2} = \frac{\text{constant}}{\mathcal{A}_c}, \tag{4-216}$$

where $\mathcal{A}_c$ is the intensity correlation area defined earlier, and the constant depends on the shape of the scattering spot, as detailed in Fig. 4.29. For a circularly-symmetric Gaussian spot, we find $n_v = 1/2\mathcal{A}_c$, or, on the average, one vortex for each two speckle intensity correlation areas.

4.8.4 The Density of Vortices for Fully Developed Speckle Plus a Coherent Background

Suppose that the observed speckle pattern is the superposition of fully developed speckle plus a mutually coherent plane wave of amplitude A_0 and intensity $I_0 = |A_0|^2$. Without loss of generality we can choose the real axis to coincide with the direction of the phasor representing the plane wave (cf. Eq. (3-22)). The effect of the plane wave is to introduce a nonzero mean to the real component of the combined field. A zero of the intensity of the combined fields now occurs when $\mathcal{R} = -A_0$ and $\mathcal{I} = 0$, and the expression for the density of vortices takes the modified form

$$n_v = \overline{\delta(\mathcal{R} + A_0)\delta(\mathcal{I})|\mathcal{R}_x \mathcal{I}_y - \mathcal{R}_y \mathcal{I}_x|}. \tag{4-217}$$

The joint probability density function of $(\mathcal{R}, \mathcal{I}, \mathcal{R}_x, \mathcal{I}_x, \mathcal{R}_y, \mathcal{I}_y)$ becomes

$$p(\mathcal{R}, \mathcal{I}, \mathcal{R}_x, \mathcal{I}_x, \mathcal{R}_y, \mathcal{I}_y) = \left(\frac{\exp\left[-\frac{(\mathcal{R}-A_0)^2 + \mathcal{I}^2}{2\sigma^2}\right]}{2\pi\sigma^2} \right)$$

$$\times \left(\frac{\exp\left[-\frac{\mathcal{R}_x^2 + \mathcal{I}_x^2}{2b_x}\right]}{2\pi b_x} \right) \left(\frac{\exp\left[-\frac{\mathcal{R}_y^2 + \mathcal{I}_y^2}{2b_y}\right]}{2\pi b_y} \right), \tag{4-218}$$

and the expression for n_v becomes

$$n_v = \frac{\exp(-I_0/\sigma^2)}{2\pi\sigma^2} \int_{-\infty}^{\infty}\int_{-\infty}^{\infty}\int_{-\infty}^{\infty}\int_{-\infty}^{\infty} |\mathcal{R}_x \mathcal{I}_y - \mathcal{R}_y \mathcal{I}_x|$$
$$\times \frac{\exp\left[-\frac{\mathcal{R}_x^2 + \mathcal{I}_x^2}{2b_x}\right]}{2\pi b_x} \frac{\exp\left[-\frac{\mathcal{R}_y^2 + \mathcal{I}_y^2}{2b_y}\right]}{2\pi b_y} d\mathcal{R}_x\, d\mathcal{I}_x\, d\mathcal{R}_y\, d\mathcal{I}_y. \qquad (4\text{-}219)$$

The integration required is identical to that performed in the previous section, and the average vortex density is given by

$$n_v = \frac{\sqrt{b_x b_y}}{2\pi\sigma^2} e^{-2\frac{I_0}{\overline{I_n}}} = \frac{b}{2\pi\sigma^2} e^{-2\frac{I_0}{\overline{I_n}}}, \qquad (4\text{-}220)$$

where, as in Section 3.3.2, we have defined $\overline{I_n} = 2\sigma^2$ to represent the average intensity of the random part of the field, and the right-hand part of the equation holds when $b_x = b_y = b$.

As might be expected, the presence of the coherent background reduces the probability that a vortex is formed, and therefore reduces the average density of vortices in a fashion that depends exponentially on the "signal-to-noise" ratio $I_0/\overline{I_n}$.

For additional properties of the optical vortices in speckle patterns, the reader may wish to consult Refs. [50] and [145]. Vortices appear in many other optical phenomena besides speckle. For a short and basic discussion, see [106], Chapter 15.

5 Optical Methods for Suppressing Speckle

In some applications, such as imaging with coherent light, speckle is a distinct annoyance and ways are sought to reduce or eliminate it. In other applications, such as speckle interferometry in nondestructive testing, speckle is used to advantage and there is no desire to suppress it. In this chapter, we take the former point of view and discuss a variety of ways that can be used to reduce the effects of speckle in coherent imaging.

There have been a number of different studies that have attempted to quantify the degree to which speckle reduces the ability of the human observer to extract details from coherent images [96] [180] [55] [95] [7] [6]. Effects on both resolution and contrast sensitivity have been studied. It is difficult to summarize the results of these studies in a few words, except to say that there is no doubt that fully-developed, unaveraged speckle seriously degrades the process of information extraction from images. The interested reader should consult the references above for details.

Before addressing how speckle can be suppressed, it is worth pointing out that speckle can be completely avoided if images are formed with incoherent light, so when such light can be used, it is wise to do so. However, many imaging modalities are fundamentally coherent, and often the objects being imaged are rough on the scale of a wavelength. In such cases, speckle must be either tolerated or dealt with. Examples are conventional holography, synthetic-aperture radar imagery, and conventional coherence tomography, to name a few. In other cases, the brightness afforded by laser sources makes the use of coherent illumination highly desirable, provided speckle can be suppressed to a satisfactory degree. An example would be high-brightness image or video projectors suitable for very large displays. Each application of interest may be amenable to different methods for suppressing speckle, so the approaches described below are not all applicable to all cases. Nonetheless, it is useful to have a discussion of many available methods in one place, and therefore we embark upon such an endeavor.

5.1 Polarization Diversity

When the diffuser or rough surface has the property that different sets of scatterers contribute to the light in the two orthogonal polarizations, or when the same set of scatterers exhibits different phases in the two orthogonal polarizations, then the speckle patterns present in the two polarization components of intensity will be independent, and, in accord with the discussion of Section 3.4, the speckle contrast will be reduced. The amount of reduction depends on the degree of polarization $\mathcal{P}$ of the light in the speckle pattern, but at maximum the reduction is by a factor of $1/\sqrt{2}$.

The fact that the scattered light is often depolarized by a rough surface is easy to understand with a simple example. With reference to Fig. 5.1, suppose that a wave linearly polarized in the $\hat{y}$ direction, traveling in the $\hat{z}$ direction out of the page, is incident on a perfectly conducting plane (representing, perhaps, the local behavior of a metallic random surface) at grazing incidence. The boundary conditions that govern reflection are that the component of reflected field parallel to the metal surface must exactly cancel the corresponding component of the electric field of the incident wave, while the component of field normal to the surface remains unchanged. As can be seen from the figure, the resultant field after reflection contains a polarization component in the $\hat{x}$ direction. The situation is more complex when the angle of incidence on the surface is more general and the tilt of the conducting plane is more general. For a more complete discussion, see Chapter 8 of [16]. For a dielectric surface, the Fresnel reflection coefficients likewise differ for polarizations parallel to and orthogonal to a dielectric interface ([139], section 6.2). A further cause of depolarization can be multiple scattering, in which an incident wave is reflected by a multitude of surface facets, usually inclined at different angles, before leaving the surface.

Whether a given diffuser or diffusely reflecting surface depolarizes the speckle can best be determined experimentally, by observing the speckle pattern through a

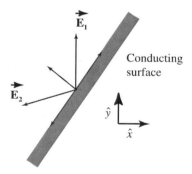

Figure 5.1 Reflection of a linearly polarized wave from a conducting surface. Components of field parallel and normal to the surface are shown but unlabeled. $\vec{E}_1$ is the incident electric field and $\vec{E}_2$ is the reflected electric field. In this simple example, the direction of propagation is out of the page, and the wave strikes the surface at grazing incidence.

polarization analyzer and determining whether the speckle pattern changes when the analyzer is rotated, and whether the average intensities of the two different speckle patterns are comparable. When this is the case, the contrast reduction factor $1/\sqrt{2}$ can be expected. If the average intensities of the two independent speckle patterns are not equal, then the degree of polarization is greater than zero, and a reduction less than $1/\sqrt{2}$ will occur.

However, in this case further reduction can usually be achieved if the polarization of the source is rotated by 90 degrees and the diffuser has the property that it produces two speckle patterns that are independent of the first two patterns. In this case, provided the incident polarization is switched between orthogonal states faster than the response time of the detector, effectively four, independent speckle patterns will be added on an intensity basis. In the special case of all four patterns being of equal average intensity, the speckle contrast will be reduced by a factor of $1/2$. In many applications, every possible reduction of contrast may be needed, and the small factor afforded by polarization diversity will still be welcome.

5.2 Temporal Averaging with a Moving Diffuser

5.2.1 Background

We consider a stationary, planar, transparent object that is illuminated through an optically rough, moving diffuser placed in close proximity to the object. The geometry is illustrated in Fig. 5.2. Part (a) of the figure shows the imaging system without an added diffuser, part (b) shows the system with an added moving diffuser, and part (c) shows the system with both a moving diffuser and a fixed diffuser. The moving diffuser is assumed to be moved with speed v in the α direction shown by the arrows. For simplicity, we assume that the magnification of the imaging system is unity.

The transparency object is assumed to be uniformly transmitting, so that we can concentrate on the properties of the speckle observed, and not on the intensity transmittance of the object. The object transparency itself may be optically smooth or optically rough on the scale of an optical wavelength. It should be noted that for an optically smooth object, there is little motivation to introduce a single moving diffuser, since when the diffuser is not present, there is no speckle. However, if the object is optically rough, then speckle will be present in any case, and the goal of introducing the second, moving diffuser is to reduce the speckle by time averaging. Note that the case depicted in part (c) of Fig. 5.2 is entirely equivalent to the case of a moving diffuser and an optically rough object. In the future we shall refer to this case as that of an "optically rough object," even though the roughness may come from a separate fixed diffuser.

As the diffuser is moved, any point on the object experiences a changing phase of illumination. Thus for any given point in the image, the region on the object contributing to that image point generates a changing random walk of amplitudes, in which the phases of the various contributions at this point are changing with time

144 CHAPTER 5 Optical Methods for Suppressing Speckle

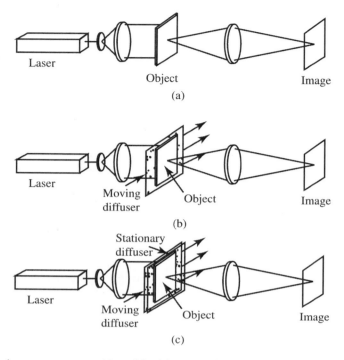

Figure 5.2 Imaging a transparent object. (a) with no moving diffuser, (b) with a moving diffuser, and (c) with both a moving diffuser and a fixed diffuser.

in a complex way. In effect, new realizations of the random walk are created as time progresses, and as a consequence the speckle intensity at any point on the image changes with time. The detected intensity measured with an integration time T is simply the integral over a number of independent speckle realizations. It should also be noted again that the case of an optically rough object and a moving diffuser is equivalent to the case of a cascade of a moving diffuser, a stationary diffuser and the smooth object. Our analysis will be general enough to cover all of these cases.

An analysis of the geometry of interest begins with a representation of the field in the image plane in terms of the field transmitted by the compound object (that is, a moving diffuser plus a rough object transparency or a moving diffuser plus a stationary diffuser plus a smooth object transparency), again under the assumption of unity magnification,

$$\mathbf{A}(x, y; t) = \iint_{-\infty}^{\infty} \mathbf{k}(x + \alpha, y + \beta) \mathbf{a}(\alpha, \beta; t) \, d\alpha \, d\beta, \tag{5-1}$$

where both **A** and **a** are time-varying, by virtue of the moving diffuser, and **k** represents the amplitude point-spread function of the imaging system[1]. For simplicity, we assume that the object transparency has an intensity transmittance that is uniform but that the phase has two independent parts, one contributed by the stationary object (ϕ_o) (or stationary diffuser) and one contributed by the moving diffuser (ϕ_d),

$$\mathbf{a}(\alpha, \beta; t) = \mathbf{a}_o e^{j\phi_o(\alpha,\beta)} e^{j\phi_d(\alpha-vt,\beta)}, \tag{5-2}$$

where $\mathbf{a}_o$ is a constant, and the diffuser is assumed to be moving in the positive α direction with speed v.

Our approach to finding the amount of speckle suppression and its dependence on diffuser motion will be to find the normalized temporal autocovariance function of the image field $\mathbf{A}(x, y; t)$, take the squared magnitude of this quantity to find the normalized autocovariance function of intensity, and then assess the number of temporal degrees of freedom M in the image plane as a function of the motion of the diffuser. Since the imaging system has been assumed to be space invariant in writing the object–image relationship above, it suffices to calculate these autocovariance functions at a single pair of image coordinates, which we take to be $(x = 0, y = 0)$.

We begin with an expression for the temporal autocorrelation function of the image fields,

$$\Gamma_\mathbf{A}(\tau) = \overline{\mathbf{A}(0,0;t)\mathbf{A}^*(0,0;t-\tau)} = \iint_{-\infty}^{\infty}\iint_{-\infty}^{\infty} \mathbf{k}(\alpha_1, \beta_1)\mathbf{k}^*(\alpha_2, \beta_2)$$

$$\times \overline{\mathbf{a}(\alpha_1, \beta_1; t)\mathbf{a}^*(\alpha_2, \beta_2; t-\tau)} \, d\alpha_1 \, d\beta_1 \, d\alpha_2 \, d\beta_2. \tag{5-3}$$

The average inside the integral sign can be rewritten

$$\overline{\mathbf{a}(\alpha_1,\beta_1;t)\mathbf{a}^*(\alpha_2,\beta_2;t-\tau)} = |\mathbf{a}_o|^2 \overline{e^{j[\phi_o(\alpha_1,\beta_1)-\phi_o(\alpha_2,\beta_2)]}} \, \overline{e^{j[\phi_d(\alpha_1-vt,\beta_1)-\phi_d(\alpha_2-vt+v\tau,\beta_2)]}} \tag{5-4}$$

where it has been assumed that the random phase processes ϕ_o and ϕ_d are statistically independent. If in addition these processes are Gaussian, zero mean, and statistically stationary, then from the relationship of these averages to characteristic functions, we have

$$\overline{e^{j[\phi_o(\alpha_1,\beta_1)-\phi_o(\alpha_2,\beta_2)]}} = e^{-\sigma_o^2[1-\mu_o(\Delta\alpha,\Delta\beta)]}$$

$$\overline{e^{j[\phi_d(\alpha_1-vt,\beta_1)-\phi_d(\alpha_2-vt+v\tau,\beta_2)]}} = e^{-\sigma_d^2[1-\mu_d(\Delta\alpha-v\tau,\Delta\beta)]}, \tag{5-5}$$

where $\Delta\alpha = \alpha_1 - \alpha_2$, $\Delta\beta = \beta_1 - \beta_2$, σ_o^2 is the variance of ϕ_o, σ_d^2 is the variance of ϕ_d, while μ_o and μ_d are the normalized autocorrelation functions of the two phase processes.

---

1. We use $(x + \alpha, y + \alpha)$ as the argument of **k** here, rather than $(x - \alpha, y - \alpha)$, to account for image inversion.

Returning to Eq. (5-3), the autocorrelation function can be simplified to

$$\mathbf{\Gamma_A}(\tau) = |\mathbf{a}_o|^2 \iint\limits_{-\infty}^{\infty} \mathbf{K}(\Delta\alpha, \Delta\beta) e^{-\sigma_o^2[1-\mu_o(\Delta\alpha, \Delta\beta)]} e^{-\sigma_d^2[1-\mu_d(\Delta\alpha-v\tau, \Delta\beta)]} \, d\Delta\alpha \, d\Delta\beta, \quad (5\text{-}6)$$

where

$$\mathbf{K}(\Delta\alpha, \Delta\beta) = \iint\limits_{-\infty}^{\infty} \mathbf{k}(\alpha_1, \beta_1) \mathbf{k}^*(\alpha_1 - \Delta\alpha, \beta_1 - \Delta\beta) \, d\alpha_1 \, d\beta_1 \quad (5\text{-}7)$$

is the deterministic autocorrelation function of the amplitude point-spread function of the imaging system.

The processes $e^{j\phi_o}$ and $e^{j\phi_d}$ have nonzero means, which can be significant if the variances of the phase processes are small. These means represent the specular components of transmittance. The effects of these means are the nonzero residual values of the diffuser and object correlation functions when the phase correlation functions μ_o and μ_d fall to zero,

$$\begin{aligned} e^{-\sigma_o^2[1-\mu_o(\Delta\alpha, \Delta\beta)]} &\to e^{-\sigma_o^2} \quad \text{when } \mu_o \to 0 \\ e^{-\sigma_d^2[1-\mu_d(\Delta\alpha-v\tau, \Delta\beta)]} &\to e^{-\sigma_d^2} \quad \text{when } \mu_d \to 0. \end{aligned} \quad (5\text{-}8)$$

Thus, if we wish to find the autocovariance function of the fields, i.e. $\mathbf{C_A} = \mathbf{\Gamma_A} - \overline{\mathbf{A}}\,\overline{\mathbf{A}^*}$, we subtract terms on the right-hand side of Eq. (5-8) from the integrand of our expression for $\mathbf{\Gamma_A}$, yielding

$$\mathbf{C_A}(\tau) = |\mathbf{a}_o|^2 e^{-(\sigma_o^2+\sigma_d^2)} \iint\limits_{-\infty}^{\infty} \mathbf{K}(\Delta\alpha, \Delta\beta)[e^{\sigma_o^2 \mu_o(\Delta\alpha, \Delta\beta)} - 1]$$

$$\times [e^{\sigma_d^2 \mu_d(\Delta\alpha-v\tau, \Delta\beta)} - 1] \, d\Delta\alpha \, d\Delta\beta. \quad (5\text{-}9)$$

The form of the function $\mathbf{K}$ can be found rather easily. Since it is the autocorrelation function of the point-spread function, its Fourier transform is the squared magnitude of the pupil function of the lens. For a lens that has a pupil function that is unity inside the pupil and zero outside the pupil, the form of $\mathbf{K}$ must be exactly the same as the form of the point-spread function $\mathbf{k}$. In addition, since we are ultimately concerned with the normalized autocovariance function, we can neglect any multiplicative constants that will be cancelled by the normalization. Therefore, for the case of a square pupil function of width L, we have

$$\mathbf{K}(\Delta\alpha, \Delta\beta) \propto \operatorname{sinc}\left(\frac{L\Delta\alpha}{\lambda z}\right) \operatorname{sinc}\left(\frac{L\Delta\beta}{\lambda z}\right),$$

where z is the distance from the entrance pupil to the object plane, and for a circular pupil of diameter D,

$$\mathbf{K}(\Delta\alpha, \Delta\beta) \propto 2 \frac{J_1\left(\frac{\pi D r}{\lambda z}\right)}{\frac{\pi D r}{\lambda z}}$$

where $r = \sqrt{\Delta\alpha^2 + \Delta\beta^2}$. The first term in the integrand represents the effects of the point-spread function of the system, the second term the effect of the object roughness or alternatively the effect of a separate fixed diffuser, and the third term represents the effect of the moving diffuser. Note that in general, the width of $\mathbf{K}$ will be much greater than the width of the two correlation functions, for we are focusing on fully developed speckle, for which there must be many correlation areas of phase within the area of the point-spread function.

Further analytical progress requires the assumption of a form for the normalized autocorrelation functions of the phases, $\boldsymbol{\mu}_o$ and $\boldsymbol{\mu}_d$. We make the assumption that these functions are Gaussian,

$$\boldsymbol{\mu}_o(\Delta\alpha, \Delta\beta) = e^{-\frac{\Delta\alpha^2 + \Delta\beta^2}{r_o^2}}$$
$$\boldsymbol{\mu}_d(\Delta\alpha, \Delta\beta) = e^{-\frac{\Delta\alpha^2 + \Delta\beta^2}{r_d^2}}, \qquad (5\text{-}10)$$

giving

$$\mathbf{C}_\mathbf{A}(\tau) = \iint_{-\infty}^{\infty} \mathbf{K}(\Delta\alpha, \Delta\beta) \left[\exp\left(\sigma_o^2 e^{-\frac{\Delta\alpha^2 + \Delta\beta^2}{r_o^2}}\right) - 1\right]$$
$$\times \left[\exp\left(\sigma_d^2 e^{-\frac{(\Delta\alpha - v\tau)^2 + \Delta\beta^2}{r_d^2}}\right) - 1\right] d\Delta\alpha \, d\Delta\beta. \qquad (5\text{-}11)$$

The normalized autocovariance function of the field $\mathbf{A}$ is given by

$$\boldsymbol{\mu}_\mathbf{A}(\tau) = \mathbf{C}_\mathbf{A}(\tau)/\mathbf{C}_\mathbf{A}(0).$$

While it appears to complicate matters, the interpretation of the expression for $\mathbf{C}_\mathbf{A}$ is helped if we normalize each of the three factors in the integrand of $\mathbf{C}_\mathbf{A}$ by their maximum values, applying the same normalizations in the numerator and denominator of $\boldsymbol{\mu}_\mathbf{A}$ so as not to change its value. Representing the normalized version of $\mathbf{C}_\mathbf{A}$ by $\hat{\mathbf{C}}_\mathbf{A}$,

$$\hat{\mathbf{C}}_\mathbf{A}(\tau) = \iint_{-\infty}^{\infty} \frac{\mathbf{K}(\Delta\alpha, \Delta\beta)}{\mathbf{K}(0,0)} \left[\frac{\exp\left(\sigma_o^2 e^{-\frac{\Delta\alpha^2 + \Delta\beta^2}{r_o^2}}\right) - 1}{e^{\sigma_o^2} - 1}\right]$$
$$\times \left[\frac{\exp\left(\sigma_d^2 e^{-\frac{(\Delta\alpha - v\tau)^2 + \Delta\beta^2}{r_d^2}}\right) - 1}{e^{\sigma_d^2} - 1}\right] d\Delta\alpha \, d\Delta\beta, \qquad (5\text{-}12)$$

where now

$$\boldsymbol{\mu}_\mathbf{A}(\tau) = \hat{\mathbf{C}}_\mathbf{A}(\tau)/\hat{\mathbf{C}}_\mathbf{A}(0).$$

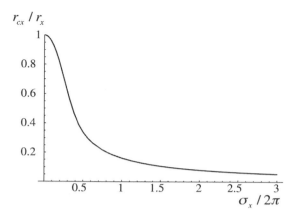

Figure 5.3 Normalized correlation radius vs. normalized standard deviation of the phase. The symbol x stands for either o or d, depending on which term is of interest.

We can gain some insight into the evaluation of these integrals by first considering the width of the bracketed factors. The result presented in Eq. (4-109), which gives the amplitude correlation area of such a factor, can be used directly. Since in the circularly symmetric case, the correlation radius r_c is related to the correlation area $\mathcal{A}_c$ through

$$r_c = \sqrt{\mathcal{A}_c/\pi},$$

we have the correlation radii of the second and third factors in the integrand being

$$r_{co} = r_o \left[\frac{\mathrm{Ei}(\sigma_o^2) - \mathcal{E} - \ln(\sigma_o^2)}{(e^{\sigma_o^2} - 1)} \right]^{1/2}$$

$$r_{cd} = r_d \left[\frac{\mathrm{Ei}(\sigma_d^2) - \mathcal{E} - \ln(\sigma_d^2)}{(e^{\sigma_d^2} - 1)} \right]^{1/2},$$

(5-13)

where, as in Eq. (4-109), $\mathrm{Ei}(x)$ represents the exponential integral and $\mathcal{E}$ is Euler's constant.

Figure 5.3 shows a plot of the normalized correlation radius (where the symbol x stands for either o or d) vs. the normalized standard deviation of the phase. From this plot we can determine the approximate width of the second and third factors in the integrand of $\hat{C}_A$.

Further insight is gained by plotting the three factors of the integrand in a typical case. Generally we are interested in the case of fully developed speckle, in which case the width of the **K** factor must be much wider than the widths of the two bracketed factors in order that many independent correlation areas contribute to the random walk that generates speckle. Figure 5.4 shows typical forms for each of the three fac-

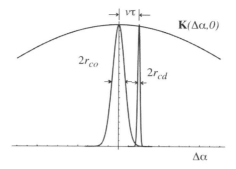

Figure 5.4 Illustration of the three factors in the integrand of the numerator.

tors in the integrand of $\hat{C}_A(\tau)$ along the $\Delta\alpha$ axis. The function $\mathbf{K}(\Delta\alpha, 0)$ is much broader than the two correlation function factors. In this example, we have assumed that the diffuser has a narrower correlation function than the object (i.e. that the object is "smoother" than the diffuser), and the diffuser correlation function is shown shifted to the right by $v\tau$. Clearly when $v\tau > r_{cd} + r_{co}$, the overlap of the two correlation functions has dropped significantly, and the value of the integral will be small. The picture appropriate for the denominator of $\hat{C}_A$ would have no shift of the correlation function of the moving diffuser.

To progress further we must look at some special cases. First we examine the case of a smooth object and a moving diffuser. Then we look at the case of a rough object and a moving diffuser.

5.2.2 Smooth Object

Consider first the case of a smooth object and a single moving diffuser (case (b) in Fig. 5.2). In such a case, $\sigma_o^2 \to 0$, and the second factor in Eq. (5-12) becomes unity, leaving

$$\mu_A(\tau) = \frac{\iint\limits_{-\infty}^{\infty} \mathbf{K}(\Delta\alpha, \Delta\beta) \left[\exp\left(\sigma_d^2 e^{-\frac{(\Delta\alpha-v\tau)^2+\Delta\beta^2}{r_d^2}}\right) - 1\right] d\Delta\alpha\, d\Delta\beta}{\iint\limits_{-\infty}^{\infty} \mathbf{K}(\Delta\alpha, \Delta\beta) \left[\exp\left(\sigma_d^2 e^{-\frac{\Delta\alpha^2+\Delta\beta^2}{r_d^2}}\right) - 1\right] d\Delta\alpha\, d\Delta\beta}. \qquad (5\text{-}14)$$

For a circular pupil in the imaging system, the factor $\mathbf{K}(\Delta\alpha, \Delta\beta)$ has a width of approximately $\lambda z/D$, whereas the diffuser correlation function has a width of approximately $2r_c$ in the $\Delta\alpha$ direction, which in general is much smaller than the width of $\mathbf{K}$. This equation can be regarded as a normalized deterministic cross-correlation function of the function $\mathbf{K}$ with the much narrower diffuser autocorrelation function, and as such, the integral will result in a close approximation to $\mathbf{K}$ itself. That is, to a good approximation,

$$\mu_A(\tau) \approx \frac{\mathbf{K}(v\tau,0)}{\mathbf{K}(0,0)} = 2\frac{J_1\left(\frac{\pi D v\tau}{\lambda z}\right)}{\frac{\pi D v\tau}{\lambda z}} \tag{5-15}$$

for a circular pupil, and

$$\mu_A(\tau) \approx \frac{\mathbf{K}(v\tau,0)}{\mathbf{K}(0,0)} = \operatorname{sinc}\left(\frac{Lv\tau}{\lambda z}\right) \tag{5-16}$$

for a square pupil.

The correlation time of the speckle intensity[2] in each of these cases is found from

$$\tau_c = \int_{-\infty}^{\infty} |\mu_A(\tau)|^2 \, d\tau, \tag{5-17}$$

which yields in these two cases

$$\text{circular pupil:} \quad \tau_c = \frac{32\lambda z}{3\pi^2 vD} \tag{5-18}$$

$$\text{square pupil:} \quad \tau_c = \frac{\lambda z}{vL}. \tag{5-19}$$

We are now prepared to address the main question of interest: how far must the diffuser move to suppress the speckle contrast by some prespecified amount, and how does that distance depend on the properties of the optical system and the diffuser? To answer this question, we must use Eq. (4-141) appropriate for temporal integration of speckle. Using this equation and Eqs. (5-15) and (5-16) with a small change of variables, we have

$$M = \left[2\int_0^1 (1-x) \frac{J_1^2\left(\pi \frac{vT}{\lambda z/D} x\right)}{\left(\pi \frac{vT}{\lambda z/D} x\right)^2} dx\right]^{-1}$$

$$M = \left[2\int_0^1 (1-x) \operatorname{sinc}^2\left(\frac{vT}{\lambda z/L} x\right) dx\right]^{-1}. \tag{5-20}$$

Figure 5.5 shows plots of the speckle contrast $C = \sqrt{1/M}$ vs. normalized distance of motion for the cases of circular and square pupils. The critical parameter in both cases can be seen to be the ratio of the distance moved by the diffuser, vT, to the approximate linear dimension of an image resolution cell on the object, that is, $\lambda z/D$ in the circular case, and $\lambda z/L$ in the square case. The asymptotic behavior of C can be found by returning to Eq. (4-141). As T grows sufficiently large, we find that $1 - \frac{\tau_c}{T} \approx 1$ over the region for which $|\mu_A|^2$ has significant value, and therefore

---

[2]. We consistently take the correlation time of speckle intensity to be the expression given, but the correlation time of the speckle amplitude to be $\int_{-\infty}^{\infty} \mu_A(\tau) \, d\tau$.

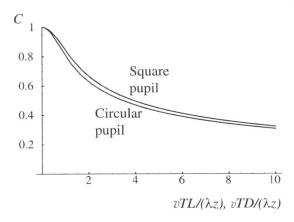

Figure 5.5 Plots of contrast vs. the ratio of distance moved by the diffuser (vT) to the approximate image resolution dimension in the object space ($\lambda z/D$ in the circular-pupil case and $\lambda z/L$ in the square-pupil case).

$$M \approx \left[\frac{1}{T}\int_{-\infty}^{\infty} |\mu_A(\tau)|^2\, d\tau\right]^{-1} = T/\tau_c. \qquad (5\text{-}21)$$

It follows that when T/τ_c is large, the contrast C behaves as

$$C \approx \sqrt{\frac{\tau_c}{T}} \approx \begin{cases} \sqrt{\dfrac{\lambda z/D}{vT}} & \text{circular pupil} \\ \sqrt{\dfrac{\lambda z/L}{vT}} & \text{square pupil.} \end{cases} \qquad (5\text{-}22)$$

To summarize the results of this subsection, when a single moveable diffuser is placed in close proximity to a smooth, transparent object, the contrast of the speckle is reduced in proportion to the square root of the number of resolution widths on the object that the diffuser is moved during the exposure period. This is a rather slow reduction in speckle, requiring comparatively large motions of the diffuser during the exposure time. In the next subsection, we shall see that if the object itself is optically rough, or if a second fixed diffuser is introduced in proximity to a smooth object, the motion required to suppress the contrast is greatly reduced.

5.2.3 Rough Object

The case of moving diffuser and a rough object, or alternatively a cascade of a moving diffuser, a fixed diffuser, and a smooth object, can be studied using Eq. (5-11) as a starting point. For simplicity of analysis, we assume here that the rough object case actually refers to a smooth object, preceded by a fixed diffuser, preceded in turn by a

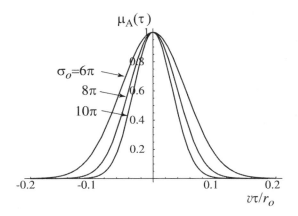

Figure 5.6 Plots of μ_A vs. $v\tau/r_o$ for three values of σ_o.

moving diffuser, and in addition, we assume that the two diffusers are statistically similar, so that $\sigma_d^2 = \sigma_o^2$ and $r_d = r_o$. Since the widths of the correlation functions of the diffusers are much narrower than the width of **K**, we can replace $\mathbf{K}(\Delta\alpha, \Delta\beta)$ by $\mathbf{K}(0,0)$, in which case the expression for $\mathbf{C_A}$ becomes

$$\mathbf{C_A}(\tau) = \iint_{-\infty}^{\infty} \left[\exp\left(\sigma_o^2 e^{-\frac{\Delta\alpha^2+\Delta\beta^2}{r_o^2}}\right) - 1\right]$$

$$\times \left[\exp\left(\sigma_o^2 e^{-\frac{(\Delta\alpha-v\tau)^2+\Delta\beta^2}{r_o^2}}\right) - 1\right] d\Delta\alpha\, d\Delta\beta. \tag{5-23}$$

Figure 5.6 shows plots of the corresponding $\mu_A(\tau)$ vs. $v\tau/r_o$ for three different values of the standard deviation of the phase imparted by either of the diffusers. As expected, the temporal correlation narrows as the standard deviation of phase increases.

The contrast C of the temporally integrated speckle is given by

$$C = \sqrt{\frac{1}{M}} = \sqrt{\frac{2}{T}\int_0^T \left(1 - \frac{\tau}{T}\right)|\mu_A(\tau)|^2\, d\tau}, \tag{5-24}$$

which can be found by numerical integration and is shown in Fig. 5.7, plotted vs. vT/r_o. If we calculate the normalized diffuser correlation radii r_c/r_o from Eq. (5-13) we find that they have values of approximately 0.05, 0.04 and 0.03 for the cases $\sigma_o = 6\pi$, 8π and 10π, respectively. A motion of one phase correlation radius r_o actually corresponds to motion over approximately 19, 25 and 31 diffuser correlation radii r_c, respectively.

The primary conclusion from this discussion of the case of two identical diffusers, one stationary and one moving, is that the speckle contrast suppression with time in-

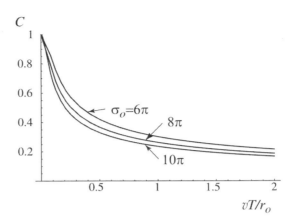

Figure 5.7 Contrast vs. normalized distance of diffuser motion for 3 values of diffuser phase standard deviation.

tegration is determined by the number of diffuser correlation cells moved, rather than by the number of object resolution elements moved. Since $r_c \ll \lambda z/D$, speckle suppression for a given movement vT is much greater with two diffusers than with one. This fact was first reported by Lowenthal and Joyeux in Ref. [103]. However, this extra speckle suppression comes at a price. The angular divergence of the light after passing through two identical diffusers is approximately twice what it is for a single diffuser. Therefore, an optical imaging system with a finite aperture will collect less of the light leaving the object plane when two diffusers are present. There is therefore a tradeoff between speckle suppression and brightness of the image. In some cases this tradeoff is acceptable, and in other cases it may not be, depending on the power of the source, the sensitivity of the detector, and the flexibility of the geometry of the imaging system.

Note that if the divergence of the two diffusers in cascade is matched to the angular acceptance of the imaging lens so that there is little or no light lost, then the correlation areas on the two diffusers are such that the moving diffuser must move by at least as much as the motion required of a single diffuser, and the advantage of using two diffusers vanishes.

5.3 Wavelength and Angle Diversity

The detailed distribution of intensity in a speckle pattern depends on the angles of illumination and observation, and on the wavelength of the incident light. In the analyses that follow, we first consider the case of free-space propagation, and then turn attention to an imaging geometry. For alternative approaches to parts of the subject matter of interest here, see [65], [125] and [56].

Throughout this section, the following assumptions will be adopted:

1. For both a transmission geometry and a reflection geometry, the scattering or diffusing surface is assumed to be rough on the scale of a wavelength for all angles of illumination and observation, and for all wavelengths. Under this condition, the specular component of transmission or reflection will be negligible, and the observed speckle fields will obey circular complex Gaussian statistics. This eliminates from consideration angles near grazing incidence, for which virtually all surfaces behave like a smooth mirror.
2. In the case of free-space propagation, the propagation of light from a plane just above the scattering spot to the observation region will be assumed to be describable by the Fresnel diffraction equation. This assumption restricts the angles of observation with respect to the z-axis (normal to the plane above the surface) to be not too large, and also restricts the size of the scattering spot to be small compared with the distance to the observation region.
3. Polarization effects will be neglected in the analysis.
4. Changes of scattering spot shape (for example, from circular at normal incidence to elliptical at off-normal incidence) will be neglected since they are not the prime factor affecting the detailed structure of the speckle.
5. Shadowing caused by oblique illumination of the surface is assumed to be insignificant, and any multiple scattering effects are excluded.

5.3.1 Free-Space Propagation, Reflection Geometry

Before embarking on analyses, we state in words what the dependencies of speckle are on angle of illumination, angle of observation, and wavelength. While these effects are discussed in the context of a reflection geometry, similar statements, slightly modified, apply to the transmission geometry as well. Consider each of these variations individually:

- Angle of illumination: An increase of the angle of illumination *with respect to the surface normal* (all other parameters remaining fixed) has two effects. First, the observed speckle pattern rotates by an amount equal and opposite to the rotation of the illumination angle, by virtue of the changes of phase of illumination arriving at the various scatterers on the surface. Second, the rotated speckle pattern undergoes internal changes due to "foreshortening" of the surface height fluctuations caused by the larger angle of illumination (which is closer to grazing incidence).
- Angle of observation: An increase in the angle of observation with respect to the surface normal (all other parameters being fixed) moves the observer through the speckle pattern, but also causes "foreshortening" of the surface height fluctuations due to the fact that the observation angle is closer to grazing incidence.
- Wavelength: An increase of wavelength (all other parameters remaining fixed) has two effects. First the observed speckle pattern undergoes a spatial scaling by an amount that depends on distance from the point where a mirror reflection of

the incident k-vector would pierce the observation plane. When wavelength increases, the speckle pattern expands about this point, and when wavelength decreases the speckle pattern shrinks about this point. This effect is caused by the fact that diffraction angles are proportional to wavelength. A second effect is upon the random phase shifts imparted to the scattered wave by the surface-height fluctuations. Since those phase shifts are proportional to h/λ, an increase in wavelength will decrease the random phase shifts, while a decrease of wavelength will increase the random phase shifts.

Hence the use of changing directions of illumination and/or observation, or changes of wavelength, can induce changes in the speckle pattern and, when such speckle patterns are superimposed on an intensity basis, thereby reduce the observed contrast. Alternatively, the use of two separate wavelengths in the illumination, can, under the right conditions to be derived, reduce speckle contrast. Finally, if two mutually incoherent sources illuminate the scattering surface from different directions, the resulting speckle patterns will add on an intensity basis and, again under the proper conditions to be derived, will reduce the speckle contrast.

To quantify these statements somewhat, consider the geometry shown in Fig. 5.8. A coordinate system $\vec{\alpha} = (\alpha, \beta, z)$ is attached to a plane just above the scattering surface, with the z axis being normal to that plane. The transverse components of $\vec{\alpha}$ are represented by the vector

$$\vec{\alpha}_t = \alpha\hat{\alpha} + \beta\hat{\beta}, \tag{5-25}$$

where $\hat{\alpha}$ and $\hat{\beta}$ are unit vectors in the α and β directions, respectively.

The average wave-vector in the illuminating beam is represented by $\vec{k}_i$, which has length $k = 2\pi/\lambda$ and direction $\hat{\imath}$. Note that this is only an "average" wave-vector, because to generate a finite spot requires a range of wave-vector angles. However, our

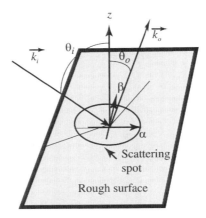

Figure 5.8 Scattering geometry, reflection case.

assumptions above assure that this angular range will be small. The wave-vector in the observation direction is $\vec{k}_o$, which has length $k = 2\pi/\lambda$ and direction $\hat{o}$.

As discussed in Section 4.5.4 (see Eq. (4-97)), the phase shift imparted to the reflected wave by the rough surface, as measured in the (α, β) plane, is

$$\phi(\alpha, \beta) = 2\pi(-\hat{i} \cdot \hat{z} + \hat{o} \cdot \hat{z})\frac{h(\alpha, \beta)}{\lambda} = ((-\vec{k}_i + \vec{k}_o) \cdot \hat{z})h(\alpha, \beta). \qquad (5\text{-}26)$$

Here $\hat{z}$ is a unit vector in the normal direction. Following Parry [125], we define a *scattering vector* $\vec{q}$ by the relation

$$\vec{q} = \vec{k}_o - \vec{k}_i. \qquad (5\text{-}27)$$

Note that

$$\vec{q} = q_\alpha \hat{\alpha} + q_\beta \hat{\beta} + q_z \hat{z} = \vec{q}_t + q_z \hat{z}, \qquad (5\text{-}28)$$

with $\vec{q}_t$ being the transverse component of $\vec{q}$,

$$\vec{q}_t = q_\alpha \hat{\alpha} + q_\beta \hat{\beta} = (\vec{k}_o)_t - (\vec{k}_i)_t \qquad (5\text{-}29)$$

with magnitude

$$|\vec{q}_t| = q_t = k|\sin \theta_o - \sin \theta_i|, \qquad (5\text{-}30)$$

and $q_z = \vec{q} \cdot \hat{z}$ being the normal component of $\vec{q}$,

$$q_z = k[\cos \theta_o + \cos \theta_i], \qquad (5\text{-}31)$$

where $k = 2\pi/\lambda$ and the angles θ_o and θ_i are the angles subtended by $\vec{k}_o$ and $-\vec{k}_i$ with respect to the z-axis. The phase shift ϕ can be written as

$$\phi(\alpha, \beta) = q_z h(\alpha, \beta). \qquad (5\text{-}32)$$

We return to discussing the scattering vector in more detail in a moment.

In this geometry, the speckle intensity is observed in the (x, y) plane, which is parallel to the (α, β) plane and distance z from it. Only free space lies between the two planes. The transverse coordinates in this plane are represented by a vector

$$\vec{x}_t = x\hat{x} + y\hat{y}.$$

Note that, in terms of the observation direction $\hat{o}$,

$$\begin{aligned} x &\approx z(\hat{o} \cdot \hat{x}) = z(\hat{o} \cdot \hat{\alpha}) \\ y &\approx z(\hat{o} \cdot \hat{y}) = z(\hat{o} \cdot \hat{\beta}), \end{aligned} \qquad (5\text{-}33)$$

where small angles have been assumed in the approximation, and the right-hand side of these equations arises from the fact that $\hat{x}$ and $\hat{\alpha}$ point in the same direction, as do also $\hat{y}$ and $\hat{\beta}$.

Our initial goal will be to find the cross-correlation between the field $\mathbf{A}_1(x_1, y_1)$ observed at coordinates (x_1, y_1) and a second field $\mathbf{A}_2(x_2, y_2)$ observed at (x_2, y_2), where the two fields differ through any or all of the following: (1) wavelength of illumination; (2) average angle of illumination of the surface; (3) average angle of observation. Thus our goal is to find

$$\Gamma_A(x_1, y_1; x_2, y_2) = \overline{\mathbf{A}_1(x_1, y_1)\mathbf{A}_2^*(x_2, y_2)}, \qquad (5\text{-}34)$$

where dependencies on $\hat{i}$, $\hat{o}$ and λ are present but not explicit in the notation.

A short digression on the scattering vector $\vec{q}$ is in order. Note that, from its definition,

$$\vec{k}_i + \vec{q} = \vec{k}_o, \qquad (5\text{-}35)$$

and thus the scattering vector $\vec{q}$ closes a k-vector triangle. How does an incident k-vector $\vec{k}_i$ get transformed into a reflected k-vector $\vec{k}_o$? The answer is that it diffracts from a grating component of the surface that has the right period to send the reflected wave in the direction $\hat{o}$. The function

$$\mathbf{f}(\alpha, \beta; q_z) = \exp[jq_z h(\alpha, \beta)] \qquad (5\text{-}36)$$

can be expressed in terms of a 2D Fourier spectrum

$$\mathbf{f}(\alpha, \beta; q_z) = \left(\frac{1}{2\pi}\right)^2 \iint_{-\infty}^{\infty} \mathbf{F}(\vec{q}) e^{j\vec{q}_t \cdot \vec{\alpha}_t} \, d\vec{q}_t. \qquad (5\text{-}37)$$

Now note that a given combination of $\vec{k}_{i1}$, $\vec{k}_{o1}$ and λ_1 requires a particular scattering vector $\vec{q}_1$, to close the k-vector triangle, while a second combination of these three parameters in general requires a different scattering vector $\vec{q}_2$ to close the triangle. Since $h(x, y)$ is a random process over an ensemble of possible surfaces, so too are the functions $\mathbf{f}$ and $\mathbf{F}$. In particular the quantities $\mathbf{F}(\vec{q}_1)$ and $\mathbf{F}(\vec{q}_2)$ have random amplitudes and phases, and their correlation determines the correlation between the observed fields (and intensities) under the two sets of conditions being considered. When $\vec{q}_1$ and $\vec{q}_2$ are close together, we can expect high correlation between the observed fields, but when they are far apart, the correlation drops or vanishes. Thus the correlation between the complex fields $\mathbf{A}_1(x_1, y_1)$ and $\mathbf{A}_2(x_2, y_2)$ is entirely determined by the correlation between the Fourier coefficients $\mathbf{F}(\vec{q}_1)$ and $\mathbf{F}(\vec{q}_2)$, and is therefore a function of $\vec{q}_1$ and $\vec{q}_2$.

The analysis required is sufficiently complex that we defer it to Appendix D, and focus here on an interpretation of the results derived there. The primary result obtained in the appendix is the following expression for the normalized cross-correlation μ_A of the two speckle fields $\mathbf{A}_1$ and $\mathbf{A}_2$ (i.e. a normalized version of Γ_A, with the normalization factor being Γ_A when $\vec{q}_1 = \vec{q}_2$). Thus

$$\mu_A(\vec{q}_1, \vec{q}_2) = \mathbf{M}_h(\Delta q_z)\mathbf{\Psi}(\Delta \vec{q}_t) \qquad (5\text{-}38)$$

where $\mathbf{M}_h(\omega)$ is the first-order characteristic function of the surface-height fluctuations h,

$$\Psi(\Delta \vec{q}_t) = \frac{\iint\limits_{-\infty}^{\infty} |S(\alpha, \beta)|^2 e^{-j\Delta \vec{q}_t \cdot \vec{x}_t} \, d\alpha \, d\beta}{\iint\limits_{-\infty}^{\infty} |S(\alpha, \beta)|^2 \, d\alpha \, d\beta}, \qquad (5\text{-}39)$$

$|S|^2$ being the intensity distribution across the scattering spot, and $\Delta\vec{q}_t$ being the transverse component of the scattering vector difference $\vec{q}_1 - \vec{q}_2$, while Δq_z is the magnitude of the normal component of the same vector difference, given by

$$\Delta q_z = \left| \frac{2\pi}{\lambda_1}[\cos\theta_{o1} + \cos\theta_{i1}] - \frac{2\pi}{\lambda_2}[\cos\theta_{o2} + \cos\theta_{i2}] \right|. \qquad (5\text{-}40)$$

As anticipated, the result is found to depend on the scattering vectors $\vec{q}_1$ and $\vec{q}_2$. In effect, the correlation of interest depends on the correlation between the complex amplitudes associated with the two scattering vectors involved in the two cases.

While it is by no means obvious at this point, each of the two terms in Eq. (5-38) is influenced by different factors:

- $M_h(\Delta q_z)$ is determined by the ratio of the rms surface-height fluctuations to a wavelength, provided the surface-height fluctuations are foreshortened by the effects of nonnormal angles of incidence and observation;
- $\Psi(\Delta\vec{q}_t)$ represents two types of change, one a pure translation of the speckle pattern with respect to the observation point when the angles of incidence or observation are changed, and the second an expansion (longer wavelengths) or contraction (shorter wavelengths) of the speckle pattern about the mirror reflection angle caused by changes in wavelength.

Discussion in following sections will clarify these points.

The cross-covariance of the speckle intensity is the squared magnitude of μ_A, since under the assumptions we started with, the surface is sufficiently rough to assure that A is a circular Gaussian random process.

Attention is now turned to various specific cases of interest.

Normal Incidence, Normal Observation, Sensitivity to Wavelength Change

The first specific case of interest assumes that the rough surface is illuminated from the normal direction, the speckle is observed in the vicinity of the normal to the surface, but the wavelength is decreased between observations. Figure 5.9 illustrates the wave-vectors in this case. Both the figure and Eq. (5-30) show that the transverse component of $\Delta\vec{q}$ is zero, while the figure and Eq. (5-31) show that the normal component is

$$\Delta q_z = |q_{z1} - q_{z2}| = |2k_1 - 2k_2| = 4\pi\left|\frac{1}{\lambda_1} - \frac{1}{\lambda_2}\right| = 4\pi\frac{|\Delta\lambda|}{\lambda_1\lambda_2} \approx 4\pi\frac{|\Delta\lambda|}{\bar{\lambda}^2}, \qquad (5\text{-}41)$$

where $\Delta\lambda = |\lambda_2 - \lambda_1|$ and $\bar{\lambda} = (\lambda_1 + \lambda_2)/2$. The correlation function can now be written

5.3 Wavelength and Angle Diversity

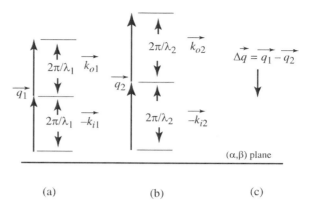

Figure 5.9 Wave-vectors for normal incidence and observation, reflection case. The wavelength is decreased from λ_1 to λ_2. (a) Wave-vectors for λ_1; (b) Wave-vectors for λ_2; (c) $\Delta\vec{q}$ is $\vec{q}_1 - \vec{q}_2$.

$$\boldsymbol{\mu}_{\mathbf{A}}(\vec{q}_1, \vec{q}_2) \approx \mathbf{M}_h\left(4\pi \frac{|\Delta\lambda|}{\bar{\lambda}^2}\right), \tag{5-42}$$

where the approximation assumes $|\Delta\lambda| \ll \bar{\lambda}$.

Since a measure of the width of a probability density function is its standard deviation σ, a measure of the width of its Fourier transform (the corresponding characteristic function) is $2\pi/\sigma$. So roughly speaking, when

$$4\pi \frac{|\Delta\lambda|}{\bar{\lambda}^2} > \frac{2\pi}{\sigma_h}, \tag{5-43}$$

where σ_h is the standard deviation of the surface-height fluctuations, the correlation will have dropped significantly. To state the condition another way, note that the change in optical frequency, $\Delta\nu$, is given by

$$\Delta\nu = c\left(\frac{1}{\lambda_1} - \frac{1}{\lambda_2}\right). \tag{5-44}$$

Then decorrelation will be reached (approximately) when the change of optical frequency satisfies

$$|\Delta\nu| > \frac{c}{2\sigma_h}, \tag{5-45}$$

where c is the speed of light. Note the larger the standard deviation of the surface-height fluctuations, the smaller the frequency change that decorrelates the fields.

If the surface-height fluctuations obey Gaussian statistics, the intensity correlation is given by

$$|\mathbf{M}_h(\Delta q_z)|^2 = \exp(-\sigma_h^2|\Delta q_z|^2) = \exp\left[-(4\pi)^2\left(\frac{\sigma_h}{\bar{\lambda}}\right)^2\left(\frac{|\Delta\lambda|}{\bar{\lambda}}\right)^2\right]. \tag{5-46}$$

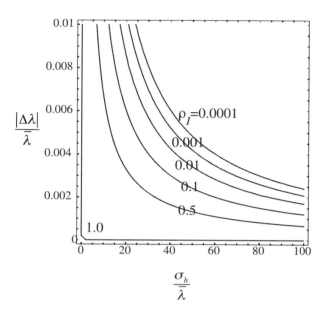

Figure 5.10 Contours of constant intensity correlation coefficient ρ_I as a function of normalized surface-height standard deviation and normalized wavelength change.

Figure 5.10 shows plots of contours of constant intensity correlation ρ_I in a plane defined by orthogonal coordinates $\sigma_h/\bar{\lambda}$ and $|\Delta\lambda|/\bar{\lambda}$. As a useful rule of thumb, reduction of the speckle intensity correlation to value $1/e^2$ or less requires

$$|\Delta\lambda| \geq \frac{1}{2\sqrt{2\pi}} \frac{\bar{\lambda}^2}{\sigma_h}, \tag{5-47}$$

or equivalently

$$|\Delta\nu| \geq \frac{1}{2\sqrt{2\pi}} \frac{c}{\sigma_h}. \tag{5-48}$$

Fixed Incidence at an Angle, Fixed Observation at the Mirror Angle, Change of Wavelength

Let the incident light have a wave-vector $\vec{k}_i$ that is not normal to the surface, and let the speckle be examined at a position where the mirror reflection of $\vec{k}_i$ would pierce the observation plane. The wave-vector diagram is shown in Fig. 5.11. As Eq. (5-30) shows, whenever $\theta_i = \theta_o$ (as is true when under the current geometry), the transverse component of $\vec{q}$ is zero. Again the wavelength has been decreased, and again the $\Delta\vec{q}$ vector is pointing in the negative $\hat{z}$ direction. The only difference with respect to the previous example is that the vector $\Delta\vec{q}$ is now shorter, due to the shallower angles with respect to the surface. If θ represents the angle of $-\vec{k}_i$ with respect to the normal

5.3 Wavelength and Angle Diversity

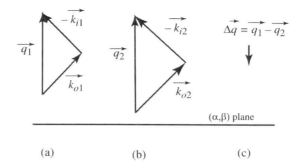

Figure 5.11 Wave-vectors for nonnormal incidence and mirror-direction observation, reflection case. The wavelength has again been decreased from λ_1 to λ_2. (a) Wave-vectors for λ_1; (b) wave-vectors for λ_2; (c) $\Delta\vec{q}$.

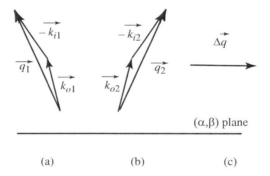

Figure 5.12 Geometry for which $\Delta\vec{q}$ is entirely transverse, reflection case; (a) geometry for first observation, (b) geometry for second observation, (c) $\Delta\vec{q}$.

to the surface, and $-\theta$ is the angle with respect to the normal for the observation vector $\vec{k}_o$, then the frequency difference that generates a $1/e^2$ decorrelation of the speckle intensities is now

$$|\Delta\nu| \geq \frac{c}{2\sqrt{2\pi}\cos\theta\sigma_h}. \tag{5-49}$$

A Case in Which $\Delta\vec{q}$ is Entirely Transverse

Consider now the geometry shown in Fig. 5.12. In this case, the wavelength is held constant, while $\vec{k}_{i1}$ and $\vec{k}_{i2}$ are reflected about a normal to the surface, as are also $\vec{k}_{o1}$ and $\vec{k}_{o2}$. The vector $\Delta\vec{q}$ is now seen to be entirely transverse, in which case

$$\mu_\mathbf{A}(\vec{q}_1, \vec{q}_2) = \Psi(\Delta\vec{q}_t). \tag{5-50}$$

Dependence of the Field Correlation on the Transverse Component of $\Delta\vec{q}$

Consider a general case in which the change in the scattering vector $\Delta\vec{q}$ need not be entirely transverse, but does have a nonzero transverse component $\Delta\vec{q}_t$. We focus here only on the portion of μ_A that depends on $\Delta\vec{q}_t$, namely $\Psi(\Delta\vec{q}_t)$. Let $|S(\alpha,\beta)|^2 = I(\alpha,\beta)$ represent the intensity associated with the scattering spot, in which case

$$\Psi(\Delta\vec{q}_t) = \frac{\iint_{-\infty}^{\infty} I(\alpha,\beta) e^{-j(\Delta q_\alpha \alpha + \Delta q_\beta \beta)}\, d\alpha\, d\beta}{\iint_{-\infty}^{\infty} I(\alpha,\beta)\, d\alpha\, d\beta}. \tag{5-51}$$

To make further progress, a specific shape for the scattering spot must be assumed. Let the spot be uniform and circular with diameter D, centered on the origin. For a circularly symmetric spot shape, the Fourier transform above can be written as a Hankel transform or a Fourier–Bessel transform,

$$\Psi(\Delta\vec{q}_t) = \frac{\int_0^\infty \rho I(\rho) J_0(\Delta q_t \rho)\, d\rho}{\int_0^\infty \rho I(\rho)\, d\rho}, \tag{5-52}$$

where $\Delta q_t = |\Delta\vec{q}_t|$. Continuing, we obtain

$$\Psi(\Delta\vec{q}_t) = \frac{\int_0^{D/2} \rho J_0(\Delta q_t \rho)\, d\rho}{\int_0^{D/2} \rho\, d\rho} = 2\frac{J_1\left(\frac{D\Delta q_t}{2}\right)}{\frac{D\Delta q_t}{2}}. \tag{5-53}$$

The first zero of this function occurs at

$$\Delta q_t = 7.66/D, \tag{5-54}$$

and thus the field correlation width associated with the transverse component of $\Delta\vec{q}$ is inversely proportional to the diameter D of the scattering spot. Remembering that

$$\Delta q_t = |\vec{q}_1 - \vec{q}_2| = |(\vec{k}_{o1} - \vec{k}_{i1}) - (\vec{k}_{o2} - \vec{k}_{i2})_t|$$
$$= \left|\frac{2\pi}{\lambda_1}(\sin\theta_{o1} - \sin\theta_{i1}) - \frac{2\pi}{\lambda_2}(\sin\theta_{o2} - \sin\theta_{i2})\right|, \tag{5-55}$$

given the set of wavelengths and angles before and after a change, the degree of decorrelation can be found.

Fixed Wavelength, Fixed Observation Angle, Sensitivity to Angle of Illumination Change

Consider now a case in which the wavelength is held constant, the speckle is observed at a normal angle to the surface, and the angle of illumination is changed. How far must the angle of illumination be changed to induce a decorrelation of the observed speckle?

The intensity covariance of the two observed speckle patterns is known to be given by

5.3 Wavelength and Angle Diversity

$$|\mu_{\mathbf{A}}(\vec{q}_1, \vec{q}_2)|^2 = |\mathbf{M}_h(\Delta q_z)|^2 |\mathbf{\Psi}(\Delta q_t)|^2. \tag{5-56}$$

In the case of constant wavelength and observation fixed in the normal direction, we have from Eqs. (5-40) and (5-55), respectively,

$$\begin{aligned} \Delta q_z &= \frac{2\pi}{\lambda} |\cos(\theta_{i1} + \Delta\theta_i) - \cos\theta_{i1}| \\ \Delta q_t &= \frac{2\pi}{\lambda} |\sin(\theta_{i1} + \Delta\theta_i) - \sin\theta_{i1}|. \end{aligned} \tag{5-57}$$

In addition, for a surface with Gaussian surface-height fluctuations,

$$|\mathbf{M}(\Delta q_z)|^2 = \exp(-\sigma_h^2 \Delta q_z^2). \tag{5-58}$$

If the scattering spot is a uniformly bright circle with diameter D, then

$$|\mathbf{\Psi}(\Delta q_t)|^2 = \left[2 \frac{J_1\left(\frac{D\Delta q_t}{2}\right)}{\frac{D\Delta q_t}{2}} \right]^2. \tag{5-59}$$

Substitution of all of the these equations into Eq. (5-56) yields the intensity covariance as a function of the various parameters. The intensity covariance is found to be dominated by $|\mathbf{\Psi}|^2$, which results because the greatest reduction of the correlation of the two patterns results from translation of one with respect to the other. If we correlate two speckle patterns that have the translation between them removed, then the intensity covariance is dominated by $|\mathbf{M}(\Delta q_z)|^2$.

Suppose first that no translation to compensate for shift between the patterns is made. If we assume values for the wavelength, the surface-height standard deviation, and the diameter of the scattering spot, we can plot contours of constant intensity correlation vs. the incidence angle θ_i and the change of incidence angle $\Delta\theta_i$. Let the wavelength be 0.5 μm and the surface-height standard deviation be 100 μm. Assume that the diameter of the scattering spot is 4 cm. With these numbers we find that to reduce the intensity covariance to 0.1, angle changes of only 2 to 4 seconds of arc are needed. Thus speckle correlation is quite sensitive to a change of angle of incidence. The primary factor governing this sensitivity is the size of the scattering spot. If the spot size is smaller, the speckle lobes will be larger, and the decorrelation angular change will be larger than in this example. This is primarily a translation effect.

Suppose on the other hand that the two speckle patterns are correlated after the translational component has been removed [99]. Then the intensity correlation will be

$$\begin{aligned} |\mu_{\mathbf{A}}|^2 = |\mathbf{M}(\Delta q_z)|^2 &= \exp\left[-\left(2\pi \frac{\sigma_h}{\lambda} (\cos(\theta_i + \Delta\theta_i) - \cos\theta_i) \right)^2 \right] \\ &\approx \exp\left[-\left(2\pi \frac{\sigma_h}{\lambda} \Delta\theta_i \sin\theta_i \right)^2 \right], \end{aligned} \tag{5-60}$$

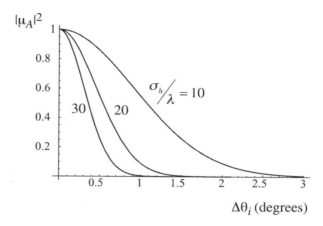

Figure 5.13 Intensity correlation between two properly translated speckle patterns as a function of illumination angle change and surface roughness.

where the approximation that $\Delta\theta_i$ is small has been made in the last step. Figure 5.13 shows the variation of intensity correlation as a function of angle change for three different values of surface roughness. If the correlation is measured, and the change of angle is known, then the surface roughness can be determined. The amount of translation needed is approximately $\Delta\theta_i z$ in the mirror (opposite) direction from the original angle change, where z is the distance between the scattering surface and the measurement plane. This result will be of use to us in Chapter 8.

5.3.2 Free-Space Propagation, Transmission Geometry

The transmission geometry is shown in Fig. 5.14. A diffuser composed of a transparent material of index n is illuminated from free space (index unity) with k-vector $\vec{k}_i$, and the scattered fields are observed in a plane normal distance z away. The direction of observation is $\hat{o}$ and the k-vector in that direction is $\vec{k}_o$. $\vec{k}_r$ is the k-vector of the refracted wave, after the incident wave enters the diffuser. For simplicity, the figure assumes that all three k-vectors lie in the (β, z) plane. Differences in this case as compared with the reflection geometry are that $\vec{k}_i$ points into the same half-space as $\vec{k}_o$, and in addition, the $\vec{k}_i$ is transformed by Snell's law into a refracted k-vector $\vec{k}_r$, as shown in the figure. This transformation preserves the length of $\vec{k}_i$ in the β direction, and increases the length of $\vec{k}_r$ in the z direction (see, for example, [78], pp. 132–134). The length of $\vec{k}_r$ is n times greater than the length of $\vec{k}_i$ and of $\vec{k}_o$.

An analysis of the transmission case is presented in Appendix D. The changes required compared with the reflection case are:

5.3 Wavelength and Angle Diversity

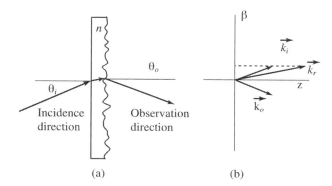

Figure 5.14 Transmission geometry. (a) Incident ray, refracted ray, and observation direction; (b) k-vectors of the incident light, the refracted light, and a k-vector in the observation direction.

- The average reflectance of the rough surface **r** must be replaced by the average transmittance of the rough interface, **t**.
- $-\hat{i}$ must be replaced by $\hat{r}$.
- Transverse components of $\vec{k}_r$ are equal to the transverse components of $\vec{k}_i$.
- The normal component of $\vec{q}$ must be replaced by an expression derived below.
- A positive value of h represents a shorter path length (positive phase shift ϕ) in the reflection case, while in the transmission case, a positive h represents a longer path length (negative value of ϕ) in the the medium of refractive index n and a shorter path length (positive value of ϕ) in free space from the diffuser to the observation point).

The phase shift in this case is given by

$$\phi(\alpha,\beta) = \frac{2\pi}{\lambda}(-n\hat{r}\cdot\hat{z} + \hat{o}\cdot\hat{z})h(\alpha,\beta), \tag{5-61}$$

where here and throughout, λ is the free-space wavelength. Again we define a scattering vector, $\vec{q}$ given by

$$\vec{q} = \vec{k}_o - \vec{k}_r. \tag{5-62}$$

The transverse component of the scattering vector is

$$\vec{q}_t = (\vec{k}_o - \vec{k}_r)_t = (\vec{k}_o - \vec{k}_i)_t, \tag{5-63}$$

where the continuity of the transverse components of $\vec{k}_i$ across the refractive boundary has been used. The z-component of the scattering vector is given by

$$q_z = k_{oz} - k_{rz} = k_{oz} - \sqrt{k_r^2 - k_{r\alpha}^2 - k_{r\beta}^2}$$

$$= k_{oz} - \sqrt{k_r^2 - k_{i\alpha}^2 - k_{i\beta}^2} = k_{oz} - \frac{2\pi}{\lambda}\sqrt{n^2 - (\hat{i}\cdot\hat{\alpha})^2 - (\hat{i}\cdot\hat{\beta})^2}$$

$$= k_{oz} - \frac{2\pi}{\lambda}\sqrt{n^2 - \sin^2\theta_i} = \frac{2\pi}{\lambda}[\cos\theta_o - \sqrt{n^2 - \sin^2\theta_i}], \tag{5-64}$$

and the phase shift suffered at the rough interface is

$$\phi(\alpha,\beta) = q_z h(\alpha,\beta). \tag{5-65}$$

The expression for the normalized correlation between two wave fields again becomes

$$\boldsymbol{\mu}_A(\vec{q}_1,\vec{q}_2) = \mathbf{M}_h(\Delta q_z)\boldsymbol{\Psi}(\Delta\vec{q}_t), \tag{5-66}$$

where the meaning of the symbols $\mathbf{M}_h$ and $\boldsymbol{\Psi}$ are unchanged, and

$$\Delta q_z = |q_{1z} - q_{2z}| = \left|\frac{2\pi}{\lambda_1}[\cos\theta_{o1} - \sqrt{n^2 - \sin^2\theta_{i1}}]\right.$$

$$\left. - \frac{2\pi}{\lambda_2}[\cos\theta_{o2} - \sqrt{n^2 - \sin^2\theta_{i2}}]\right|$$

$$= \left|\frac{2\pi}{\lambda_2}[\sqrt{n^2 - \sin^2\theta_{i2}} - \cos\theta_{o2}] - \frac{2\pi}{\lambda_1}[\sqrt{n^2 - \sin^2\theta_{i1}} - \cos\theta_{o1}]\right| \tag{5-67}$$

where λ_1 and λ_2 are free-space wavelengths. The transverse component of $\Delta\vec{q}$ is (as before)

$$\Delta\vec{q}_t = (\vec{k}_{o1} - \vec{k}_{i1})_t - (\vec{k}_{o2} - \vec{k}_{i2})_t, \tag{5-68}$$

with magnitude

$$\Delta q_t = \left|\frac{2\pi}{\lambda_1}(\sin\theta_{o1} - \sin\theta_{i1}) - \frac{2\pi}{\lambda_2}(\sin\theta_{o2} - \sin\theta_{i2})\right|. \tag{5-69}$$

We now turn attention to an example.

Normal Incidence, Normal Observation, Wavelength Change

Figure 5.15 shows the k-vectors for this example. Again the wavelength λ_2 is assumed to be smaller than the wavelength λ_1. In this case the incidence direction and observation direction are both normal, implying that θ_i, θ_r and θ_o are all zero. Under this condition the quantity Δq_z becomes

$$\Delta q_z = 2\pi(n-1)\left|\frac{1}{\lambda_2} - \frac{1}{\lambda_1}\right|. \tag{5-70}$$

Referring to Eq. (5-44), we see that

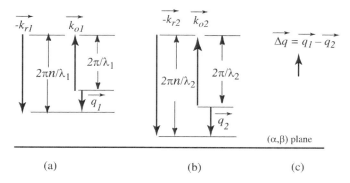

Figure 5.15 Wave-vectors for normal incidence and observation, transmission case. The wavelength is decreased from λ_1 to λ_2. (a) Wave-vectors for λ_1; (b) Wave-vectors for λ_2; (c) $\Delta \vec{q}$ is $\vec{q}_1 - \vec{q}_2$.

$$\Delta q_z = 2\pi(n-1)|\Delta\nu|/c. \qquad (5\text{-}71)$$

where c is the vacuum speed of light. For a surface with a Gaussian height distribution with standard deviation σ_h, following the development for the similar reflection case, the intensity correlation will fall to $1/e^2$ when

$$\sigma_h^2 \Delta q_z^2 = 2,$$

and decorrelation will occur when

$$|\Delta\nu| \geq \frac{c}{\sqrt{2}\pi(n-1)\sigma_h}. \qquad (5\text{-}72)$$

For a glass diffuser, with $n \approx 1.5$, the required change of frequency is about 4 times greater for a diffuser in transmission than for a rough surface in reflection, assuming that the standard deviations of the surface height fluctuations are the same.

Other examples, similar to those presented in the reflection case, can also be considered, but will not be dealt with here, since they are quite straightforward.

5.3.3 Imaging Geometry

Attention is now turned to the imaging geometry shown in Fig. 5.16. The case shown is for a reflection geometry, but at the end of this section the changes needed for the transmission geometry will be discussed. See [57] for an alternative discussion of this topic.

For simplicity, we assume that the distances from the scattering surface to the lens and from the lens to the image plane are both z, thus guaranteeing unity

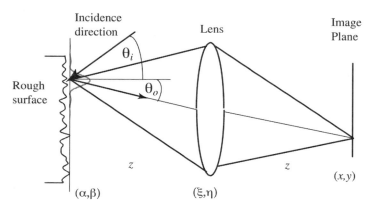

Figure 5.16 Imaging in a reflection geometry.

magnification. The imaging system can be described by its point-spread function $\mathbf{k}(x, y; \alpha, \beta)$, which is given by (see [71], p. 110)

$$\mathbf{k}(x, y; \alpha, \beta) = \frac{1}{\lambda^2 z^2} e^{j\frac{\pi}{\lambda z}(\alpha^2 + \beta^2)} \iint_{-\infty}^{\infty} \mathbf{P}(\xi, \eta) e^{-j\frac{2\pi}{\lambda z}[\xi(\alpha+x)+\eta(\beta+y)]} \, d\xi \, d\eta \qquad (5\text{-}73)$$

and the image field $\mathbf{A}(x, y)$ is related to the scattered field $\mathbf{a}(\alpha, \beta)$ by

$$\mathbf{A}(x, y) = \iint_{-\infty}^{\infty} \mathbf{k}(x, y; \alpha, \beta) \mathbf{a}(\alpha, \beta) \, d\alpha \, d\beta. \qquad (5\text{-}74)$$

In writing Eq. (5-73), we have dropped a quadratic phase factor in $x^2 + y^2$ since only the intensity of the speckle pattern in the (x, y) plane will be of interest.

Our goal again will be to find the cross-correlation

$$\mathbf{\Gamma}_A(x_1, y_1; x_2, y_2) = \overline{\mathbf{A}_1(x_1, y_1)\mathbf{A}_2^*(x_2, y_2)}, \qquad (5\text{-}75)$$

where $\mathbf{A}_1$ and $\mathbf{A}_2$ again represent the observed fields under two different conditions of observation, the differences lying in any or all of (1) wavelength of illumination, (2) angle θ_i of illumination, and (3) angle θ_o of observation, as defined by the angle of the central ray through the lens to the observation point.

The effective roughness of the surface is again influenced by the direction of illumination and the average direction of observation, and again it is helpful to define a vector $\vec{q}$ that closes the k-vector triangle,

$$\vec{q} = \vec{k}_o - \vec{k}_i = \vec{q}_t + q_z \hat{z}, \qquad (5\text{-}76)$$

where $|\vec{q}_t|$ and q_z are again given by Eqs. (5-30) and (5-31). The phase shift suffered by the illumination scattered from the surface can again be written as[3]

$$\phi(\alpha, \beta) = q_z h(\alpha, \beta). \tag{5-77}$$

Again the details of the analysis are carried out in Appendix D, the results of which we shall use here. The normalized cross-correlation function μ_A is found to be

$$\mu_A(\Delta x, \Delta y) = \mathbf{M}_h(\Delta q_z)\mathbf{\Psi}(\Delta x, \Delta y), \tag{5-78}$$

where

$$\mathbf{\Psi}(\Delta x, \Delta y) = \frac{\iint\limits_{-\infty}^{\infty} |\mathbf{P}(\xi, \eta)|^2 e^{-j\frac{2\pi}{\lambda_2 z}(\xi \Delta x + \eta \Delta y)} d\xi\, d\eta}{\iint\limits_{-\infty}^{\infty} |\mathbf{P}(\xi, \eta)|^2 d\xi\, d\eta}, \tag{5-79}$$

$\mathbf{P}$ represents the (possibly complex-valued) pupil function of the imaging system, and $\mathbf{M}_h$ is again the characteristic function of the surface-height fluctuations. The quantity Δq_z has the same meaning as in previous sections. In a reflection geometry it is given by

$$\Delta q_z = \left| \frac{2\pi}{\lambda_1}[\cos\theta_{o1} + \cos\theta_{i1}] - \frac{2\pi}{\lambda_2}[\cos\theta_{o2} + \cos\theta_{i2}] \right|, \tag{5-80}$$

while in a transmission geometry it is given by

$$\Delta q_z = \left| \frac{2\pi}{\lambda_1}[\cos\theta_{o1} - \sqrt{n^2 - \sin^2\theta_{i1}}] - \frac{2\pi}{\lambda_2}[\cos\theta_{o2} - \sqrt{n^2 - \sin^2\theta_{i2}}] \right|. \tag{5-81}$$

In closing this section we make special note of the fact that, in an imaging geometry, within the approximations we have made, the sole effect of a change of wavelength, angle of incidence or angle of observation is through the change of the effective surface-height fluctuations, as represented by the term $\mathbf{M}_h(\Delta q_z)$. The squared magnitude of the normalized Fourier transform of the pupil function determines the lateral extent of the speckle correlation, and within our approximations, is not appreciably affected by the changes under consideration.

---

3. Clearly there is a spread of scattered directions collected by the lens. However, the approximation is made that the direction of the central ray can be used in calculating the phase shift ϕ. Such an approximation has already been used for the direction of illumination, which for a finite spot, must consist of a range of angles. Both approximations are valid if the angular spreads involved are small.

5.4 Temporal and Spatial Coherence Reduction

In optics, speckle is most frequently encountered with fully coherent light, such as provided by a CW laser. However, it can also occur with less coherent sources, and even, under some circumstances, with white light. Nonetheless, reducing the coherence of the light being used is another strategy for reducing the contrast of speckle. In this section, we begin with an introduction to some of the common concepts of coherence used in optics, and compare these concepts with corresponding ones used in the description of speckle. That discussion is followed by descriptions of several possible approaches to coherence reduction and discussions of their effectiveness. For an alternative discussion of the effects of coherence on speckle, see [125].

5.4.1 Coherence Concepts in Optics

The fundamental concepts of coherence were introduced in optics by Wolf [176], [177]. For a more detailed discussion of coherence concepts than can be presented here, see [70], Chapter 5. Central to these concepts is the *mutual coherence function*, which for complex-valued scalar optical waves $\mathbf{u}_1(t) = \mathbf{u}(P_1, t)$ and $\mathbf{u}_2(t) = \mathbf{u}(P_2, t)$ is defined by

$$\tilde{\Gamma}_{12} = \tilde{\Gamma}(P_1, P_2; \tau) = \langle \mathbf{u}_1(t)\mathbf{u}_2^*(t+\tau) \rangle, \tag{5-82}$$

where P_1 and P_2 are two points in space, τ is a relative time delay between the two beams, the angle brackets signify an infinite time average,

$$\langle \mathbf{g}(t) \rangle = \lim_{T \to \infty} \frac{1}{T} \int_{-T/2}^{T/2} \mathbf{g}(t)\, dt,$$

and in order to avoid possible future confusion, we have used a tilde over quantities that are defined by time averages. The mutual coherence function is thus seen to be a temporal cross-correlation function between the two optical signals $\mathbf{u}_1$ and $\mathbf{u}_2$. A normalized version of $\tilde{\Gamma}$ is known as the *complex degree of coherence* and is defined by

$$\tilde{\gamma}_{12}(\tau) = \frac{\tilde{\Gamma}_{12}(\tau)}{[\tilde{\Gamma}_{11}(0)\tilde{\Gamma}_{22}(0)]^{1/2}}, \tag{5-83}$$

which obeys the inequality

$$0 \leq |\tilde{\gamma}_{12}(\tau)| \leq 1. \tag{5-84}$$

Note that $\tilde{\Gamma}_{11}(0)$ and $\tilde{\Gamma}_{22}(0)$ are simply the time-average intensities $\tilde{I}(P_1)$ and $\tilde{I}(P_2)$, respectively.

Two additional concepts are useful if the light is narrowband, that is, if $\Delta \nu \ll \nu_o$, where ν_o is the mean frequency of the waves. In this case

$$\mathbf{u}_1(t) = \mathbf{A}_1(t)e^{-j2\pi\nu_o t} \quad \text{and} \quad \mathbf{u}_2(t) = \mathbf{A}_2(t)e^{-j2\pi\nu_o t}, \tag{5-85}$$

where $\mathbf{A}_1(t)$ and $\mathbf{A}_2(t)$ represent the time-varying complex envelopes of the two disturbances. We then have

$$\tilde{\mathbf{\Gamma}}_{12}(\tau) \approx \tilde{\mathbf{J}}_{12}e^{-j2\pi v_o \tau} \quad \text{and} \quad \tilde{\gamma}_{12}(\tau) \approx \tilde{\mu}_{12}e^{-j2\pi v_o \tau}, \tag{5-86}$$

where

$$\tilde{\mathbf{J}}_{12} = \langle \mathbf{A}_1(t)\mathbf{A}_2^*(t) \rangle$$

$$\tilde{\mu}_{12} = \frac{\tilde{\mathbf{J}}_{12}}{[\tilde{I}(P_1)\tilde{I}(P_2)]^{1/2}}, \tag{5-87}$$

are called, respectively, the *mutual intensity* and the *complex coherence factor* of the two disturbances. The magnitude of the complex coherence factor always lies between 0 and 1.

Clearly there is a close relationship between these quantities describing the coherence of light and the cross-correlation functions of speckle amplitudes [69]. The primary difference lies in the character of the averages being taken. In the case of conventional coherence concepts, the averages are over time, although when the random fields are *ergodic* and their statistics over time are known, the time averages can sometimes be evaluated with the help of a statistical averages. In the case of speckle, the averages of interest are statistical averages over an ensemble of rough surfaces that could cause the speckle pattern. Even when changing speckle patterns are time averaged, the quantities of interest are ensemble averages of the residual fluctuations.

At this point it is helpful to distinguish between *temporal* and *spatial* coherence. If the points P_1 and P_2 merge to the same point, say P_1, then

$$\tilde{\gamma}_{12}(\tau) \to \tilde{\gamma}_{11}(\tau),$$

which is the normalized autocorrelation function of the fields at point P_1, and can be referred to as the *complex degree of temporal coherence*. By the Wiener–Khinchin theorem (see [70], Section 3.4), $\tilde{\gamma}_{11}(\tau)$ is related to the power spectral density $\mathcal{G}_{11}(v)$ of $\mathbf{u}_1(t)$ through[4]

$$\tilde{\gamma}_{11}(\tau) = \int_{-\infty}^{0} \hat{\mathcal{G}}_{11}(v)e^{j2\pi v\tau}\,dv, \tag{5-88}$$

where

$$\hat{\mathcal{G}}_{11}(v) \equiv \frac{\mathcal{G}_{11}(v)}{\int_{-\infty}^{0} \mathcal{G}_{11}(v)\,dv} \tag{5-89}$$

is the power spectral density normalized to have unity area. The coherence time of the wave amplitude is conventionally defined to be the equivalent width (see [20], Chapter 8) of the complex degree of coherence,

$$\tau_c = \int_{-\infty}^{\infty} \tilde{\gamma}_{11}(\tau)\,d\tau. \tag{5-90}$$

---

4. Since $\mathcal{G}_{11}(v)$ is the power spectral density of an analytic signal, formed from a real signal by doubling the negative frequency Fourier amplitudes and suppressing the positive frequencies, this function has value only at negative frequencies.

With regard to spatial coherence, the mutual intensity $\tilde{\mathbf{J}}_{12}$ and the complex coherence factor $\tilde{\mu}_{12}$ are functions strictly of P_1 and P_2, and are independent of τ. They are therefore quantities that describe the *spatial* coherence of the light. For the case of free-space propagation, these quantities can be calculated with the help of the van Cittert–Zernike theorem, which we encountered already for the ensemble-average coherence of speckle amplitude (see Eq. (4-55)). When the light in an initial plane with coordinates (α, β) is delta-correlated in the time-average sense, we say that this light is *spatially incoherent*. Propagation of this light to a parallel plane distance z away, having transverse coordinates (x, y), yields a complex coherence factor given by[5]

$$\tilde{\mu}_{12}(\Delta x, \Delta y) = \frac{\iint_{-\infty}^{\infty} \tilde{I}(\alpha, \beta) e^{-j\frac{2\pi}{\lambda z}(\alpha \Delta x + \beta \Delta y)} \, d\alpha \, d\beta}{\iint_{-\infty}^{\infty} \tilde{I}(\alpha, \beta) \, d\alpha \, d\beta}, \tag{5-91}$$

where $\tilde{I}(\alpha, \beta)$ is the time-averaged intensity distribution of the initial incoherent light, quasi-monochromatic light with mean wavelength λ has been assumed, and $(\Delta x, \Delta y) = (x_1 - x_2, y_1 - y_2)$. The coherence area of the wave amplitude is conventionally defined as the equivalent area of the complex coherence factor,

$$\mathcal{A}_c = \iint_{-\infty}^{\infty} \tilde{\mu}_{12}(\Delta x, \Delta y) \, d\Delta x \, d\Delta y. \tag{5-92}$$

When the geometry is that of an imaging system, the approximation introduced by Zernike can again be used, namely the exit pupil of the imaging system can be considered to be an incoherent source, and the van Cittert–Zernike theorem can be applied to the intensity distribution across that pupil.

5.4.2 Moving Diffusers and Coherence Reduction

In Section 5.2, the properties of speckle in the presence of moving diffusers were studied extensively. In this subsection, we consider this subject again, but this time focusing on the effects of such diffusers on the time-averaged coherence of the light they transmit.

Previously, in Eq. (5-2), a moving diffuser was modeled as a structure which, when illuminated with a simple, normally incident plane wave, produces a transmitted field of the form

$$\mathbf{a}(\alpha, \beta) = \mathbf{a}_o e^{j\phi_d(\alpha - vt, \beta)}, \tag{5-93}$$

---

[5]. We have neglected a phase factor dependent on x and y, which will be immaterial since we will ultimately be concerned only with $|\tilde{\mu}_{12}|^2$.

where v is the speed of movement. The mutual coherence function for such light is

$$\tilde{\Gamma}(\alpha_1,\beta_1;\alpha_2,\beta_2;\tau) = |\mathbf{a}_o|^2 \langle e^{j\phi_d(\alpha_1-vt,\beta_1)} e^{-j\phi_d(\alpha_2-v(t+\tau),\beta_2)} \rangle. \quad (5\text{-}94)$$

As mentioned previously, it is sometimes convenient to evaluate infinite time averages by means of an ensemble average. Assuming that the diffuser phase is a stationary and ergodic Gaussian random process, we can evaluate the time average as (see Eq. (5-5))

$$\tilde{\gamma}_{12}(\tau) = \overline{e^{j[\phi_d(\alpha_1-vt,\beta_1)-\phi_d(\alpha_2-vt-v\tau,\beta_2)]}} = e^{-\sigma_\phi^2[1-\mu_\phi(\Delta\alpha+v\tau,\Delta\beta)]}, \quad (5\text{-}95)$$

where $\Delta\alpha = \alpha_1 - \alpha_2$, $\Delta\beta = \beta_1 - \beta_2$, σ_ϕ is the standard deviation of the diffuser phase, and μ_ϕ is the normalized autocorrelation function of the phase.

Further progress requires the adoption of a specific form for μ_ϕ. As before, we choose a Gaussian shape for this correlation function,

$$\mu_\phi(\Delta\alpha,\Delta\beta) = e^{-\frac{\Delta\alpha^2+\Delta\beta^2}{r_\phi^2}}, \quad (5\text{-}96)$$

where r_ϕ is the correlation radius of the phase. The normalized coherence function becomes

$$\tilde{\gamma}_{12}(\tau) = e^{-\sigma_\phi^2\left[1-e^{-\frac{(\Delta\alpha+v\tau)^2+\Delta\beta^2}{r_\phi^2}}\right]}. \quad (5\text{-}97)$$

The coherence time and coherence area can now be found, with one subtle change to our previous expressions. The Gaussian phase screen always has both a specular and a diffuse component of transmittance. The specular component is extremely small when $\sigma_\phi \gg 1$, but even an infinitesimally small specular component, when integrated over infinite limits, will cause the integral to diverge. For this reason, when evaluating the coherence time and the coherence area of the light transmitted by the moving diffuser, we must subtract the "DC" component of transmittance, which represents the specular component, and renormalize so that the integrand has value unity at the origin. This can be done by subtracting from $\tilde{\gamma}_{12}(\Delta\alpha,\Delta\beta;\tau)$ a term $\lim_{\Delta\alpha,\Delta\beta,\tau\to\infty} \tilde{\gamma}_{12}(\Delta\alpha,\Delta\beta;\tau)$, or in the specific case of interest, subtracting $e^{-\sigma^2}$ from $\tilde{\gamma}_{12}(\tau)$. In addition, we divide the integrand by $(1-e^{-\sigma^2})$ to assure an integrand with unity value at the origin. Thus the coherence time can be found as

$$\tau_c = \int_{-\infty}^{\infty} \left(\frac{\tilde{\gamma}_{11}(\tau) - \tilde{\gamma}_{11}(\infty)}{1 - \tilde{\gamma}_{11}(\infty)}\right) d\tau = \int_{-\infty}^{\infty} \frac{e^{-\sigma_\phi^2\left(1-e^{-\left(\frac{v\tau}{r_\phi}\right)^2}\right)} - e^{-\sigma_\phi^2}}{1 - e^{-\sigma_\phi^2}} d\tau$$

$$= \frac{e^{-\sigma_\phi^2}}{1 - e^{-\sigma_\phi^2}} \int_{-\infty}^{\infty} \left(e^{\sigma_\phi^2 e^{-\left(\frac{v\tau}{r_\phi}\right)^2}} - 1\right) d\tau. \quad (5\text{-}98)$$

Defining $\tau_0 = \frac{\sqrt{\pi r_\phi^2}}{v}$, which is the time it takes the diffuser to move a linear distance equal to the square root of the correlation area of the diffuser phase, the integral can be reduced to

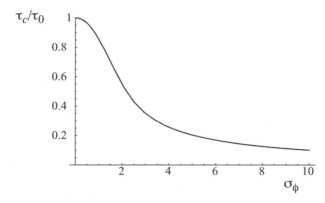

Figure 5.17 Normalized coherence time vs. standard deviation of the diffuser phase.

$$\frac{\tau_c}{\tau_0} = \frac{e^{-\sigma_\phi^2}}{1-e^{-\sigma_\phi^2}} \int_{-\infty}^{\infty} (e^{\sigma_\phi^2 e^{-\pi t^2}} - 1)\, dt. \quad (5\text{-}99)$$

While this integral does not seem to be expressible in terms of tabulated functions, nonetheless it can be evaluated numerically with the results shown in Fig. 5.17. As can be seen from that plot, as the standard deviation σ_ϕ of the diffuser phase increases, the coherence time falls to smaller and smaller fractions of the time τ_0.

In a similar manner, the coherence area is given by

$$\mathcal{A}_c = \frac{e^{-\sigma_\phi^2}}{1-e^{-\sigma_\phi^2}} \iint_{-\infty}^{\infty} \left(e^{\sigma_\phi^2 e^{-\left(\frac{\sqrt{\Delta\alpha^2+\Delta\beta^2}}{r_\phi}\right)^2}} - 1 \right) d\Delta\alpha\, d\Delta\beta$$

$$= 2\pi \frac{e^{-\sigma_\phi^2}}{1-e^{-\sigma_\phi^2}} \int_0^\infty r \left(e^{\sigma_\phi^2 e^{-\left(\frac{r}{r_\phi}\right)^2}} - 1 \right) dr. \quad (5\text{-}100)$$

This integral has been seen before, and was evaluated in Eq. (4-109) and Fig. 4.18. Note that the result is entirely independent of the speed v of the diffuser.

One subtlety that needs mention is the state of coherence when the diffuser remains stationary. Recall that the coherence concepts introduced here are all defined in terms of infinite time averages. Assuming for the moment that the light illuminating the diffuser is perfectly monochromatic, the light leaving a stationary diffuser must by definition be perfectly coherent. As soon as the diffuser begins to move, however slowly, the infinite time average involved in its definition causes the coherence area to immediately drop to the value given above. Likewise when the diffuser is stationary and the source is monochromatic, the coherence time is by definition infinite, as evidenced by the fact that τ_0 is infinite. If the illumination is non-monochromatic and the diffuser is stationary, then the coherence properties of the illumination (rather than the properties of the diffuser) determine the coherence area and coherence time.

5.4.3 Speckle Suppression by Reduction of Temporal Coherence

In the previous subsection, the temporal and spatial coherence reducing properties of moving diffusers were examined. In this subsection, we take a broader view and examine the speckle reduction properties when the coherence of the light is less than perfect, regardless of the origin of these coherence properties. Often the source itself may have less than perfect coherence, and it is of interest to know how reduced source coherence reduces the contrast of speckle. The discussion will be framed in terms of a free-space reflection geometry, but the corresponding results for a free-space transmission geometry and for an imaging geometry will also be presented.

It is convenient to examine the case of reduced temporal coherence in the spectral domain. The spectrum of the light incident on the scattering surface is assumed to be constant across that surface and will be represented by the power spectral density $\mathcal{G}(\nu)$, where we have dropped the subscripts as unnecessary in this context.

The analysis of Section 5.3 has shown that the correlation between speckle amplitudes at two different wavelengths can be found from Eq. (5-38), which we repeat here for convenience:

$$\boldsymbol{\mu}_\mathbf{A}(\vec{q}_1, \vec{q}_2) = \mathbf{M}_h(\Delta q_z)\boldsymbol{\Psi}(\Delta \vec{q}_t). \tag{5-101}$$

In this subsection, we assume that the illumination and observation angles do not change, and focus our complete attention on the changes caused by wavelength or frequency shift. In general, both factors on the right of this equation can contribute to a drop of correlation due to frequency change. However, if the incidence and observation angles are approximately equal, i.e. $\theta_i \approx \theta_o$, then $\mathbf{M}_h(\Delta q_z)$ is by far the dominant factor. We assume that this is the case, and in what follows we neglect the factor $\boldsymbol{\Psi}(\Delta \vec{q}_t)$.

Let $I(x, y; \nu)$ represent the speckle intensity at (x, y) when the frequency of the illumination is ν. Then the total intensity at this point is found by integrating $I(x, y; \nu)$ over frequency, with the normalized power spectral density function as a weighting factor. Thus

$$I(x, y) = \int_0^\infty \hat{\mathcal{G}}(-\nu) I(x, y; \nu) \, d\nu. \tag{5-102}$$

For fully developed, polarized speckle, $I(x, y; \nu)$ must obey negative exponential statistics, and the quantity $I(x, y)$ must obey the statistics of integrated speckle. Referring back to Section 4.6, we conclude that the number of degrees of freedom associated with speckle integrated over frequency is

$$M = \left[\int_{-\infty}^\infty K_{\hat{\mathcal{G}}}(\Delta \nu) |\mathbf{M}_h(\Delta q_z)|^2 \, d\Delta \nu \right]^{-1} \tag{5-103}$$

where Δq_z is a function of $\Delta\nu$, and

$$K_{\hat{\mathcal{G}}}(\Delta\nu) = \int_{-\infty}^{0} \hat{\mathcal{G}}(\xi)\hat{\mathcal{G}}(\xi - \Delta\nu)\,d\xi \tag{5-104}$$

is the autocorrelation function of the normalized power spectrum. It follows that the speckle contrast associated with an extended normalized power spectral density $\hat{\mathcal{G}}(\nu)$ is

$$C = \sqrt{\int_{-\infty}^{\infty} K_{\hat{\mathcal{G}}}(\Delta\nu)|\mathbf{M}_h(\Delta q_z)|^2\,d\Delta\nu}. \tag{5-105}$$

Surface Scattering

Assume that the surface height fluctuations are Gaussian. Then

$$|\mathbf{M}_h(\Delta q_z)|^2 = \exp(-\sigma_h^2 \Delta q_z^2), \tag{5-106}$$

where

$$\Delta q_z = \left|\frac{2\pi}{\lambda_1} - \frac{2\pi}{\lambda_2}\right|(\cos\theta_o + \cos\theta_i) = \frac{2\pi|\Delta\nu|}{c}(\cos\theta_o + \cos\theta_i). \tag{5-107}$$

To illustrate the results presented above, attention is turned to a specific example of a spectrum. We assume a reflection geometry. Let the source have a Gaussian-shaped spectrum, and assume that the light is fully polarized both before and after scattering.

The one-sided normalized power spectrum of the light is taken to be of the form

$$\hat{\mathcal{G}}(\nu) \approx \frac{2}{\sqrt{\pi}\delta\nu}\exp\left[-\left(\frac{\nu + \bar{\nu}}{\delta\nu/2}\right)^2\right], \tag{5-108}$$

where $\bar{\nu}$ is the negative of the center frequency of the spectrum (defined, customarily, to be a positive number), $\delta\nu$ represents the $1/e$ width of the spectrum (a positive number), and the assumption that $\delta\nu \ll |\bar{\nu}|$ has been made.

The autocorrelation function of the normalized power spectrum is found to be

$$K_{\hat{\mathcal{G}}}(\Delta\nu) = \sqrt{\frac{2}{\pi\delta\nu^2}}\exp\left(-\frac{2\Delta\nu^2}{\delta\nu^2}\right). \tag{5-109}$$

When these results are substituted in the expression for contrast, the result becomes

$$C = \sqrt{\frac{1}{\sqrt{1 + 2\pi^2\left(\frac{\delta\nu}{\bar{\nu}}\right)^2\left(\frac{\sigma_h}{\bar{\lambda}}\right)^2(\cos\theta_o + \cos\theta_i)^2}}}. \tag{5-110}$$

Equivalently, the fractional bandwidth needed to achieve a reduction of contrast to level C is given by

$$\delta\nu/\bar{\nu} = \delta\lambda/\bar{\lambda} = \frac{\bar{\lambda}}{\sigma_h}\sqrt{\frac{\frac{1}{C^4} - 1}{8\pi^2}}. \tag{5-111}$$

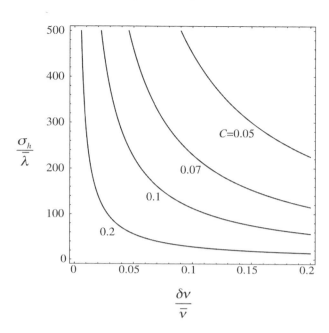

Figure 5.18 Contours of constant contrast C as a function of fractional bandwidth $\delta v/\bar{v}$ and surface height standard deviation normalized by the mean wavelength, $\sigma_h/\bar{\lambda}$. Normal incidence and observation angles are assumed.

Figure 5.18 shows contours of constant contrast in a plane with axes $\delta v/\bar{v}$ and $\sigma_h/\bar{\lambda}$, under the assumption of normal incidence and normal observation. As can be seen from this figure, if the surface height has a standard deviation of as much as 200 wavelengths, achieving a contrast ratio of 0.1 or less requires a fractional bandwidth of approximately 5% or more. To achieve a contrast ratio of 0.05 or less requires a fractional bandwidth of 20% or more. The conclusion of importance here is that when surface scattering is responsible for the speckle, the frequency (or wavelength) sensitivity of the speckle is not great. As we shall see now, the sensitivity can be much greater when volume scattering is the mechanism generating the speckle.

Volume Scattering

As discussed previously in Sect. 4.5.6, when volume scattering is the mechanism generating speckle, the path delays can be far longer than those encountered with surface scattering. As was discussed, the probability density function of path lengths, $p_l(l)$, can in some cases exceed the thickness of the scattering medium by large factors. The factor $\mathbf{M}_h(\Delta q_z)$ that dominates the frequency sensitivity of speckle in the surface scattering case must now be replaced by the factor $\mathbf{M}_l(\Delta q_z)$, which is the characteristic function of the path-length fluctuations.[6] For normal incidence and observation, the values of Δq_z are now given by

$$\Delta q_z = 2\pi \left| \frac{1}{\lambda_2} - \frac{1}{\lambda_1} \right| \approx \frac{2\pi \Delta \nu}{c} \qquad \text{reflection geometry}$$

$$\Delta q_z = 2\pi(n-1) \left| \frac{1}{\lambda_2} - \frac{1}{\lambda_1} \right| \approx \frac{2\pi(n-1)\Delta \nu}{c} \qquad \text{transmission geometry.}$$

(5-112)

The characteristic function $\mathbf{M}_l(\Delta q_z)$ corresponding to the path-length probability density function $p_l(l)$ is given by

$$\mathbf{M}_l(\Delta q_z) = \int_0^\infty p_l(l) e^{j\Delta q_z l} \, dl. \qquad (5\text{-}113)$$

For fully developed speckle, if we wish to determine the minimum frequency change that will decorrelate the two speckle intensities, we must find the $\Delta \nu$-width of $|\mathbf{M}_l(\Delta q_z)|^2$. In doing so, we can make use of Parseval's theorem of Fourier analysis, which states that

$$\frac{1}{2\pi} \int_{-\infty}^{\infty} |\mathbf{M}_l(\Delta q_z)|^2 \, d\Delta q_z = \int_0^\infty p_l^2(l) \, dl. \qquad (5\text{-}114)$$

In the reflection case, the equivalent width in $\Delta \nu$-space is then expressible as

$$W_{\Delta \nu} = \frac{\int_{-\infty}^{\infty} \left| \mathbf{M}\left(\frac{2\pi \Delta \nu}{c} \right) \right|^2 d\Delta \nu}{|\mathbf{M}(0)|^2} = c \int_0^\infty p_l^2(l) \, dl, \qquad (5\text{-}115)$$

where the fact that $\mathbf{M}(0) = 1$ has been used.

Thus once $p_l(l)$ is determined, either theoretically or experimentally, an estimate of the frequency decorrelation interval can be obtained. Experimental results reported in [172] show, for a 12 mm thick sample, an average impulse response duration on the order of 0.5 nsec, and a frequency de-correlation interval of less than 5 GHz were observed. Such frequency sensitivity is far greater than might be expected from surface scattering from a planar rough surface.

5.4.4 Speckle Suppression by Reduction of Spatial Coherence

In the previous subsection, we considered only temporal coherence. Attention is now turned to considering spatial coherence effects under the assumption that temporal coherence effects are negligible. Less than perfect spatial coherence can arise not only in the case of a moving diffuser, but also when an incoherent source illuminates the scattering surface through free space or through an optical system. In what follows we consider two cases in which the spatial coherence of the illumination is reduced and study the effects on speckle contrast.

---

6. Care must be taken with the fact that on reflection, l replaces $2h$.

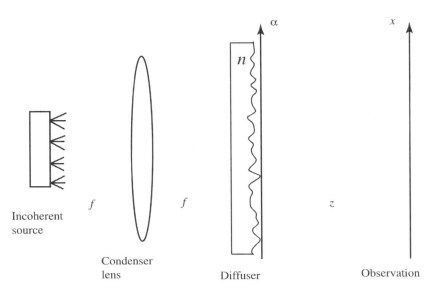

Figure 5.19 Diffuser illuminated by an incoherent source.

Example—Illumination by a Circular Incoherent Source

A common geometry, known as Köhler's illumination, is shown in Fig. 5.19. On the left, an ideal quasi-monochromatic incoherent source (i.e. a source for which all radiating points are temporally uncorrelated) illuminates a condenser lens with a circular pupil (diameter D), the function of which is to provide efficient illumination of the object, which in this case is a scattering surface. The incoherent source is in the front focal plane of the condenser lens, and the diffuser is in the rear focal plane. The light from the incoherent source is assumed to fill the condenser lens. A free-space transmission geometry is again assumed from the diffuser to the observation plane, with modifications for other geometries mentioned later. The intensity of the transmitted light is examined a distance z beyond the diffuser. It is assumed that the correlation area of the diffuser amplitude transmittance is sufficiently small that light from all points on the diffuser contributes to the light at a measurement point in the observation plane.

By the van Cittert–Zernike theorem, using Zernike's approximation, the complex coherence factor $\tilde{\mu}_{12}$ incident on the diffuser is as expressed by Eq. (5-91), that is, as a Fourier transform of the intensity distribution across the pupil of the condenser lens. The appropriate Fourier transform has already been found in Eq. (4-67),

$$\boldsymbol{\mu}_{12}(r) = \left[2 \frac{J_1\left(\frac{\pi D}{\lambda f} r\right)}{\frac{\pi D}{\lambda f} r} \right], \qquad (5\text{-}116)$$

where $r = \sqrt{\Delta\alpha^2 + \Delta\beta^2}$. The coherence area of the light incident on the diffuser is thus

$$A_c = 2\pi \int_0^\infty r \left[2 \frac{J_1\left(\frac{\pi D}{\lambda f} r\right)}{\frac{\pi D}{\lambda f} r} \right] dr = \frac{4}{\pi} \left(\frac{\lambda f}{D}\right)^2 \qquad (5\text{-}117)$$

Assume that the diffuser is uniformly illuminated over an area A_D. Then, under the assumptions we have made, there will be approximately

$$M = \frac{A_D}{A_c}$$

coherence areas on the diffuser, each of which will contribute an independent speckle intensity pattern to the observation plane.[7] It follows that the contrast of the speckle will be

$$C \approx \begin{cases} 1 & A_c > A_D \\ \sqrt{\dfrac{A_c}{A_D}} & A_c < A_D \end{cases}. \qquad (5\text{-}118)$$

When a rough surface is illuminated by a similar optical system in a reflection geometry, the results are unchanged.

When there is an imaging system between the scattering surface and the observation region, the results change. In this case the parameter M is the number of coherence areas within the equivalent area of the amplitude point-spread function (or weighting function) of the system, as seen at the scattering surface. If the imaging system has a lens with a circular pupil function $\mathbf{P}$ of diameter P, then the spread function (neglecting phase factors) is given by Eq. (4-114),

$$\mathbf{k}(\alpha,\beta) = \frac{1}{(\lambda z)^2} \iint_{-\infty}^{\infty} \mathbf{P}(\xi,\eta) \exp\left[j \frac{2\pi}{\lambda z}(\alpha\xi + \beta\eta) \right] d\xi\, d\eta = \frac{\pi P^2}{(\lambda z)^2} \left[2 \frac{J_1\left(\frac{\pi P r}{\lambda z}\right)}{\frac{\pi P r}{\lambda z}} \right], \qquad (5\text{-}119)$$

where r has the same meaning as above, and z is the distance from the lens to the scattering plane. The equivalent area of this spread function is

$$A_s = \frac{4}{\pi}\left(\frac{\lambda z}{P}\right)^2, \qquad (5\text{-}120)$$

and the speckle contrast becomes

---

[7]. This statement assumes that the spatial coherence area incident on the diffuser is large enough to cover many correlation areas of the diffuser structure.

5.4 Temporal and Spatial Coherence Reduction

$$C \approx \begin{cases} 1 & \mathcal{A}_c > \mathcal{A}_s \\ \sqrt{\dfrac{\mathcal{A}_c}{\mathcal{A}_s}} = \dfrac{fP}{zD} & \mathcal{A}_c < \mathcal{A}_s \end{cases}. \quad (5\text{-}121)$$

A generalization of this result, valid even for large angles, would be

$$C \approx \begin{cases} 1 & \mathrm{NA}_{\mathrm{illum}} < \mathrm{NA}_{\mathrm{image}} \\ \dfrac{\mathrm{NA}_{\mathrm{image}}}{\mathrm{NA}_{\mathrm{illum}}} & \mathrm{NA}_{\mathrm{illum}} > \mathrm{NA}_{\mathrm{image}} \end{cases}. \quad (5\text{-}122)$$

Here, NA stands for numerical aperture, which is defined as $n \sin \theta$, where n is the refractive index, and θ is the half-angle subtended by a pupil. In this case, NAs are measured in the scattering plane; $\mathrm{NA}_{\mathrm{illum}}$ is the numerical aperture of the condenser lens, as seen from the scattering plane, and $\mathrm{NA}_{\mathrm{image}}$ is the numerical aperture of the imaging lens, again as seen from the scattering plane. Achieving suppression of the speckle requires that $\mathrm{NA}_{\mathrm{illum}} > \mathrm{NA}_{\mathrm{image}}$.

It is worth noting that the degree of speckle suppression depends on the properties of the illumination and imaging systems, but not on properties of the scattering surface, other than the usual assumptions that it is rough on the scale of a wavelength and its angular scattering overfills the imaging optics (for all source points).

Comparing the two expressions for C, one for the free-space geometry and one for the imaging geometry, it is clear that it is much more difficult to suppress speckle in an imaging geometry using spatial coherence, since in general $\mathcal{A}_s \ll \mathcal{A}_D$; that is, the area of the diffuser will generally be far larger than the equivalent area of the point-spread function, and therefore the light illuminating the scattering surface must be much more incoherent in the imaging case to achieve a given reduction of contrast.

Example—Illumination by Two Mutually Incoherent Point Sources

As a second example, consider the illumination and imaging optics shown in Fig. 5.20. Assume that the illumination arises from two equal-intensity, mutually-incoherent point sources in the same plane as the previous extended incoherent source, with those two point sources being equally spaced above and below the optical axis. Again assume Köhler's illumination optics, but this time insert an imaging lens, which we assume to have unity magnification (object and image distances both z). Let the angles of the sources subtended with the optical axis be $+\zeta$ and $-\zeta$, respectively, and consider the image at coordinates $(0, 0)$ in the (x, y) plane. The center frequencies of the two sources are assumed to be identical, but they arise from two different sources which are totally uncorrelated.

Because they lie in the front focal plane of the condenser lens, each of the point sources generates a plane wave incident on the diffuser, with angles of inclination $+\zeta$ and $-\zeta$. Our goal here is to explore the minimum angular separation $2\zeta_{\min}$ of the two sources required to reduce the contrast of the speckle by $1/\sqrt{2}$. Note that, because

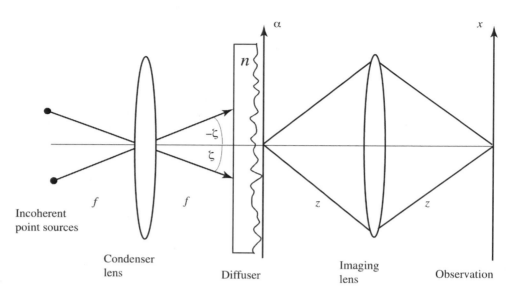

Figure 5.20 Illumination and imaging geometry with two mutually incoherent point sources.

the two sources are mutually incoherent, the two image speckle patterns will add on an intensity basis. However, this does not necessarily mean that the speckle contrast will be reduced, for if the angular separation of the sources is too small, the two speckle patterns produced may be correlated.

The analysis required is related to that performed in Appendix D.2, but different in detail. This time we wish to find the cross-correlation between two image fields $\mathbf{A}_1(x, y)$ and $\mathbf{A}_2(x, y)$ arising from the illuminations at angles $+\zeta$ and $-\zeta$. Under the assumption that the speckle is well-developed, the fields have complex circular Gaussian statistics and the correlation of the intensities $I_1(x, y)$ and $I_2(x, y)$ will be the squared magnitude of the correlation of the fields.

Following the outline of Appendix D.2, we again have that the image field $\mathbf{A}(x, y)$ is related to the field $\mathbf{a}(\alpha, \beta)$ in the plane in close proximity with the scattering surface through

$$\mathbf{A}(x, y) = \iint\limits_{-\infty}^{\infty} \mathbf{k}(x + \alpha, y + \beta) \mathbf{a}(\alpha, \beta) \, d\alpha \, d\beta, \tag{5-123}$$

where $\mathbf{k}$ is the point-spread function of the imaging system. Thus the cross-correlation of fields $\mathbf{A}_1$ and $\mathbf{A}_2$ at image point (x, y), which we will represent by $\mathbf{\Gamma}_{12}(x, y; \zeta)$, can be found from

5.4 Temporal and Spatial Coherence Reduction

$$\Gamma_{12}(x,y;\zeta) = \iint_{-\infty}^{\infty} \iint_{-\infty}^{\infty} \mathbf{k}(x+\alpha_1, y+\beta_1)\mathbf{k}^*(x+\alpha_2, y+\beta_2)$$
$$\times \overline{\mathbf{a}_1(\alpha_1,\beta_1)\mathbf{a}_2^*(\alpha_2,\beta_2)}\, d\alpha_1\, d\beta_1\, d\alpha_2\, d\beta_2, \quad (5\text{-}124)$$

where $\mathbf{a}_1$ is the field generated by the upper point source and $\mathbf{a}_2$ is the field generated by the lower point source. The fields $\mathbf{a}_1$ and $\mathbf{a}_2$ can in turn be written

$$\mathbf{a}_1(\alpha_1, \beta_1) = \mathbf{r}\,\mathbf{S}(\alpha_1,\beta_1) e^{j2\pi\frac{\sin\zeta}{\lambda}\beta_1} e^{j2\pi\frac{h(\alpha_1,\beta_1)\cos\zeta}{\lambda}}$$
$$\mathbf{a}_2(\alpha_2, \beta_2) = \mathbf{r}\,\mathbf{S}(\alpha_2,\beta_2) e^{-j2\pi\frac{\sin\zeta}{\lambda}\beta_2} e^{j2\pi\frac{h(\alpha_2,\beta_2)\cos\zeta}{\lambda}}, \quad (5\text{-}125)$$

where again $\mathbf{r}$ is the average amplitude reflectivity of the surface, $\mathbf{S}$ is the amplitude distribution incident on the surface (excluding the phase tilt due to the inclined illumination angle), the third terms represent the effect of the inclined illumination, while the last terms represent the effect of the surface roughness. It follows that

$$\overline{\mathbf{a}_1(\alpha_1,\beta_1)\mathbf{a}_2^*(\alpha_2,\beta_2)} = |\mathbf{r}|^2 \mathbf{S}(\alpha_1,\beta_1)\mathbf{S}^*(\alpha_2,\beta_2)$$
$$\times e^{j2\pi\frac{\sin\zeta}{\lambda}(\beta_1+\beta_2)} \overline{e^{j2\pi\frac{\cos\zeta}{\lambda}[h(\alpha_1,\beta_1)-h(\alpha_2,\beta_2)]}}$$
$$= \kappa|\mathbf{r}|^2 |\mathbf{S}(\alpha_1,\beta_1)|^2 e^{j4\pi\frac{\sin\zeta}{\lambda}\beta_1}\delta(\alpha_1-\alpha_2, \beta_1-\beta_2), \quad (5\text{-}126)$$

where we have adopted the δ-correlated assumption for the $\mathbf{a}$'s, and the term to be averaged has become equal to unity as a result.[8] Substituting Eq. (5-126) in Eq. (5-124) yields

$$\Gamma_{12}(x,y;\zeta) = \kappa|\mathbf{r}|^2 \iint_{-\infty}^{\infty} |\mathbf{S}(\alpha,\beta)|^2 e^{j4\pi\frac{\sin\zeta}{\lambda}\beta} |\mathbf{k}(x+\alpha, y+\beta)|^2 \, d\alpha\, d\beta, \quad (5\text{-}127)$$

where we have dropped the subscripts on (α,β) because they are no longer needed. For a good imaging system, $|\mathbf{k}|^2$ is a very narrow function compared with $|\mathbf{S}|^2$, and the approximation

$$\Gamma_{12}(x,y;\zeta) \approx \kappa|\mathbf{r}|^2 |\mathbf{S}(-x,-y)|^2 \iint_{-\infty}^{\infty} |\mathbf{k}(x+\alpha, y+\beta)|^2 e^{j4\pi\frac{\sin\zeta}{\lambda}\beta} \, d\alpha\, d\beta \quad (5\text{-}128)$$

is accurate. A normalized version of this cross-correlation can now be written, the normalization being by $\Gamma_{12}(x,y;0)$, that is, by the value of the cross-correlation when the two point sources coincide. Thus

---

8. κ is again a constant that is needed when making the δ-correlated assumption in order to achieve a dimensionally correct result.

$$\mu_{12}(x, y; \zeta) = \frac{\iint\limits_{-\infty}^{\infty} |\mathbf{k}(x+\alpha, y+\beta)|^2 e^{j4\pi\frac{\sin\zeta}{\lambda}\beta}\, d\alpha\, d\beta}{\iint\limits_{-\infty}^{\infty} |\mathbf{k}(x+\alpha, y+\beta)|^2\, d\alpha\, d\beta}. \tag{5-129}$$

A change of variables of integration will simplify the result: let $\alpha' = x + \alpha$ and $\beta' = y + \beta$. The expression for μ_{12} becomes

$$\mu_{12}(x, y; \zeta) = \frac{e^{-j4\pi\frac{\sin\zeta}{\lambda}y} \iint\limits_{-\infty}^{\infty} |\mathbf{k}(\alpha', \beta')|^2 e^{j4\pi\frac{\sin\zeta}{\lambda}\beta'}\, d\alpha'\, d\beta'}{\iint\limits_{-\infty}^{\infty} |\mathbf{k}(\alpha', \beta')|^2\, d\alpha'\, d\beta'}. \tag{5-130}$$

Since we are interested only in the modulus of μ_{12}, the exponential factor before the integral can be dropped, and we are left with

$$|\mu_{12}(\zeta)| = \left| \frac{\iint\limits_{-\infty}^{\infty} |\mathbf{k}(\alpha', \beta')|^2 e^{j4\pi\frac{\sin\zeta}{\lambda}\beta'}\, d\alpha'\, d\beta'}{\iint\limits_{-\infty}^{\infty} |\mathbf{k}(\alpha', \beta')|^2\, d\alpha'\, d\beta'} \right|. \tag{5-131}$$

Since $|\mathbf{k}|^2$ is the *intensity* point-spread function of the imaging system, it follows directly that $|\mu_{12}|$ can be expressed in terms of the *optical transfer function* (OTF) of the imaging system,[9]

$$|\mu_{12}(\zeta)| = \left| \mathcal{H}\left(0, -\frac{2\sin\zeta}{\lambda}\right) \right|, \tag{5-132}$$

where the OTF is defined by

$$\mathcal{H}(v_x, v_y) = \frac{\iint\limits_{-\infty}^{\infty} |\mathbf{k}(x, y)|^2 e^{-j2\pi(v_x x + v_y y)}\, dx\, dy}{\iint\limits_{-\infty}^{\infty} |\mathbf{k}(x, y)|^2\, dx\, dy}.$$

For an imaging system with a circular pupil of diameter D, the OTF has the form

$$\mathcal{H}(\rho) = \begin{cases} \dfrac{2}{\pi}[\arccos(\rho/\rho_0) - (\rho/\rho_0)\sqrt{1 - (\rho/\rho_0)^2}] & \rho < \rho_0 \\ 0 & \text{otherwise,} \end{cases} \tag{5-133}$$

---

9. For a discussion of optical transfer functions, see [71], Section 6.3.

where $\rho = \sqrt{v_x^2 + v_y^2}$ and $\rho_0 = \frac{D}{\lambda z}$ is the frequency cutoff of the OTF. From this result it is clear that the correlation μ_{12} will fall to zero when

$$\sin \zeta \geq \frac{D}{2z} \approx \text{NA}_{\text{image}}, \quad (5\text{-}134)$$

where NA_{image} is again the numerical aperture of the imaging system. Again the ability to suppress speckle depends on the character of the illumination and imaging systems, and not directly on the properties of the scattering surface.

If two mutually incoherent sources separated according to this criterion are used, the speckle will be reduced by a factor $1/\sqrt{2}$. However, it is possible to form a two-dimensional array of M mutually incoherent sources, with separations satisfying this criterion in both dimensions, and provided the scattering angle of the diffuser is sufficiently great to fill the imaging lens aperture for all such sources, the contrast of the speckle will be reduced by a factor $1/\sqrt{M}$. The smaller the numerical aperture of the imaging system, the easier it is to obtain speckle suppression by this method, but such a condition comes at the expense of resolution in the image.

5.5 Use of Temporal Coherence to Destroy Spatial Coherence

In some applications, a single spatially coherent source with a relatively short coherence length may be available. To reduce speckle, it may be desirable to reduce the spatial coherence of the source. This can be done by splitting the source into an array of M separate beams, and delaying each beam by a different time. If each delay differs from every other delay by more than the coherence time, then the spatial array of sources lacks spatial coherence. An example of one way to do this is shown in Fig. 5.21 (cf. U.S. patent #6,924,891 B2). Light is equally split into each of M fibers in a bundle of fibers; the bundle is separated, with each fiber entering a different delay loop. The incremental delay between fibers is made greater than the coherence time of the light source; the fibers are then brought back to a bundle and carried to the

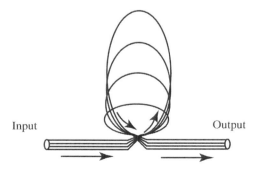

Figure 5.21 Fiber bundle with different delays for each fiber.

desired location of the effective source. The end of the bundle now contains M different mutually incoherent sources, which constitute a new source with limited spatial coherence. Provided the conditions described in Section 5.4.4 are satisfied, some suppression of speckle can be expected from the use of such a source. The maximum amount of speckle contrast suppression would be a factor $1/\sqrt{M}$, but a smaller reduction of speckle will occur if the conditions spelled out in Section 5.4.4 regarding the required separation of such sources are not satisfied.

Many other methods for introducing different delays into different portions of a light beam can be envisioned, including methods using waveguides, methods using gratings, and methods using parallel partially reflecting/transmitting plates (cf. U.S. patent #6,801,299 B2).

5.6 Compounding Speckle Suppression Techniques

In this chapter we have discussed a variety of methods for reducing speckle contrast. Each method may be viewed as introducing a certain number of degrees of freedom. For example, polarization diversity can provide two degrees of freedom, while temporal averaging, angular and wavelength diversity, and coherence reduction all can contribute their own degrees of freedom. In most cases,[10] if we have N independent mechanisms for introducing new degrees of freedom, then the total number M of degrees of freedom is simply

$$M = \prod_{n=1}^{N} M_n, \qquad (5\text{-}135)$$

and the total contrast of the resulting speckle will be

$$C = \frac{1}{\sqrt{M}}. \qquad (5\text{-}136)$$

If reduction of contrast is the goal, then as many different approaches as possible should be tried simultaneously in order to achieve the greatest amount of overall speckle suppression.

---

10. An exception occurs for simultaneous temporal and spatial averaging when a changing diffuser is imaged onto a fixed diffuser. See the section beginning on p. 214.

6 Speckle in Certain Imaging Applications

6.1 Speckle in the Eye

There is an interesting experiment that can be performed with a group of people in a room using a CW laser (even a small laser pointer will do if the beam is expanded a bit and the lights are dimmed). Shine the light from the laser on the wall or on any other planar rough scattering surface and ask the members of the group to remove any eyeglasses they may be wearing (this may be difficult to do for people with contact lenses, in which case they can continue to wear their lenses). Ask all members of the group to look at the scattering spot and to move their heads laterally left to right and right to left several times. Now ask if the speckles in the spot moved in the same direction as their head movement or in the opposite direction from their head movement. The results will be as following:

- Individuals who have perfect vision or who are wearing their vision correction will report that speckle motion was hard to detect. In effect, the speckle structure appears fixed to the surface of the scattering spot, and does not appear to move *with respect to the spot*, but does undergo some internal change that is not motion.
- Individuals who are farsighted and uncorrected will report that the speckle moved in the same direction as their head moved, translating through the scattering spot in that direction.
- Individuals who are nearsighted will report that the speckles move through the scattering spot in the opposite direction to their head movement.

Our goal in this section is to give a brief explanation of the results of this experiment. For alternative but equivalent approaches to explaining this phenomenon, see [11], [113], or [49], p. 140.

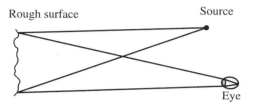

Figure 6.1 Source, rough surface and the eye of the observer.

Let the object consist of an illuminated scattering spot on a planar rough surface, as shown in Fig. 6.1. Let the direction of illumination and the direction of observation, with respect to the surface, be near normal. Note that movement of the eye in one direction when the source and scattering surface are fixed is equivalent to motion of the source and scattering surface in the opposite direction when the eye is fixed.

As the pupil of the eye moves, it intercepts different contributions of scattered light from the rough surface, with a complete change of the intercepted light caused by the movement of one eye pupil diameter. As a consequence, as the eye moves, speckle motion is always accompanied by a certain degree of internal changes of the structure of the speckle pattern, which we refer to here as *boiling* of the speckle. Note in addition that, as the eye moves side to side, the eye also rotates (in its socket) to hold the scattering spot in the center of the field of view.

Figure 6.2 shows the geometry assumed, and illustrates the cases of a well-corrected eye, a farsighted eye, and a nearsighted eye. Here we show a movement of Δx of the scattering surface downward with respect to the eye, equivalent to a movement of the pupil upward by Δx. Note that the difference between the location of the retina and the true image plane for both the farsighted and the nearsighted eye can arise due to elongation or shortening of the eyeball and/or decreased or increased power of the lens as compared with the normal eye.

It is important to distinguish between the retinal image of the scattering spot and the true image of the scattering spot. In general these two images lie in different planes. An important fact that is needed to understand this phenomenon is that, regardless of where the scattering spot is imaged, whether on the retina, behind the retina, or in front of the retina, there is always speckle incident on the retina, and that speckle moves on the retina *by the same distance that the true image of the scattering spot has moved*, which in general can be a different distance than the retinal image of the spot has moved.

Consider first the well-corrected eye. As the scattering spot moves by Δx, the image of the scattering spot, and the speckle, fall on the retina and both move in the opposite direction by $(z_i/z_o)\Delta x$. But the eye also rotates to hold the retinal image of the scattering spot at a fixed location on the retina. As a consequence the speckle does not appear to move at all, but does boil since the pupil is capturing a changing portion of the light leaving the scattering surface.

6.1 Speckle in the Eye 189

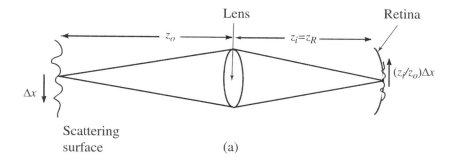

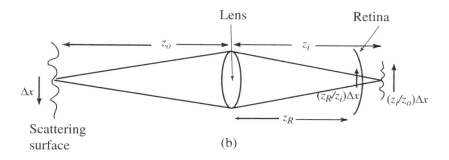

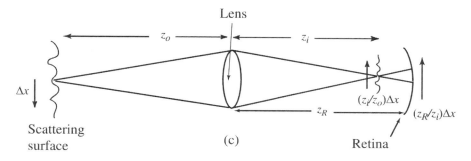

Figure 6.2 Looking down on (a) a well-corrected eye, (b) a farsighted eye, and (c) a nearsighted eye.

Now consider the farsighted (hypermetropic) eye shown in part (b) of the figure. If the scattering surface moves by Δx, the true image of the scattering spot and the speckle move by $(z_i/z_o)\Delta x$, while the retinal image moves by $(z_R/z_o)\Delta x$. Since $z_R < z_i$, the speckle moves a greater distance than the retinal image of the spot. If the eye rotates to hold the center of the spot at a constant position on the retina, the speckle appears to move relative to the spot in the same direction the spot has moved.

Consider next the nearsighted (myopic) eye. The retina is farther from the lens than the image plane. The true image of the scattering spot and the speckle move by $(z_i/z_o)\Delta x$, while the retinal image moves by $(z_R/z_o)\Delta x$. Since $z_R > z_i$, the speckle moves a smaller distance than the retinal image of the spot. If the eye rotates to hold the center of the spot at a constant position on the retina, the speckle appears to move relative to the spot in the opposite direction to the motion of the spot.

The previous discussion may raise the question as to why speckle patterns are not more widely used in ophthalmic testing for eyeglass prescriptions. One answer is that a corrective lens should assure the best possible image on the retina averaged over all wavelengths in the visible spectrum of light, rather than at the single wavelength characteristic of a laser. This problem can be overcome by using several different lasers with wavelengths distributed throughout the visible spectrum, or by using a single widely tunable laser. The application of laser speckle patterns to ophthalmic testing has received considerable attention in the literature; some classic publications are [79], [25], [24] and [10].

6.2 Speckle in Holography

Speckle suppression is generally most difficult with imaging techniques that rely fundamentally on the coherence of light to form images. Holography is one such technique (for background on holography, see, for example, [71], Chapter 9). While methods for forming holograms with incoherent light do exist, they are complex and do not produce images of the same quality as can be obtained with coherent light.

6.2.1 Principles of Holography

A hologram is formed by interfering the light transmitted by or reflected from an object of interest with light in a mutually coherent reference wave. Many recording geometries exist, but we illustrate with the one shown in Fig. 6.3. Light transmitted by a transparent object interferes with a mutually coherent tilted plane wave that serves as the reference wave. Usually the object transparency is preceded by a diffuser in order to spread the light evenly over the hologram plane and to allow the image to be viewable from all parts of the hologram. In other cases the object may be a reflective 3D structure, with optically rough surfaces. In both cases, the object light at the hologram consists of the sum of the reference wave, which we represent by $\mathbf{R}(x, y)$, and a speckled object field represented by $\mathbf{O}(x, y)$. The interference of these waves generates an intensity pattern

$$\begin{aligned} I(x, y) &= |\mathbf{R}(x, y) + \mathbf{O}(x, y)|^2 \\ &= |\mathbf{R}(x, y)|^2 + |\mathbf{O}(x, y)|^2 + \mathbf{R}(x, y)\mathbf{O}^*(x, y) + \mathbf{R}^*(x, y)\mathbf{O}(x, y). \end{aligned} \quad (6\text{-}1)$$

6.2 Speckle in Holography

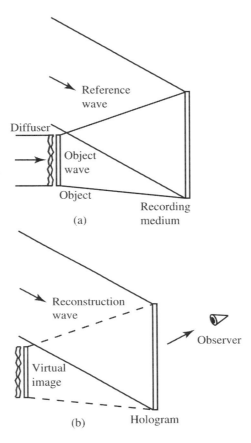

Figure 6.3 (a) Recording a hologram of a diffuse object, and (b) reconstructing the virtual image.

If the reference wave is a simple plane wave, tilted in the vertical direction and inclined at angle θ to the normal to the recording medium, then

$$\mathbf{R}(x, y) = \mathbf{R}_0 e^{-j2\pi \frac{\sin\theta}{\lambda} y}, \tag{6-2}$$

and

$$I(x, y) = |\mathbf{R}_0|^2 + |\mathbf{O}(x, y)|^2 + \mathbf{O}^*(x, y) e^{-j2\pi \frac{\sin\theta}{\lambda} y} + \mathbf{O}(x, y) e^{j2\pi \frac{\sin\theta}{\lambda} y}. \tag{6-3}$$

This intensity pattern is recorded on the photosensitive medium, and after appropriate processing, forms a hologram. The amplitude transmittance $\mathbf{t}(x, y)$ of the hologram is made proportional to exposure, with the result

$$\mathbf{t}(x, y) = \mathbf{t}_0 + \beta|\mathbf{O}(x, y)|^2 + \beta\mathbf{O}^*(x, y) e^{-j2\pi \frac{\sin\theta}{\lambda} y} + \beta\mathbf{O}(x, y) e^{j2\pi \frac{\sin\theta}{\lambda} y}, \tag{6-4}$$

where β is a proportionality constant and $\mathbf{t}_0$ is a constant bias transmittance. The spatial frequency

$$\alpha = \frac{\sin\theta}{\lambda} \tag{6-5}$$

is often referred to as the "carrier frequency" of the hologram.

Let the hologram be illuminated by a duplicate of the original reference wave. We call this the "playback" wave, and represent it by

$$\mathbf{P}(x, y) = \mathbf{P}_0 e^{-j2\pi\frac{\sin\theta}{\lambda}y}, \tag{6-6}$$

where the wavelength λ is assumed to be the same during both recording and playback. The fourth term of the product of $\mathbf{P}$ with $\mathbf{t}$ yields the reconstructed wave component of interest,

$$\mathbf{A}_4(x, y) = \beta \mathbf{P}_0 \mathbf{O}(x, y) \tag{6-7}$$

which, aside from proportionality constants, is a duplicate of the original object wavefront incident on the hologram. This wave component propagates to the right, where a portion of it is intercepted by the observer's eye. If the angular offset of the reference wave has been chosen large enough, the wave components generated by other terms in $\mathbf{t}(x, y)$ propagate in different directions and are not intercepted by the observer. Since the wave component that travels to the observer is a duplicate of the original object wavefront, the observer sees a virtual image of the object behind the hologram. Because of the diffuser through which the object was illuminated, the image of the object contains speckle.

To be more complete, we mention that if the hologram is illuminated by a playback wave propagating at an angle $-\theta$ rather than θ, the playback wave takes the form

$$\mathbf{P}(x, y) = \mathbf{P}_0 e^{j2\pi\frac{\sin\theta}{\lambda}y}, \tag{6-8}$$

and in this case the *third* term of the transmitted fields becomes

$$\mathbf{A}_3(x, y) = \beta \mathbf{P}_0 \mathbf{O}^*(x, y), \tag{6-9}$$

which is the conjugate of the original object wave. This wave component propagates to the right of the hologram and forms a real image of the object in space. It too contains speckle.

Note that this has been the most basic introduction to holography. Many other geometries and types of holograms are known, but will not be covered here, since the main interest is speckle.

6.2.2 Speckle Suppression in Holographic Images

Unfortunately, because holography is fundamentally a coherent imaging technique, suppression of speckle is a particularly difficult problem. The difficulty is compounded by the fact that the recording and playback steps are usually distinct and separated in time. While there are a few exceptions, usually speckle can be suppressed only with an accompanying reduction in image resolution. In what follows we discuss several specific methods.

Angle and Wavelength Diversity in Playback

Suppose that during the reconstruction or playback step, the hologram is illuminated by two mutually incoherent plane waves traveling in slightly different directions. Each of the two incident plane waves creates an image of the original object, and those images add on an intensity basis. However, the locations of the two images are different. The central locations of the virtual and real images are determined by the grating equation, as applied to the grating period associated with the carrier frequency, in particular the period

$$\Lambda = \frac{\lambda}{\sin \theta}. \tag{6-10}$$

According to the grating equation (see Ref. [71], Appendix D), if θ_1 represents the angle with respect to the normal (measured clockwise) of the incident playback wave, then the angle of diffraction of that wave with respect to the normal (again measured clockwise) is

$$\sin \theta_2 = \frac{\lambda}{\Lambda} + \sin \theta_1. \tag{6-11}$$

Thus a change of angle of incidence of the playback wave clearly results in a shift of the images produced by the hologram. The superposition of two shifted images reduces the speckle contrast by the factor $1/\sqrt{2}$, provided the shift is by more than a single speckle correlation area, but at the same time it destroys image detail. In the more general case of a continuous range of angles of independent plane-wave playback beams, the image is continuously blurred, speckle is further suppressed, and image detail is again lost due to the resulting image blur.

A similar phenomenon happens when different wavelengths of illumination are used in a single plane-wave playback beam. The grating equation again implies that the images produced by a different wavelengths are shifted with respect one another.[1] The superposition of two images again reduces speckle, but at the price of blurring the resulting image.

The image blur that results from these methods may be of little consequence in some applications, for the resolution may be even more severely limited by the eye. However, in other applications, particularly those that involve electronic detection of the real image, the loss of resolution can be objectionable. It should also be kept in mind that for a blur-free image containing strong speckle, the image information that can be extracted is limited by the speckle itself, so the tradeoff between speckle

---

1. The images at different wavelengths are also displaced from one another in axial distance.

contrast and image resolution may, over certain limits, be one that leaves the information extractable from the image unchanged.

Time-Varying Hologram Masks

Another method that can be used to trade speckle contrast for resolution involves the use of changing masks in the hologram plane. This method can also be used with nonholographic coherent imaging systems, and has been explored in this context by Dainty and Welford in Ref. [31] and by McKechnie in Ref. [110]. During playback the holographic aperture can be broken up into separate subapertures, and the subapertures can be opened in time sequence, with one open at a time. Alternatively a single subaperture can be translated or rotated in the hologram plane. Separate images of the object will be formed by each subaperture, and those images will have different speckle patterns. If several images are obtained during the time integration of the eye or an electronic detector being used, the superposition of these images will reduce the speckle contrast observed.

The price paid by this method is that the resolution obtainable from a single subaperture is smaller than that obtainable from the full hologram aperture. This may be of little consequence if we are viewing the virtual image with the eye, for the pupil of the eye is then the limiting aperture, but if the real image is being detected, the resolution obtainable from the image is determined by the diffraction limit of the subaperture, not the diffraction limit of the entire hologram. Thus again we encounter the tradeoff between speckle contrast and resolution. In some applications, a small amount of blurring with an accompanying drop in speckle contrast may be quite acceptable.

One relative of this method is the earlier method of Martienssen and Spiller (Ref. [108]), in which a sequence of separate holograms is recorded, with the diffuser that precedes the object being moved or changed between exposures. When this set of holograms is illuminated, one at a time in time sequence, the speckle in each holographic image is different and the speckle contrast is reduced by an amount that depends on how many holographic images were displayed in one detector resolution time. For a discussion of another method for speckle suppression used in holographic microscopy that also uses a form of masking, see the work of van Ligten (Ref. [167]).

Multiplexed Images in Thick Holograms

Just as Martienssen and Spiller used a sequence of separate holograms, each taken with a different diffuser preceding the object, and displayed the resulting images in rapid sequence, so too the multiplexing properties of thick holograms can be used to achieve similar results. It is possible to multiplex a multitude of different holograms in a single thick holographic recording, using different reference wave angles for each separate recording. If a different diffuser precedes the object for each such recording, then sequentially illuminating the hologram by each of the reference waves in turn will create a sequence of images that differ only in the speckle patterns present. If the integration time is long compared to the time a single image is present, the speckle will be reduced accordingly without loss of image resolution.

A number of other methods for suppressing speckle can be found in the literature, but the above discussion gives a reasonable sample of available methods.

6.3 Speckle in Optical Coherence Tomography

Optical coherence tomography, widely referred to as OCT, is, like holography, an imaging technique that depends fundamentally on the coherence of the light used in the imaging process. For this reason, speckle is a significant issue in OCT. In what follows, the simplest form of the OCT imaging process will be described, following which some methods for reducing speckle will be discussed. For a comprehensive book on this imaging technique, see Ref. [19]. Particularly relevant is Chapter 7 by Schmitt, Xiang and Yung on speckle reduction techniques. Additional general references on OCT are Refs. [142] and [43].

The OCT imaging technique is rather young as imaging methods go, and has undergone rapid progress, which will no doubt continue into the future.

6.3.1 Overview of the OCT Imaging Technique

OCT is a method for measuring the optical backscattering from a sample as a function of both depth and transverse coordinates. Often the samples are biological in nature. The resolutions obtainable are in the range 1 to 15 μm, better than that obtainable with ultrasound by between one and two orders of magnitude.

With the development of lasers that can emit pulses with durations of only a few femtoseconds (1 fsec = 10^{-15} sec), the idea of achieving depth resolution by pulse-echo timing becomes particularly attractive. However, electronic circuitry capable of responding in fsecs to returning optical signals does not exist, so another method for isolating a very short time interval of the returning signal is needed. Such isolation is accomplished in OCT by the use of broadband sources and interferometry, taking advantage of the very short coherence times of such sources. A typical source of this type might be a super-luminescent diode, with a bandwidth of 20–40 nm at 1310 nm wavelength. If the source has a short coherence time, then depth regions where the coherence with respect to a reference beam is high can be distinguished from depth regions where that coherence has vanished. Figure 6.4 illustrates one possible realization of an OCT system. The system is a fiber-based Michelson interferometer. It operates by linearly scanning the reference mirror in the axial direction to scan the region of high coherence through the depth of the object, and scanning the mirror in the object arm to move the region of measurement in the transverse direction across the object. In this fashion a 2D scan of the object is obtained.[2] The linear

---

2. Several different modes of scanning are possible. By analogy with ultrasound imaging modes, a single vertical scan in the depth direction is referred to as an "A-scan," while the combination of scanning in depth and one transverse direction is referred to as a "B-scan."

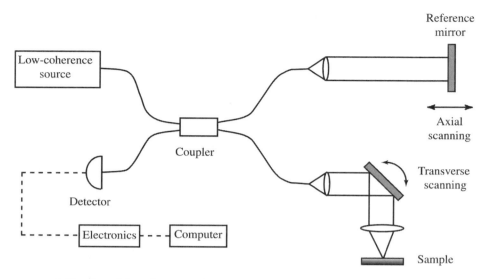

Figure 6.4 A fiber-based interferometer for use in OCT. The axial scanning mirror changes the path-length delay in the reference arm to select the axial depth and the transverse scanning mirror selects the transverse coordinates being imaged.

motion of the reference mirror in effect Doppler shifts the reference light, and when the reference and object beams are incident on the detector, a beat note is observed when the light coming from the object is coherent with the light coming from the reference mirror. Thus the amplitude of scattering from the region of high coherence can be measured by measuring the strength of the beat note. As the reference mirror scans, the scattering amplitudes from the corresponding depth regions within the object are obtained.

6.3.2 Analysis of OCT

To understand the operation of OCT in more detail, we embark on a short analysis. Incident on the detector are a reference wave and an object wave, which we represent by analytic signals $\mathbf{E}_r(t)$ and $\mathbf{E}_o(x, z, t)$, respectively,

$$\mathbf{E}_r(t) = \alpha_r \mathbf{L}\left(t - \frac{2l_r}{c}\right)$$

$$\mathbf{E}_o(x, z; t) = \alpha_o(x, z)\mathbf{L}\left(t - \frac{2l_o}{c} - 2\bar{n}\frac{z}{c}\right). \tag{6-12}$$

Here $\mathbf{L}(t)$ represents the analytic signal representation of the low-coherence light emitted by the source, assumed to be independent of transverse coordinate x. The quantity α_r represents the amplitude attenuation of the light as it passes through the coupler to the reference mirror and back through the coupler to the detector. The quantity $2l_r$ is the path length traveled *in free space* in the upper (reference) arm of the interferometer, that is, from the end of the fiber to the reference mirror and back to the fiber, while c is the free-space speed of propagation of light. The quantity $2l_o$ is the distance traveled by the light in the lower arm from the end of the fiber to the top surface of the object and back to the fiber. The quantity z represents depth into the object and $\bar{n}$ is the average refractive index within the object.[3] It is assumed that the optical path lengths traveled in fiber in the upper and lower arms are the same. $\alpha_o(x,z)$ represents the amplitude attenuation of the field returned from position z within the object of interest.

The intensity incident on the photodetector is given by

$$I(x,z) = \langle |\mathbf{E}_r(t) + \mathbf{E}_o(x,z;t)|^2 \rangle$$
$$= \langle |\mathbf{E}_r(t)|^2 \rangle + \langle |\mathbf{E}_o(x,z;t)|^2 \rangle + \langle \mathbf{E}_r(t)\mathbf{E}_o^*(x,z;t) \rangle + \langle \mathbf{E}_r^*(t)\mathbf{E}_o(x,z;t) \rangle, \quad (6\text{-}13)$$

where $\langle \cdot \rangle$ signifies an infinite time average. Substituting Eqs. (6-12) into Eqs. (6-13) and defining $I_r = \langle |\mathbf{E}_r(t)|^2 \rangle = \alpha_r^2 \langle |\mathbf{L}(t)|^2 \rangle$ and $I_o(x,z) = \langle |\mathbf{E}_o(x,z;t)|^2 \rangle = \alpha_o^2(x,z)\langle |\mathbf{L}(t)|^2 \rangle$, we obtain

$$I(x,z) = I_r + I_o(x,z) + \tilde{\Gamma}_\mathbf{E}(x,z) + \tilde{\Gamma}_\mathbf{E}^*(x,z) \quad (6\text{-}14)$$

where

$$\tilde{\Gamma}_\mathbf{E}(x,z) = \langle \mathbf{E}_r(t)\mathbf{E}_o^*(x,z;t) \rangle = \alpha_r \alpha_o(x,z) \left\langle \mathbf{L}\left(t - \frac{2l_r}{c}\right) \mathbf{L}^*\left(t - \frac{2l_o}{c} - 2\bar{n}\frac{z}{c}\right) \right\rangle$$
$$= \alpha_r \alpha_o(x,z) \tilde{\Gamma}_\mathbf{L}\left(\frac{2l_r}{c} - \frac{2l_o}{c} - 2\bar{n}\frac{z}{c}\right), \quad (6\text{-}15)$$

and

$$\tilde{\Gamma}_\mathbf{L}(\tau) = \langle \mathbf{L}(t)\mathbf{L}^*(t+\tau) \rangle. \quad (6\text{-}16)$$

It is somewhat more convenient to work with a normalized version of $\tilde{\Gamma}_\mathbf{L}(\tau)$, namely

$$\tilde{\gamma}_\mathbf{L}(\tau) = \frac{\tilde{\Gamma}_\mathbf{L}(\tau)}{\tilde{\Gamma}_\mathbf{L}(0)} = \frac{\tilde{\Gamma}_\mathbf{L}(\tau)}{I_L}, \quad (6\text{-}17)$$

where

$$I_L = \langle |\mathbf{L}(t)|^2 \rangle. \quad (6\text{-}18)$$

---

[3]. We make the simplifying approximation that the average refractive index of the object does not change by a significant fraction of its mean value over the depth of the object. In addition, wavelength dispersion introduced during propagation through the object has been neglected for simplicity. Dispersion can change the time structure of the signal returning in the object arm of the interferometer, reducing its correlation with the reference wave. The reader interested in understanding the effects of dispersion should consult Ref. [76], Section 2.2.4.

If $\bar{v}$ represents the center frequency of the laser output, then $\tilde{\gamma}_L$ can be rewritten in the form

$$\tilde{\gamma}_L(\tau) = |\tilde{\gamma}_L(\tau)|e^{-j(2\pi\bar{v}\tau - \phi_L(\tau))} = \tilde{\mu}_L(\tau)e^{-j2\pi\bar{v}\tau}, \tag{6-19}$$

where $|\tilde{\mu}_L(\tau)| = |\tilde{\gamma}_L(\tau)|$ and $\arg[\tilde{\mu}_L(\tau)] = \phi_L(\tau)$. The intensity pattern can then be written

$$I(x,z) = I_r + I_o(x,z) + 2\alpha_r\alpha_o(x,z)I_L\left|\tilde{\mu}_L\left(\frac{2l_r}{c} - \frac{2l_o}{c} - 2\bar{n}\frac{z}{c}\right)\right|$$

$$\times \cos\left[\frac{4\pi}{\bar{\lambda}_0}(l_r - l_o - \bar{n}z) - \phi_L\left(\frac{2l_r}{c} - \frac{2l_o}{c} - 2\bar{n}\frac{z}{c}\right)\right], \tag{6-20}$$

where $\bar{\lambda}_0 = c/\bar{v}$ is the mean wavelength in free space.

Consider now the result of scanning the reference mirror in the axial direction with speed V_r. For simplicity, let the transverse coordinate of interest be $x = 0$ and consider how image resolution is obtained in the vertical object dimension. Let $2l_r - 2l_o = 2V_r t'$ and consider the response from a single point scatterer at depth z_k. The response is[4]

$$I(0,z_k,t') = I_L\left[\alpha_r^2 + \alpha_o^2(0,z_k) + 2\alpha_r\alpha_o(0,z_k)\left|\tilde{\mu}_L\left(2\frac{V_r}{c}t' - 2\bar{n}\frac{z_k}{c}\right)\right|\right.$$

$$\left.\times \cos\left[\frac{4\pi}{\bar{\lambda}_0}(V_r t' - \bar{n}z_k) - \phi_L\left(2\frac{V_r}{c}t' - 2\bar{n}\frac{z_k}{c}\right)\right]\right]. \tag{6-21}$$

Figure 6.5 shows a plot of a simulated fringe pattern obtained by scanning the reference mirror. Various parameters of this fringe pattern are labeled.

A remaining unanswered question concerns the width of the fringe pattern in the time dimension. The answer to this question (see [70], p. 164) lies in the Fourier transform relationship that exists between the the function $\tilde{\mu}_L(\tau)$ and the power spectral density of the light, $\hat{\mathcal{G}}(v)$, where the circumflex over $\mathcal{G}$ indicates that the function has been normalized to have unit area,

$$\tilde{\mu}_L(\tau) = \int_{-\infty}^{0} \hat{\mathcal{G}}(v)e^{j2\pi v\tau}\,dv, \tag{6-22}$$

where

$$\hat{\mathcal{G}}(v) = \frac{\mathcal{G}(v)}{\int_{-\infty}^{0}\mathcal{G}(v)\,dv}. \tag{6-23}$$

---

4. The reader may wonder how an intensity, which is defined as an infinite time average, can have temporal variations. The answer is that the detector is assumed to average over a time period that is long compared with the fluctuations of the source, but short compared with the intensity variations induced by movement of the reference mirror. To avoid confusion, we have used the symbol t' to represent time changes induced by mirror motion.

6.3 Speckle in Optical Coherence Tomography

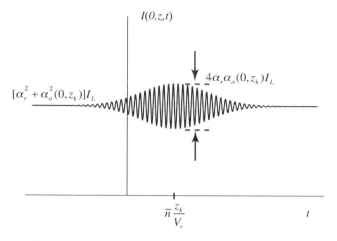

Figure 6.5 Temporal fringe pattern detected as a result of reference mirror motion.

As an example, a Gaussian power spectrum, normalized to have unit area and with full-width-half-maximum bandwidth Δv and center frequency $\bar{v}$ (again taken to be a positive number), has the form

$$\hat{\mathcal{G}}(v) = \frac{2\sqrt{\ln 2}}{\sqrt{\pi}\Delta v} \exp\left[-\left(2\sqrt{\ln 2}\,\frac{v+\bar{v}}{\Delta v}\right)^2\right], \tag{6-24}$$

and the corresponding forms of $|\tilde{\mu}_{\mathbf{L}}(\tau)|$ and $\phi_{\mathbf{L}}(\tau)$ are

$$|\tilde{\mu}_{\mathbf{L}}(\tau)| = \exp\left[-\left(\frac{\pi \Delta v \tau}{2\sqrt{\ln 2}}\right)^2\right] \tag{6-25}$$

$$\phi_{\mathbf{L}}(\tau) = 0. \tag{6-26}$$

The depth resolution can be found by calculating the depth Δz that corresponds to the full width of $|\tilde{\mu}_{\mathbf{L}}(\tau)|$ at half its maximum value, yielding

$$\Delta z = \frac{2c \ln 2}{\pi \bar{n} \Delta v}. \tag{6-27}$$

With the substitution $\Delta v \approx (c/\bar{n})\Delta\lambda/\bar{\lambda}^2$, an equivalent expression is

$$\Delta z \approx \frac{2 \ln 2}{\pi} \frac{\bar{\lambda}^2}{\Delta\lambda}, \tag{6-28}$$

where $\bar{\lambda} = c/(\bar{n}\bar{v})$ and $\Delta\lambda$ is the change in wavelength that corresponds to the change in frequency Δv.

The resolution in the z (or axial) direction is now known. Resolution in the transverse direction x is determined by the usual resolution equations for microscopy. An often-used expression ([19], p. 7) is

$$\Delta x = \frac{4\bar{\lambda}}{\pi}\left(\frac{f}{d}\right), \tag{6-29}$$

where d is the the spot size on the objective lens and f is the focal length.

6.3.3 Speckle and Speckle Suppression in OCT

The response of an OCT system to a point scatterer has been analyzed in the previous section. We now turn to a discussion of how speckle arises in such a system and how it can in principle be at least partially suppressed.

Origin of Speckle in OCT

Consider what happens when N scatterers lie within a single depth resolution cell, again with $x = 0$. The field $\mathbf{E}_o(0;t)$ arriving at the detector from this one resolution cell is

$$\mathbf{E}_o(0;t) = \sum_{k=1}^{N} \alpha_o(0, z_k) \mathbf{L}\left(t - \frac{2l_o}{c} - 2\bar{n}\frac{z_k}{c}\right). \tag{6-30}$$

Neglecting the interference of these contributions with themselves, which generates the I_o term, focus instead on the fringe pattern or reference–object interference term $I_\Delta(t')$, which generates the beat frequency at the detector. The form of the interference term is, from Eq. (6-21),

$$I_\Delta(t') = I_L \alpha_r \sum_{k=1}^{N} \alpha_o(0, z_k) \left|\tilde{\mu}_{\mathbf{L}}\left(2\frac{V_r}{c}t' - 2\bar{n}\frac{z_k}{c}\right)\right|$$
$$\times \cos\left[\frac{4\pi}{\bar{\lambda}_0}(V_r t' - \bar{n}z_k) - \phi_{\mathbf{L}}\left(2\frac{V_r}{c}t' - 2\bar{n}\frac{z_k}{c}\right)\right]. \tag{6-31}$$

To simplify further, suppose that, as in the case of a Gaussian spectrum, the phase $\phi_{\mathbf{L}}$ is zero, yielding

$$I_\Delta(t') = I_L \alpha_r \sum_{k=1}^{N} \alpha_o(0, z_k) \left|\tilde{\mu}_{\mathbf{L}}\left(2\frac{V_r}{c}t' - 2\bar{n}\frac{z_k}{c}\right)\right| \cos\left[\frac{4\pi}{\bar{\lambda}_0}(V_r t' - \bar{n}z_k)\right]. \tag{6-32}$$

It is now clear that the intensity $I_\Delta(t')$ is the result of a random walk. The cosines are entirely analogous to the complex exponentials found in the usual random walk in the complex plane. The amplitudes of the cosines vary with z_k, but most importantly the *phases* of the cosines are random, with values $4\pi\bar{n}z_k/\bar{\lambda}_0$. These randomly phased contributions, as well as other randomly phased contributions from the entire transverse resolution cell, add together to generate speckle.[5]

---

5. The mathematical representation of the field as a superposition of contributions from individual randomly phased scatterers implicitly assumes single scattering. When multiple scattering takes place, it is impossible to associate individual scatterers with unique field contributions. However, the field at a given point can in principle still be expressed as a sum of various path integrals with different delays.

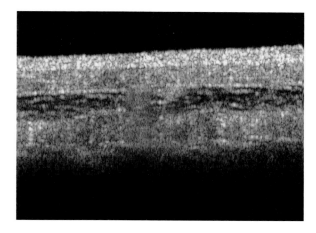

Figure 6.6 Image of hamster skin and underlying tissue obtained by OCT. Image courtesy of Prof. Jennifer Kehlet Barton of the University of Arizona; reprinted from [15] with permission of the Optical Society of America.

The detected photocurrent is typically quadrature detected. The two quadrature components are squared and added together, and the square root of the sum is taken. This yields a signal proportional to the magnitude of the optical field incident from the object. If the depths of the scatterers are random over a distance comparable with a wavelength, the resulting statistics of the measured signal will be very close to Rayleigh distributed when the light is fully polarized. Note only the signal component that is co-polarized with the reference wave will generate an AC current. Methods for detecting both polarization components by using two orthogonally polarized reference signals do exist.

Figure 6.6 shows an image of the skin and underlying tissue of a hamster, obtained by OCT. From the top of the image to the bottom, layers of connective tissue, muscle, fat, and dermis/epidermis can be seen. The small region in the center (interrupting the fat layer) with a higher spatial frequency speckle is the location of a blood vessel with flowing blood. This was obtained with a 1300 nm OCT system with 15 μm axial and lateral resolution. Speckle is clearly seen in the image.

Suppression of Speckle in OCT

We have now seen how speckle arises in OCT. Speckle in this imaging modality is a fluctuation in amplitude, rather than in intensity. Nonetheless, the presence of these fluctuations in the images obtained can hinder the detection of small structures that are near the resolution limit. It is therefore important to consider ways to suppress speckle. The problem is complicated by the fact that OCT, like holography, is fundamentally a coherent imaging technique, so methods for destroying the spatial coherence of the source can not be used. In what follows, we discuss three different methods for suppressing speckle. Others exist, but we limit ourselves to three here. For discussion of a few additional techniques, see [19], Chapter 7.

- **Polarization Diversity.** In many cases the speckle observed in two different polarizations of the signal from the object may be uncorrelated. If the signals from the two polarizations are detected separately and superimposed on a *magnitude* basis, speckle suppression by a factor of at most $1/\sqrt{2}$ can be realized. For some systems it may be possible to perform two scans, each with an orthogonal polarization illuminating the object. If the two returning polarizations are detected separately for each incident polarization, then for some objects, four uncorrelated speckle patterns can be obtained. If these patterns are superimposed on a magnitude basis, speckle suppression by at most $1/2$ can be realized.
- **Frequency or Wavelength Diversity.** We have seen in Chapter 5 that speckle structure is dependent on optical frequency, and under some conditions, the speckle at two different frequencies will be uncorrelated. This fact raises the possibility that two or more scans of the object can be performed, each with a different frequency, and provided those frequencies are separated sufficiently to guarantee uncorrelated speckle patterns, speckle suppression will result when the separate images are added on a magnitude basis. The condition for decorrelation is, roughly speaking, that the frequency change be at least the reciprocal of the dispersion time for a single resolution element on the object. The depth of a single resolution cell is given by Eq. (6-27), and therefore the dispersion time is approximately

$$\delta t = 2\bar{n}\Delta z/c = \frac{4 \ln 2}{\pi \Delta \nu}. \qquad (6\text{-}33)$$

The frequency change that decorrelates the speckle from one resolution cell is therefore[6]

$$\delta \nu = \frac{1}{\delta t} = \frac{\pi \Delta \nu}{4 \ln 2} \approx \Delta \nu. \qquad (6\text{-}34)$$

Thus at first glance it appears relatively easy to obtain uncorrelated speckle patterns, the frequency translation required being essentially the source bandwidth. But recall that the depth resolution depends inversely on $\Delta \nu$, and therefore large source bandwidths are desirable if high depth resolution is to be achieved. As a consequence, while the required frequency shift is $\Delta \nu$, that may in fact be a large frequency shift in absolute terms.
- **Aperture Masks.** In the case of holography, we saw that it is possible to suppress speckle at the price of lower transverse resolution by using time-varying aperture masks. Similar approaches are possible in OCT. Suppose a separate scan is performed with the aperture of the imaging lens divided up into several smaller subregions, only one of which is open on a single scan. On successive scans, different

---

6. Remember that $\Delta \nu$ represents the spectral width of the source, while $\delta \nu$ represents the amount the center frequency of the source has been changed. The particular numerical constant in this equation holds for a Gaussian-shaped spectrum.

subapertures are opened. Since different speckle fields are incident on each subaperture, the image speckle produced by one subaperture will be uncorrelated with the speckle produced by the others. If the images obtained by M different subapertures are combined on a modulus basis, the speckle contrast will be reduced by $1/\sqrt{M}$. Unfortunately, the transverse resolution will also be reduced by $1/\sqrt{M}$. In addition, note that if a fiber-based interferometer is used, each subaperture must be large enough to focus the light into the collecting fiber with little light loss. For another technique that is related to the method described here and can be accomplished with a single scan, see [141].

In closing the discussion of speckle in OCT, we note that we have not covered the subject of digital image processing to suppress speckle, since it is beyond the scope of this book. A reference on this subject in the context of OCT can be found in Ref. [19], Chapter 7.

6.4 Speckle in Optical Projection Displays

Projection displays have found widespread use in large-screen TVs, large computer monitors, advertising, simulators, and other applications. When these displays utilize lasers as the sources of the three primary colors, speckle appears as a disturbing noise that must be dealt with for a satisfactory viewing experience. In addition, as we shall see, speckle can be a problem even when the sources are not lasers. The source of this speckle is the screen onto which the images are projected, which must be optically rough if viewing is to be possible over a wide range of angles. This section is devoted to speckle in projection displays and some of the means for reducing its disturbing effects.

At the start it is important to distinguish between three types of projection displays: (1) full-frame displays that optically project an entire 2D frame at a time; (2) line-scan displays that scan a 1D optical line over a 2D frame; and (3) raster-scanned displays, which scan an optical spot over a 2D frame. Each type of display requires different treatment here, because the opportunities for suppressing speckle differ between display types.

It is also possible to distinguish between front-projection displays, for which the image is viewed in reflection from a screen, and back-projection displays, for which the image is viewed in transmission through a diffusing screen. For our purposes there is little or no difference between these two types. Both will suffer from speckle when the illumination is coherent. Our illustrations will be for front-projection displays.

Projection displays have been used for video display for some years, but only recently has interest been increasing in the use of lasers as the sources in such displays. This increasing interest has been sparked in part by the advent of semiconductor lasers that emit in the red and the blue. A green semiconductor source has not yet been realized, and therefore green light is usually obtained by use of nonlinear optics, starting with an IR source. There are four different advantages that lasers afford, as compared with other light sources: (1) higher contrast, (2) improved color gamut, (3) potentially greater brightness, and (4) in some configurations projecting a very narrow raster-scanned beam, nearly infinite depth of focus. A price paid for the use of lasers is the need to actively suppress speckle, because fully developed speckle, if not suppressed, provides too low a signal-to-noise ratio for viewer satisfaction. Note that semiconductor lasers generally have line widths in the range of 2 to 4 nm, which for most screens is too small a bandwidth for speckle suppression to occur due to wavelength diversity.

While there exist many different standards for display of video and computer information, we focus here on the original NTSC standard for broadcast video. This display has a 4:3 horizontal/vertical format with 525 scan lines per image, of which 480 are used to display the active area of the image, and 45 are used for vertical blanking. For an analog display, the number of pixels per line is approximately 640. A full image is displayed (in interlaced or progressive format) every 1/30th of a second, meaning that, in a line-scanned display, an individual scan line is present for about 70 μsec, and in a raster display, an individual pixel is present for somewhat more than 100 nsec. For comparison purposes, the HDTV 1080i standard, with a 16:9 horizontal/vertical format and a 1/30th second frame time, has 1080 lines and 1920 pixels per line, implying that a scan line lasts approximately 30 μsec and a pixel is displayed about every 15 nsec.

6.4.1 Anatomies of Projection Displays

Full-Frame Displays

Figure 6.7 shows a simplified block diagram of a typical full-frame display. The beams from red, green and blue sources[7] are expanded to a size appropriate for a full-frame, 2D spatial light modulator, and each passes through such a modulator. Each modulator imposes on the transmitted intensity an image appropriate for that particular color and frame. The three color images are then combined and projected onto the viewing screen. The viewer sees the light scattered off of the rough screen.

If the projection display does not use lasers, then a single white light source can be filtered to produce the three colored sources. However, we will be considering primarily laser-based projection displays here.

Different kinds of 2D spatial light modulators have been used, but most common are the Digital Light Projection (DLP) device, and the Liquid Crystal on Silicon (LCOS) device. The DLP device, developed by Texas Instruments, consists of a 2D silicon micro-mirror array with a mirror for each pixel to be projected (see Ref. [35] for an excellent overview of this device). Each mirror tilts under voltage control to allow varying amounts of light to be transmitted through the optical system that fol-

---

7. For a method for producing red, green and blue coherent sources from a single infrared laser, see U.S. Patent #5,740,190.

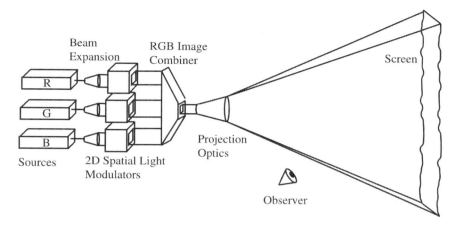

Figure 6.7 Schematic layout of a full-frame projection display. Various optical elements have been simplified or omitted for simplicity.

lows. CMOS drivers can be integrated on the same substrate as the mirrors. The mirrors can be driven either in a digital (on/off) fashion or in an analog fashion, but most common is the digital mode of operation, in which the brightness of the picture in a given color is controlled by the amount of time a pixel is in the "on" state during a single frame time. Various digital time codes are chosen for different intensity levels.

We will assume that three separate DLP chips are used, one for each color. However, it is possible to use a single chip and a rotating color filter wheel to sequentially display three colors per frame time, but such an approach makes most sense when a single white-light source is used. In the case of three laser sources, a similar approach could be taken by use of an electro-optic or micro-electro-mechanical (MEMs) switch, which changes the color incident on the DLP chip every 1/3rd of a frame time. Since the DLP device is capable of switching frames at 10 kHz rates, this approach is possible.

A second technology that allows display of a frame at a time is the Liquid Crystal on Silicon (LCOS) spatial light modulator (see Ref. [112] for an overview). For this approach, a silicon chip containing an array of independently controlled liquid crystal pixels is made, again with drivers on the chip. Due to a metallic layer under the liquid crystal cells, the chip operates in reflection rather than in transmission. An LCOS device is an analog device, in which each liquid crystal pixel controls the intensity transmitted by means of a voltage controlled polarization rotation induced by the liquid crystal, followed by a polarizer. A frame is loaded into the chip, and appropriate voltages are applied to the liquid crystal pixels to rotate the polarization by the right amount to make the polarizer that follows transmit an intensity that is appropriate for each pixel. Three modulators can be used, one for each color. While LCOS devices can be wasteful of optical power if an ordinary non-laser source is used, due to the fact that the half of the optical power in the wrong polarization component must be thrown away before the light reaches the device, nonetheless for laser sources, which often are well polarized at the start, this is not as much of a problem. While we will assume that the system uses three LCOS chips, it is possible to use a single chip and a sequential display of the three colors, as with the DLP.

In a system with three LCOS devices, a frame is displayed for 1/30 sec, and any mechanical action to reduce speckle must take place in this time. In a system with three DLP chips, the situation is more complex due to the use of binary pulse-width modulation to represent intensity. The image to be projected is displayed by sequentially projecting bit planes, with the duration of the kth bit plane proportional to 2^k. The available time is minimum for the minimum intensity that is greater than 0, while the maximum time is available for the highest intensity, for which the pixel is in the "on" state for the full frame time. If the display has 10 bits of intensity resolution, as would be required to achieve a 1000:1 contrast ratio, one would ideally like the speckle intensity standard deviation to be smaller than the intensity associated with the least significant bit. However, in practice, speckle is most objectionable in the brighter areas of the image and reducing speckle for those areas may suffice. As an approximation, we assume that the minimum time available for speckle manipulation during a frame is about 1/8th of a frame time, or about 4 msec.

Line-Scanned Displays

Figure 6.8 shows a simplified block diagram of a line-scan projection system. Again red, green and blue lasers are utilized. The beams from these lasers are shaped into lines, and each passes to a separate 1D spatial light modulator, which imposes the intensity information appropriate for a particular line in that color. The three color

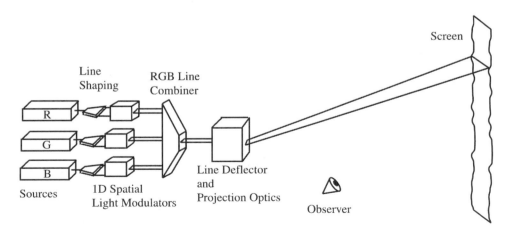

Figure 6.8 Schematic layout of a line-scan projection system. Various optical elements have been simplified or omitted for clarity.

lines are merged and a line deflector plus projection optics project the color-merged line onto the viewing screen.

The most common 1D spatial light modulator used in projection displays is the so-called Grating Light Valve (GLV) developed by Silicon Light Machines (see Ref. [165]), the rights to which in projection TV applications are owned by Sony Corporation. The GLV is also a silicon-based micro-mechanical mirror device, in which the small electrostatically deflectable ribbons are interlaced with static ribbons. Initially all the ribbons are in a "raised" position. Under this condition, there is no grating encountered by the light, and a mirror reflection from the device takes place. When the deflectable ribbons are lowered, a grating pattern exists, and a portion of the light is sent into first diffraction orders. The optical system that follows blocks the undiffracted light and passes the diffracted light, with the amount of light in the diffracted orders being a function of the voltage applied locally to the deflectable ribbons. In this way, the intensity of the transmitted light in any one color is modulated in an analog fashion across an entire line at a time. The scanning of the line is slow enough that a scanning mirror can be used for this purpose. The ribbons can be deflected in times as short as about 1 μsec.

Raster-Scanned Displays

A common type of projection display is the raster-scanned display. The elements of such a display are illustrated in Fig. 6.9. Red, green and blue sources are shown on the left, followed by a modulator for each color. The three color beams are then combined and sent to a beam deflector, which provides a slow scan vertically (over the various lines in the display) and a fast scan horizontally (over the pixels in each line). Usually the deflector consists of two different deflectors, one for fast scan over

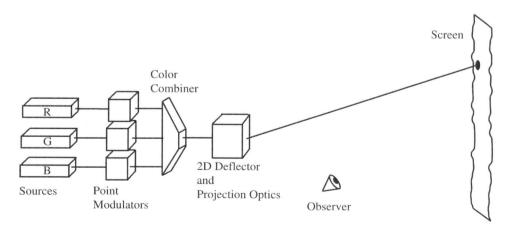

Figure 6.9 Schematic layout of an optical raster-scanned projection display. Various optical imaging elements have been omitted for simplicity.

pixels and one for slower scan over lines. Different technologies can be used for the fast scan and the slow scan. Projection optics are not shown in this schematic, but may be required. The slow scan over the lines can again be achieved with a rotating faceted mirror. The fast scan along a line requires a different technology, for example acousto-optic, electro-optic, or a very fast micro-mechanical mirror.

6.4.2 Speckle Suppression in Projection Displays

In considering the possible approaches to speckle suppression in projection displays, it is helpful to make a list of all the approaches that come to mind, and then to assess the usefulness of each approach for each of the three display types described above. The list of methods to be considered is as follows:

1. Introduce polarization diversity;
2. Introduce a moving screen;
3. Introduce a specially designed screen that minimizes the generation of speckle;
4. For each color, broaden the spectrum of the sources or use multiple lasers at slightly different frequencies, thereby achieving wavelength diversity in the illumination;
5. For each color, use multiple independent lasers separated spatially, thereby achieving angle diversity in the illumination;
6. Overdesign the projection optics as compared with the resolution of the eye (this approach requires some explanation, to follow);
7. Image a changing diffuser with random phase cells onto the screen; and
8. Image a changing diffuser with deterministic orthogonal phase codes onto the screen.

Each of these approaches and its limitations will be discussed in what follows.

6.4.3 Polarization Diversity

Polarization diversity can gain at most a small factor in speckle contrast reduction, so it does not provide much relief from speckle. Nonetheless, every factor that contributes to reducing speckle is welcome, and indeed many approaches can and should be used together.

If a polarized source illuminates the reflecting screen, and the screen has the property that it generates two equal-strength, independent speckle patterns in each of two orthogonal components of polarization, then the speckle contrast will be reduced by a factor $1/\sqrt{2}$. In some cases, a reduction of contrast by a factor of $1/2$ rather than $1/\sqrt{2}$ can be obtained using polarization. This case requires that the screen depolarize the light into two independent, equal-strength speckle patterns, but further that when the polarization of the light illuminating the screen changes between two orthogonal states, the resulting speckle patterns in both reflected orthogonal polarizations change as well. If all four of these reflected speckle patterns are independent, then an illuminating source that rotates the incident polarization rapidly (faster than the response time of the eye) will produce an observed intensity that is the average of four independent speckle patterns, resulting in a reduction of the speckle contrast by a total factor of 1/2. This effect will not be observed if the two incident polarization states are present simultaneously and are mutually coherent, because then similarly polarized reflected speckle patterns will add on an amplitude basis rather than an intensity basis, and contrast will be reduced only by a factor $1\sqrt{2}$. A reduction by 1/2 can also be accomplished if, rather than illuminating the screen with two orthogonal polarizations *sequentially*, the two polarizations arise *simultaneously* from independent laser sources, or two light beams from the same laser that have been made uncorrelated by delaying one beyond its coherence length.

Unfortunately, many of the projection systems described above depend on the use of polarization for other functions, such a modulation or beam combination. In such cases, polarization diversity can not be used as a speckle reduction method. In any case, there is more interest in methods that have the potential to reduce speckle by larger factors than offered by polarization diversity.

6.4.4 A Moving Screen

A simple method for reducing speckle in this application is to rapidly move the screen back and forth by a large enough amount to result in significant speckle suppression. Movement of the screen results in similar movement of the speckle and a change of the speckle pattern intercepted by the eye. The net effect is what we have called "boiling" of the speckle sensed by the observer, and time integration of the boiling speckle results in a reduction of contrast.

How far must the screen move in order to decorrelate the speckle? By analogy with the discussions of Sect. 5.2, the screen must move by an entire resolution cell of the eye on the screen in order to completely decorrelate the speckle. The size of a resolution element of the eye on the screen depends on both the state of correction of the eye (we will assume the observer is well corrected), and the darkness of the environment, which determines the pupil size. Taking a typical pupil size of 3.0 mm, the angular width of the line-spread function of the eye, measured between the points where that function has fallen to 1/10th of its maximum value, is about 4 arc mins (see [168], p. 30).[8]

For near-normal viewing, the diameter d of a eye resolution cell on the screen can be expressed as

$$d = 2z_e \tan(a/2), \quad (6\text{-}35)$$

---

8. Note that larger pupil size generally results in poorer resolution, due to increased aberrations of the eye.

where z_e is the distance from the observer to the screen, and a is the angular resolution of the eye. At 3 m distance from the screen, this corresponds to a resolution cell width of 3.5 mm, while at 10 m from the screen (a larger scale display) the corresponding width is 1.2 cm. For the smaller display, the screen must be moved by some multiple of 3.5 mm. If the goal is to reduce the speckle contrast by a factor of 10, and linear motion is used, then the screen must be moved by 100 resolution cells of the eye, or by the uncomfortably large distance of 35 cm, while the larger display requires more than 3 times that movement. If instead the screen is moved in a spiral motion, with gradually increasing radius, an equivalent number of resolution cells can be covered in a circle of about 3.5 cm diameter. To further make the task difficult, motion by the required amount must be accomplished in one frame time.

An approach with more promise, but one that is still difficult, is to rotate the screen slightly about a central vertical or horizontal axis. Rotation of the screen by angle φ results in a rotation of the speckle pattern by angle 2φ. To decorrelate the speckle seen by the eye, the rotation must be sufficiently large to introduce a phase change of 2π across the resolution cell of the eye on the screen, from the resolution-cell edge farthest from the center of rotation to the resolution-cell edge closest to the center of rotation. To suppress the speckle contrast by a factor $1/\sqrt{M}$, the required rotation angle of the screen is

$$\varphi = \frac{M\lambda}{2d} = \frac{M\lambda}{2z_e a}, \qquad (6\text{-}36)$$

where small angle approximations have been used, again d is the diameter of the eye resolution cell on the screen, a is the angular resolution of the eye and z_e is the distance from the eye to the particular resolution cell on the screen being considered. To suppress speckle by a factor of 10, with a screen at 3 m distance from the observer and a wavelength of 0.5 μm, the required angle of rotation of the screen is about 0.4 degree.

For raster-scanned and line-scanned displays, the short time that the scanning spot or line dwells on a single eye resolution cell on the screen makes it impossible to rotate the screen fast enough to achieve any suppression of the speckle during the dwell time. However, both the spot and line scan systems revisit the same eye resolution element 30 times per second, while the full-frame system spends that same time displaying a single frame. Thus, rotation of the screen such that several independent speckle patterns appear within 1/30 second is the requirement for some speckle reduction in all 3 systems. In the example above, this requires a screen rotation rate of about 12 degrees/sec. Note that, since a given eye resolution element is visited about 30 times a second in the scanning systems, the maximum possible suppression of speckle contrast will be $\approx \sqrt{1/30}$. Greater amounts of suppression are possible in principle with the full-frame system.

We conclude that, indeed, speckle can be reduced by moving the screen onto which the image is projected, provided the rotation rate of the screen can be made large enough. The ability to rotate the screen by a small angle rapidly depends, of course, on the size and mass of the screen.

6.4.5 Wavelength Diversity

Wavelength diversity can occur or can be introduced in a projection display through two different means: (1) the three laser sources may have line widths that are wide enough to impact the contrast of the speckle, or (2) the line widths of the sources may be intentionally spread to achieve the same result.

Unfortunately, even for near-normal illumination and near-normal viewing, screens that function mainly by surface scattering do not have a high degree of frequency sensitivity, and therefore rather large line widths are required to suppress speckle by any significant amount. Using Eq. (5-111) we can show that even for a screen with a 250 μm surface-height standard deviation and a center wavelength of 0.5 μm, about a 2% bandwidth is needed to reduce the speckle contrast to 1/10 (10 nm bandwidth at 500 nm) and a reduction of the speckle contrast to 1/100 is not achievable at all, since a fractional bandwidth greater than 1 would be required.

When volume scattering, rather than surface scattering, is the dominant scattering mechanism, much greater differential path delays are present and the frequency sensitivity of the speckle can be significantly enhanced. However, this improvement must not be accompanied by a transverse spreading of the beam to the extent that a projected spot on the screen becomes larger than the resolution element of the eye, thereby blurring the projected image. As mentioned in Chapter 5, a 5 GHz frequency decorrelation interval has been demonstrated experimentally in transmission through a 12 mm thick diffuser [172]. However, the results of transmission through such a thick diffuser are probably not directly applicable to the projection display, due to the lateral spreading that accompanies strong multiple scattering, although there is no doubt that frequency sensitivity could be enhanced by exploiting milder versions of this phenomenon.

6.4.6 Angle Diversity

Suppose that the illumination source for one of the three color components consists of an array of semiconductor lasers, each oscillating independently. Because the elements of the array are spatially separate, with a suitably designed system, they can illuminate the screen from slightly different angles, but with identical modulations. To what extent can such an arrangement suppress the speckle observed in that particular color component? In particular, under what conditions will the use of an array of M different sources result in a speckle contrast of $1/\sqrt{M}$?

Consider the rather generic full-frame projection arrangement shown in Fig. 6.10, with several angularly separated sources, each of wavelength λ. The transparency labeled "object" is projected onto the observing screen, magnified by a factor m. Magnification of the image by a factor m is accompanied by a demagnification of the angles in the image space by a factor of m.

The discussions associated with Fig. 5.20 contain the answer to our question. Each source in the array generates a separate speckle intensity pattern, and because the sources are mutually incoherent, those intensity patterns add. For near normal

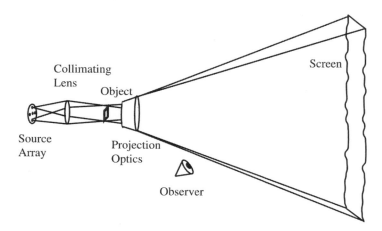

Figure 6.10 Generic full-frame projection arrangement with angularly separated sources.

observation, the dominant effect of the different angles of illumination is an angular shift of the speckle patterns, with little internal change of structure. In accord with the discussion of Fig. 5.20, the superimposed speckle patterns will be uncorrelated when the change of illumination angle exceeds the angular subtense of the eye pupil. This is equivalent to requiring that the speckle pattern incident on the pupil of the eye be spatially shifted by an amount equal to the pupil diameter. Remembering that the angles incident on the screen are smaller than the angles incident on the object by the magnification factor m, the angular spacing of the sources must be a factor of m larger than the angular subtense of the eye.

The discussion above is appropriate for a full-frame display. For a line-scan display, the speckle incident on the pupil of the eye is much wider in the direction orthogonal to the scan line than it is in the direction parallel to the scan line. For such a display, the sources should be separated in a direction parallel to the scan line if minimum separation is desired. For a raster-scanned display, equal separations in both directions can be used.

We conclude that angle diversity can in principle reduce speckle contrast for all three types of display. The chief practical challenge is associated with the angular separations required of the sources, which must be m times the angular diameter of the eye pupil of the closest observer. If the projector is, for example, projecting an image of a 10 mm chip onto a 3 m screen, the magnification m of the projector is 300. Then the angular separation of the sources must be 300 times the angular subtense of the pupil, which we take to be 3 mm over a viewing distance of perhaps 6 m, or 5×10^{-4} radians. The angular separations of the sources must therefore be about 0.15 radians or 8.6 degrees. If the focal length of the collimating lens is 1 cm, then the sources must be separated by 1.5 mm. This is simply an example of the considerations that must be gone through by the system designer.

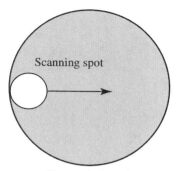

Figure 6.11 Small projected spot scanning across a larger eye resolution element on the screen.

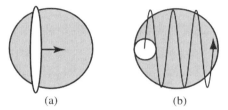

Figure 6.12 Two variants on the spot-scanning format.

6.4.7 Overdesign of the Projection Optics

Speckle suppression can be achieved in line-scan and raster-scan displays by overdesign of the projector optics. Such an approach does not yield speckle suppression for a full-frame display. By overdesign of the projector optics, we mean projecting a line width or spot size that is smaller than the resolution of the eye.

The basic idea is illustrated in Fig. 6.11. A small scanning spot moves through the resolution element of the eye as it is scanned across the screen. Each distinct nonoverlapping position of the spot within the eye resolution element will generate a new and independent speckle pattern, and since the spot crosses the eye resolution element in only a tiny fraction of a frame time, the eye integrates these different speckle patterns, thereby reducing the perceived contrast of the speckle. If the diameter of the eye resolution element on the screen is d and the diameter of the projector spot is s, then the perceived speckle contrast will be

$$C = \sqrt{s/d}. \tag{6-37}$$

Two variants of this approach are illustrated in Fig. 6.12. In part (a) of the figure, the spot is elongated in the vertical direction, with the speckle suppression being again given by Eq. (6-37). In part (b) of the figure, a more complex scanning path is used that sends the projector spot over most of the available area under the eye resolution element. Since d^2/s^2 independent spots are covered in this scan, the speckle reduction will be improved to

$$C = s/d. \tag{6-38}$$

See also Ref. [169], in which a diffractive optical element is used to fill the resolution spot of the eye with a variety of circles, all of which scan across the screen.

For line-scan systems a similar approach can be taken. In this case a narrow illuminated line scans across the wider resolution cell of the eye on the screen. The amount of speckle contrast reduction possible is given again by Eq. (6-37), where s is now the narrow width of the scanning line.

There is, of course, extra cost associated with overdesigned projection optics. However, this approach does provide a means for achieving significant contrast reduction when such reduction is at a premium.

6.4.8 Changing Diffuser Projected onto the Screen

Attention is now turned to the use of a changing diffuser for speckle suppression. Consider the geometry shown in Fig. 6.13. In the case shown, an object, which can be taken to be a full frame generated by a spatial light modulator, is imaged onto a changing diffuser. The cases of line projection and raster-scan projection will be considered briefly later. The diffuser is assumed to be thin in the sense that it does not distort the image, but does add a random phase to each pixel. Such diffusers can be made by etching glass according to a lithographically defined pattern. They can be

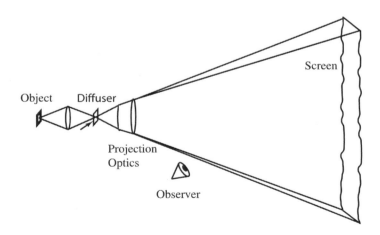

Figure 6.13 Simplified geometry for imaging a frame onto a moving diffuser and projecting the resulting image.

moved back and forth by a linear transducer, or rotated. For simplicity, we assume that the image frame to be projected is entirely white, with no gray-scale information, in order to concentrate on the speckle rather than the structure of the image to be projected. In addition, we assume that all fields are linearly polarized, although polarization effects can be taken into account in accord with the earlier subsection on this topic.

As the diffuser changes, for example by motion, the random phase of each diffuser pixel changes. The transmitted image is then projected onto the rough screen and the light reflected from that screen is intercepted by the eye of the observer. Two different conditions are considered: (1) the diffuser has been designed so that the light it transmits uniformly fills the projection lens but does not overfill it; and (2) the diffuser overfills the projection lens, with some light being lost as a result. If the diffuser does not overfill the projection lens, then that lens simply images the random phase cells of the diffuser onto the screen without introducing intensity fluctuations in each such cell. If the diffuser overfills the projection lens, then each projector resolution cell undergoes a random walk of amplitudes scattered from the diffuser, resulting in an independent Rayleigh-distributed amplitude and uniformly distributed phase for each such projection-lens resolution cell incident on the screen.

Figure 6.14 forms a geometrical basis for our approximate analysis. For effectiveness of diffuser motion in suppressing speckle, the projector lens should have higher resolution on the screen than the eye. The large circle of diameter d is a resolution element of the eye on the screen, corresponding to the area contributing light to one point on the retina. The smaller circles, of diameter s, are the resolution elements of the projector lens on the screen. There are

$$K = \left(\frac{d}{s}\right)^2 \tag{6-39}$$

resolution elements of the projector lens within one eye resolution element.

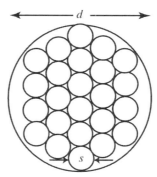

Figure 6.14 Regions on the screen. The large circle of diameter d represents one resolution element of the eye, while the smaller circles of diameter s represent resolution elements of the projector lens. Partial small circles are omitted.

Each resolution element of the projector lens is assumed to have incident either a random phase when the diffuser just fills the projection lens, or a random walk in amplitude when the diffuser over-fills the projector lens. The screen is, of course, optically rough, so each resolution element of the projector lens on the screen will generate an independent speckle field at the point on the retina corresponding to this particular eye resolution element. As the diffuser moves, the K different speckle field contributions change their relative phases (projector lens just filled) or both their amplitude and phase (projector lens over-filled).

We assume that when the diffuser changes, it yields a diffuser phase pattern that is statistically independent of the previous diffuser phase pattern. If the diffuser is in motion, then it must move the linear distance subtended by an eye resolution element to yield statistical independence. The total integrated intensity incident on the retina at point (x, y) will be

$$I(x, y) = \sum_{m=1}^{M} I_m(x, y) \qquad (6\text{-}40)$$

where each $I_m(x, y)$ is generated by an independent diffuser phase pattern. If the measurement time is T and the time it takes the diffuser to assume an independent realization of phases is τ, then $M = T/\tau$.

To find the contrast of the speckle that results, it is necessary to perform two different averages. One is an average over the ensemble of diffuser phase patterns, for which we assume that each projected cell of the diffuser has either a random phase with uniform statistics on $(-\pi, \pi)$ (projector lens just filled) or a speckle field with Rayleigh amplitude and uniform phase statistics (projector lens over-filled). The second is an average over an ensemble of rough screens, for which we assume that, in the absence of a diffuser, each projector-lens resolution cell yields at the retina the usual random walk in the complex plane that is associated with fully developed speckle. Since the screen is not moving, the speckle amplitudes associated with the screen do not change in the absence of a diffuser.

The theoretical derivations of expressions for contrast are complex and are deferred to Appendix E.

Diffuser Does Not Overfill the Projection Lens

In this case, as shown in Appendix E, the speckle contrast is found to be

$$C = \sqrt{\frac{M + K - 1}{MK}}, \qquad (6\text{-}41)$$

where again, M is the number of independent diffuser realizations averaged, and K is the number of projector-lens resolution elements lying under a single eye resolution element on the screen. Note that M and K can never be smaller than unity.

Several special cases should be explored to see whether the result makes physical sense. First consider the case $K = 1$, which was mentioned above. In this case, there is only one projector resolution cell in an eye resolution cell. The contrast becomes

$$C = \sqrt{\frac{M}{M}} = 1. \tag{6-42}$$

Thus, in this case no suppression of speckle is obtained by a changing diffuser, which agrees with physical intuition, since changing the phase of the one projector resolution element that contributes to the given point on the retina will not change the intensity statistics.

Next, consider what happens when K becomes very large and M is fixed. The contrast becomes

$$C \approx \sqrt{\frac{K}{MK}} = \sqrt{\frac{1}{M}}. \tag{6-43}$$

Thus for large K, the contrast drops as one would expect for integration over M different time-correlation cells. Again, this makes good physical sense.

Finally, consider the case of finite K and $M \to \infty$, that is, a very long time average with a finite K. The result becomes

$$C \approx \sqrt{\frac{M}{MK}} = \sqrt{\frac{1}{K}}. \tag{6-44}$$

This result may be a bit unexpected. It says that, for a finite number of resolution elements of the projector lens lying within one eye resolution element, the contrast never goes to zero with increased time averaging. Instead it approaches an asymptote $\sqrt{1/K}$ as integration time is increased. This effect has been observed experimentally,[9] and it has important implications for speckle observed in projection systems that use non-laser sources, to be discussed shortly.

Figure 6.15(a) illustrates contours of constant speckle contrast from Eq. (6-41). Figure 6.15(b) shows plots of contrast vs. M for various values of K. The asymptotic behavior of C when M grows large is evident.

Diffuser Overfills the Projection Lens

In this case, as shown in Appendix E, the contrast of the speckle is found to be

$$C = \sqrt{\frac{M + K + 1}{KM}}. \tag{6-45}$$

Note that this is the same speckle contrast found in Eq. (3-108) for speckled speckle, which is indeed the same case encountered here.

Again in the limit of very large K the contrast as a function of M behaves as

$$C \approx \sqrt{\frac{1}{M}}. \tag{6-46}$$

---

9. Private communication, J.I. Trisnadi.

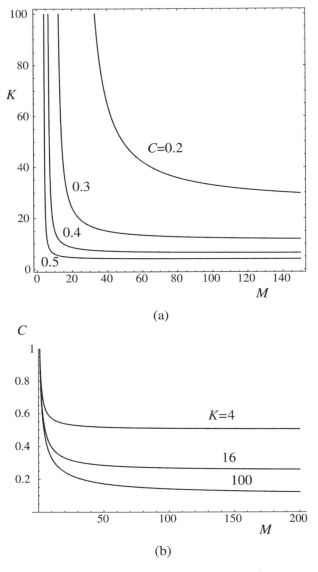

Figure 6.15 (a) Contours of constant speckle contrast, from Eq. (6-41); (b) contrast vs. M for various values of K, showing the asymptotic behavior of Eq. (6-44).

In the limit of large M, for fixed and finite K, the contrast approaches an asymptote,

$$C \approx \sqrt{\frac{1}{K}}, \qquad (6\text{-}47)$$

as it did before. The primary difference of the overfilled case is what happens when the projector resolution-cell size is the same as the eye resolution-cell size. In this case,

$$C = \sqrt{\frac{M+2}{M}} = \sqrt{1 + \frac{2}{M}}. \tag{6-48}$$

For $M = 1$ (only one realization of the diffuser), the speckle contrast is $\sqrt{3}$, which is larger than unity! This is a consequence of the doubly stochastic nature of the problem, with fluctuating illumination intensity incident on a screen that itself induces fluctuations. As M grows, the contrast asymptotically approaches unity, the same value encountered in the previous section.

Plots of contrast contours vs. K and M, and of contrast vs. M for fixed values of K, do not differ appreciably from those shown in Fig. 6.15, and will not be repeated.

Approximate Expressions for the Parameter K

It is possible to find approximate expressions for the parameter K under the condition that there are many projector resolution-cells lying under one point-spread function of the eye on the screen. Using results obtained in Chapter 4, in particular Eq. (4-132) with notation changed to fit the current problem, the parameter K is given exactly by

$$K = \left[\frac{1}{\mathcal{A}_e^2} \iint\limits_{-\infty}^{\infty} \mathcal{P}_e(\Delta x, \Delta y) |\boldsymbol{\mu}_p(\Delta x, \Delta y)|^2 \, d\Delta x \, d\Delta y \right]^{-1}, \tag{6-49}$$

where $\boldsymbol{\mu}_p$ is the normalized correlation function of the light amplitude arriving at the screen from the projector lens, $\mathcal{P}_e$ is the autocorrelation function of the intensity point-spread function of the eye at the screen (we use the symbol p_e to represent that point-spread function), and $\mathcal{A}_e$ is the area under the point-spread function p_e. While it has been difficult to obtain an exact expression for K from this equation, due to the particular form of the point-spread function of the eye (to be discussed), a result valid when the diameter of the projector-lens point-spread function is much narrower than the diameter of the point-spread function of the eye can be found. For $\boldsymbol{\mu}_p$ very narrow, $\mathcal{P}_e(\Delta x, \Delta y)$ can be replaced by $\mathcal{P}_e(0,0)$, yielding

$$K \approx \left[\frac{\mathcal{P}_e(0,0)}{\mathcal{A}_e^2} \iint\limits_{-\infty}^{\infty} |\boldsymbol{\mu}_p(\Delta x, \Delta y)|^2 \, d\Delta x \, d\Delta y \right]^{-1}. \tag{6-50}$$

The autocorrelation function $\mathcal{P}_e(\Delta x, \Delta y)$ evaluated at the origin yields

$$\mathcal{P}_e(0,0) = \iint\limits_{-\infty}^{\infty} p_e^2(x, y) \, dx \, dy. \tag{6-51}$$

Note also that

$$\mathcal{A}_e = \iint_{-\infty}^{\infty} p_e(x, y)\, dx\, dy. \tag{6-52}$$

and therefore

$$K \approx \left[\frac{\iint_{-\infty}^{\infty} p_e^2(x, y)\, dx\, dy}{\left(\iint_{-\infty}^{\infty} p_e(x, y)\, dx\, dy\right)^2} \left(\iint_{-\infty}^{\infty} |\boldsymbol{\mu}_p(\Delta x, \Delta y)|^2\, d\Delta x\, d\Delta y \right) \right]^{-1} \tag{6-53}$$

Finding the intensity point-spread function $p_e(x, y)$ of the eye requires some work. A starting point is a model for the line-spread function of an eye with a 3 mm pupil, fit to experimental data by Westheimer [174][10]. He models the line-spread function $L_e(\alpha)$ as a symmetric function given by

$$l_e(\alpha) = 0.47 e^{-3.3\alpha^2} + 0.53 e^{-0.93|\alpha|}, \tag{6-54}$$

where α is the visual angle in arc minutes. Note that the eye is not diffraction limited. Finding the point-spread function of a circularly symmetric system from the line-spread function is a classic problem in optics; a solution proceeds as follows. First find the Fourier transform $L_e(\nu)$ of the line-spread function, which in this case is

$$L_e(\nu) = 0.46 \exp(-3\nu^2) + \frac{1}{0.88 + 40\nu^2}. \tag{6-55}$$

Since the line-spread function is a projection through the point-spread function, the projection-slice theorem of Fourier analysis implies that the radial profile of the 2D Fourier transform of p_e is

$$P_e(\rho) = L_e(\rho) = 0.46 \exp(-3\rho^2) + \frac{1}{0.88 + 40\rho^2}, \tag{6-56}$$

where in this case $\rho = \sqrt{v_X^2 + v_Y^2}$ and has the dimensions cycles per minute of arc. A 2D inverse Fourier transform of this function yields the point-spread function of the eye, which we find to be

$$\tilde{p}_e(s) = 0.48 \exp[-3.3 s^2] + 0.16 K_0(0.93 s), \tag{6-57}$$

where $K_0(\cdot)$ is a modified Bessel function of the second kind, order 0 and the quantity s is angular radius in minutes of arc.

Picking a common space for the integrations, we choose to express the point-spread function in screen coordinates, which, taking into account the conversion from arc minutes into radians, requires the substitution $s \to 3438 r/z_e$, yielding

---

10. This is not the only model of the eye response that appears in the literature. For an alternative model, see, for example, [175].

$$p_e(r) = 0.48 \exp(-3.9 \times 10^7 r^2/z_e^2) + 0.16 K_0(32 \times 10^2 r/z_e). \quad (6\text{-}58)$$

The square of the volume under $p_e(r)$ is equal to $1.88 \times 10^{-14} z_e^4$. The volume under $p_e^2(r)$ is equal to $3.18 \times 10^{-8} z_e^2$.

To complete the calculation, we must find the volume under $|\boldsymbol{\mu}_p(\Delta x, \Delta y)|^2$. For the case of a projection lens that is overfilled by the diffused light, this quantity can be calculated using Eq. (5-91) under the assumption that the exit pupil of the projector lens is circular with diameter D and it is located distance z_p from the screen, with the result

$$|\boldsymbol{\mu}_p(r)|^2 = \left| 2 \frac{J_1\left(\frac{\pi D}{\lambda z_p} r\right)}{\frac{\pi D}{\lambda z_p} r} \right|^2, \quad (6\text{-}59)$$

where $r = \sqrt{\Delta x^2 + \Delta y^2}$. The volume under this function is given by $1.27 \lambda^2 z_p^2 / D^2$. Finally, we find that

$$K \approx 4.7 \times 10^{-7} \frac{D^2 z_e^2}{\lambda^2 z_p^2}. \quad (6\text{-}60)$$

Noting that $\tilde{D} = (z_e/z_p)D$ is the diameter of the projection optics when projected down a cone of rays converging towards the screen, but only to the z-plane in which the eye resides, we can equivalently present the result as

$$K \approx 4.7 \times 10^{-7} \frac{\tilde{D}^2}{\lambda^2}. \quad (6\text{-}61)$$

This result is valid only for $K \gg 1$; as K shrinks, it must asymptotically approach unity.[11]

For the case of a random phase diffuser that just fills or underfills the projection optics, we assume that the random phase cells of the diffuser are projected with magnification onto the screen (i.e. we neglect diffraction effects by the finite aperture of the projection lens). Since the eye looks at the screen with a random location with respect to the grid of dividing lines between cells, it is not difficult to show that

$$|\boldsymbol{\mu}_p(\Delta x, \Delta y)|^2 = \left(1 - \frac{|\Delta x|}{mb}\right)^2 \left(1 - \frac{|\Delta y|}{mb}\right)^2, \quad (6\text{-}62)$$

where the diffuser phase cells have been assumed to be square with size b on each edge, and the magnification of the projection system has been represented by m. Using Eqs. (6-53) and (6-58), the volume under this function is $0.44 m^2 b^2$. This result then yields an expression for K valid in the just-filled or underfilled cases,

---

11. If we replace the eye by an electronic detector and a diffraction-limited imaging lens, the corresponding result becomes $K \approx \Omega_p / \Omega_d$, where Ω_p is the solid angle subtended by the projector lens aperture and Ω_d is the solid angle subtended by the detector lens aperture, with both solid angles measured from the screen. See also Eq. (5-122).

$$K \approx 1.3 \times 10^{-6} \frac{z_e^2}{b^2 m^2}. \qquad (6\text{-}63)$$

Again, this result is valid only when there are many projected phase cells within the impulse response of the eye on the screen.

Changing Diffuser with Deterministic Orthogonal Codes

In this subsection, we derive the basic result obtained by Trisnadi (Ref. [166]) in designing phase masks with certain orthogonality properties that help reduce speckle. We begin with Eq. (E-3) for the total intensity incident on one point on the retina, which we repeat below:

$$I = \sum_{m=1}^{M} \sum_{k=1}^{K} \sum_{l=1}^{K} \mathbf{A}_k \mathbf{A}_l^* \mathbf{B}_k^{(m)} \mathbf{B}_l^{(m)*}, \qquad (6\text{-}64)$$

where $\mathbf{B}_k^{(m)}$ represents the field projected onto the screen by one projector-lens resolution element during the mth diffuser realization, and $\mathbf{A}_k$ is the random speckle field that would be projected onto the retina by that one projector lens resolution element on the screen if the field $\mathbf{B}_k^{(m)}$ were unity. Unlike the previous cases of random diffusers considered, in this case the diffuser contributions are entirely deterministic. If the summation over m is performed first, then we have

$$I = \sum_{k=1}^{K} \sum_{l=1}^{K} \mathbf{A}_k \mathbf{A}_l^* \sum_{m=1}^{M} \mathbf{B}_k^{(m)} \mathbf{B}_l^{(m)*}. \qquad (6\text{-}65)$$

Now suppose that we are able to find illumination conditions and a set of M diffuser structures such that

$$\sum_{m=1}^{M} \mathbf{B}_k^{(m)} \mathbf{B}_l^{(m)*} = \beta \delta_{kl}, \qquad (6\text{-}66)$$

where β is a real and positive constant and δ_{kl} is a Kronecker delta,

$$\delta_{kl} = \begin{cases} 1 & k = l \\ 0 & k \neq l. \end{cases} \qquad (6\text{-}67)$$

Then we obtain a total intensity given by

$$I = \beta \sum_{k=1}^{K} |\mathbf{A_k}|^2. \qquad (6\text{-}68)$$

That is, the various projector resolution elements on the screen and within one eye resolution element will add on an *intensity* basis rather than an amplitude basis, and a contrast of $C = \sqrt{1/K}$ will result. Note that M is not an independent variable in this case, but rather must be chosen to make the orthogonality condition true. The minimum value of M is K, for to make the patterns add on an intensity basis requires summation of at least K patterns. Note that $C = \sqrt{1/K}$ can be achieved in K steps.

Trisnadi demonstrated this kind of speckle reduction by using phase masks etched in glass and a non-overfilled projector, for which $\mathbf{B}_k^{(m)} = \exp(j0) = +1$ or $\mathbf{B}_k^{(m)} = \exp(j\pi) = -1$. He found that masks related to a Hadamard matrix had the required properties, and successfully demonstrated speckle reduction as predicted. For details see Ref. [166]. Note that, whether the diffusers are random or deterministically orthogonal, it is impossible to reduce speckle using changing diffusers below the value $C = \sqrt{1/K}$. However, with properly chosen orthogonal diffusers, it is possible to reach this limit in a finite number of diffuser changes, whereas with random diffusers this limit is approached only asymptotically as $M \to \infty$.

Line-Scan and Raster-Scan Displays

The previous discussions have assumed full-frame display of images. To what extent are the changing-diffuser techniques applicable to line-scan and raster-scan displays?

In principle, changing random diffusers can be applied to all three display types. While the dwell time of a scanning spot or line in a single eye resolution element at the screen is too short to allow much change of speckle by a changing diffuser, both the scanning line and the scanning spot revisit that resolution element every 1/30 sec. Thus for all three display types, changes of the diffuser that can be accomplished in 1/30 sec. will help suppress speckle. However, for raster-scanned and line-scanned displays, the scanning spot or line revisits a given eye resolution element about 30 times a second, and therefore the maximum possible value of M is 30, limiting the reduction of contrast that can be achieved. For a full-frame display, larger reductions may be possible. However, in all these cases, the contrast reduction also depends on the parameter K.

In addition, in a raster-scanned display, the weighting by the eye response of the projector resolution cells within the eye resolution element will be tapered in a way that will vary in detail between observers at different distances and under different lighting conditions, and therefore the application of orthogonal codes on a pixel basis will not be practical.

We conclude that changing diffusers can be used in all three types of systems, but with contrast reductions that are limited by the eye resolution time and the value of the parameter K.

Speckle in Projection Displays that use Nonlaser Sources

To this point our discussions have been focused on projection systems that use lasers, for with such systems the existence of speckle is most obvious. In this subsection, the appearance of speckle in non-laser based systems will be discussed.

Suppose that the projector lens is again overdesigned, so that it projects K of its resolution cells onto the screen within one resolution cell of the eye. Neglect the finite bandwidth of the source for the moment, assuming that each of the three colors has been filtered to produce a relatively narrow line width. Consider just one of these three colors.

A slight digression into the temporal properties of nonlaser light, or more precisely, "thermal" light produced by most nonlaser sources, is needed. Such light is typically generated when a very large number of atoms or molecules are excited to high energy states by thermal, electrical, or other means, and then randomly and independently drop to lower energy states by the process known as spontaneous emission. Let such light pass through a polarizer, so that only one linearly polarized component is of concern. If we were able to examine the properties of this rapidly changing light at a single point in space, we would observe a light describable by a rapidly varying complex analytic signal. If $\mathbf{u}(t)$ represents the analytic signal, then

$$\mathbf{u}(t) = \mathbf{A}(t)\exp(-j2\pi\bar{\nu}t), \qquad (6\text{-}69)$$

where $\bar{\nu}$ is the center frequency of the signal, and $\mathbf{A}(t)$ is the time varying complex envelope, with

$$\mathbf{A}(t) = |\mathbf{A}(t)|\exp[j\theta(t)] = A(t)\exp[j\theta(t)]. \qquad (6\text{-}70)$$

The real part of the analytic signal is the real-valued scalar amplitude of the wave at the chosen point.

Now the complex envelope is itself the sum of a very large number of "elementary" complex envelopes generated by each of the atomic or molecular events involved in spontaneous emission,

$$\mathbf{A}(t) = \sum_{\text{all atoms}} \mathbf{a}(t) = \sum_{\text{all atoms}} a(t)\exp[j\phi(t)]. \qquad (6\text{-}71)$$

Equation (6-71), for any fixed t, will be recognized to be a random walk in the complex plane, with independent steps. As a consequence, we can conclude that the statistics of the resultant amplitude $A(t)$ at fixed t is Rayleigh distributed and the phase $\theta(t)$ is uniformly distributed on $(-\pi, \pi)$. It follows that the intensity $I(t) = |A(t)|^2$ obeys negative exponential statistics. For each correlation time of the optical wave (which is approximately the reciprocal of the bandwidth of the light), there will be a new realization of the random walk, and uncorrelated values of $I(t)$ and $\theta(t)$ will be assumed. Thus we see that *the temporal statistics of polarized thermal light are the same as the ensemble statistics of speckle*!

If the source producing the light is spatially incoherent, then from Eq. (5-91) we know that the complex coherence factor is given by a Fourier transform of the intensity distribution across the pupil of the projection lens. Therefore, if we were to freeze time to an interval much shorter than the reciprocal of the bandwidth of the light, incident on the screen would be a wave with both intensity and phase fluctuations, each with spatial scale sizes of the order of the diffraction limit of the projector lens. The temporal intensity statistics of the light in each projected cell are negative-exponential, and the temporal phase statistics are uniform on $(-\pi, \pi)$. Different cells fluctuate independently of each other.

In accord with these facts, as time progresses the screen is illuminated by a very rapidly changing speckle pattern, the time between independent realizations being approximately the reciprocal of the bandwidth of the light. For typical bandwidths, there will be an enormous number M of different realizations of this speckle within one temporal response time of the eye. However, as we have seen, when the number K of projector resolution cells lying within one eye resolution cell is finite, the enormous number of independent realizations of the speckle incident on the screen is not capable of suppressing the speckle contrast below the level $\sqrt{1/K}$. The finite bandwidth of the source can suppress the contrast more through frequency diversity, but as we have seen, speckle is relatively insensitive to changes of frequency when surface scattering is the primary mechanism at the screen.

This effect is especially noticeable with micro-displays, which are small display systems often designed to accommodate a single observer. For such displays, the projector pupil size is typically comparatively small, perhaps on the order of 5 cm or less. If the screen is a meter away, by the van Cittert–Zernike theorem the coherence area on the screen is of the order of 10 μm in diameter for visible light. Thus coherence extends over many correlation areas of the screen, and speckle will be observed. If the eye averages over K coherence areas on the screen, the contrast of the speckle will be reduced by $\sqrt{1/K}$, as discussed above. The smaller the projector lens, the smaller K and the more speckle will be a problem.

Thus it is possible to observe speckle in displays that use nonlaser sources. The degree to which it is a problem will depend to a large extent on the value of the parameter K, that is, by how much the projector optics have been overdesigned compared with the resolution of the eye.

6.4.9 Specially Designed Screens

An approach that is capable of eliminating speckle under appropriate conditions is to create a special viewing screen with properties to be described. Examples of this type of screen are the *Engineered Diffusers* of RPC Photonics, Inc. To be effective for full-frame and line-scan displays, these screens must be used with projected moving diffusers, or some other means for reducing the spatial coherence of the light incident on the viewing screen. They are also effective with nonlaser sources under conditions when speckle might otherwise be generated by a random rough screen (see last subsection of Section 6.4.8).

Consider a screen consisting of an array of small convex spherical or aspherical reflectors, as pictured in Fig. 6.16. Further suppose that the height above a plane is varied randomly for these reflectors, and assume that their sizes are approximately equal. An individual reflector spreads the light over an angle determined by its radius of curvature, allowing the screen to be viewed from a range of angles.

Assume that the various geometrical factors at the screen are as shown in Fig. 6.17. The large circle is a representation of the area on the screen that is imaged to a single point on the retina. The small circle is a representation of the area covered by the point-spread function of the projector lens. The middle-sized circle is a representation of the average size of the reflective elements on the screen. The relative sizes of these factors must be in the order shown to achieve speckle suppression. The screen element size must be smaller than the eye resolution element in order to

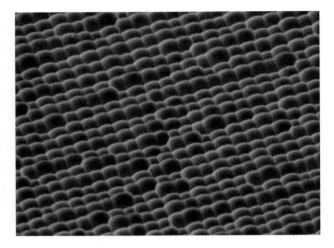

Figure 6.16 Microstructured array of reflecting elements. Courtesy of RPC Photonics, Inc.

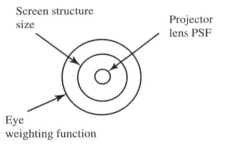

Figure 6.17 Geometrical factors at the screen.

avoid disturbing effects associated with the observer resolving the screen structure. The projector lens point-spread function must be smaller than the screen element size in order to achieve speckle suppression, as will be seen in the arguments to follow. Note that this is again a case requiring overdesign of the projector lens, or a projector lens resolution that is better than the eye resolution.

Consider first the case of a raster-scan display. As the projected spot scans across a line of the display, the vast majority of the time it illuminates only a single screen element, the only exception being the brief instants when the spot scans across a boundary between screen elements. As a consequence of the fact that only a single element of the screen is illuminated the vast majority of the time, there is no opportunity for interference between contributions from different screen elements, and the display will be free of speckle.

Next consider the cases of line-scan and full-frame displays. In both of these cases, multiple elements of the screen are illuminated simultaneously, so the screen alone will not eliminate speckle. However, if such screens are combined with the projection of changing diffusers, as discussed in the previous section, suppression of speckle can be achieved. Assume that the diffuser is changing fast enough, compared with the line time (for line-scan displays) or the frame time (for full-frame displays) that the light striking the screen from the projector lens can be considered to be partially coherent, with a coherence area that is approximately the size of the point-spread function of the projection lens, and a coherence time that is determined by the rate at which the diffuser is changing. Since the coherence area is small compared with the size of a screen element, there is no opportunity for interference between the light contributions from different screen elements, and speckle will be suppressed.

If K represents the number of coherence areas within the weighting function of the eye, and M represents the number of temporal correlation cells that are averaged during a line time or a frame time, we can again start with Eq. (E-3), writing

$$I = \sum_{m=1}^{M} \sum_{k=1}^{K} \sum_{l=1}^{K} \mathbf{A}_k \mathbf{A}_l^* \mathbf{B}_k^{(m)} \mathbf{B}_l^{(m)*}. \tag{6-72}$$

Again $\mathbf{B}_k^{(m)}$ represents the field projected onto the screen by one projector-lens resolution element, while $\mathbf{A}_k$ represents the the field that would be projected onto the retina by the kth projector lens resolution element on the screen if the field $\mathbf{B}_k$ were unity. For the screens considered in this section, the fields $\mathbf{A}_k$ are approximately of the form

$$\mathbf{A}_k = e^{j\phi_p} \tag{6-73}$$

where ϕ_p is the phase imparted to the field reflected from the screen by the pth screen element. Making appropriate changes to the analysis of Appendix E, namely replacing $J_A = \overline{|\mathbf{A}_k|^2}$ by unity and $\overline{|\mathbf{A}_k|^4}$ by unity, we find the contrast of the speckle to be

$$C = \sqrt{\frac{K-1}{KM}} \tag{6-74}$$

when the changing diffuser just fills the projection lens, and

$$C = \sqrt{\frac{1}{M}} \tag{6-75}$$

when the changing diffuser overfills the projection lens. Note that when the diffuser just fills the projector lens and there is only one projector resolution cell within the eye response function, the speckle vanishes, but the screen structure may be noticeable and objectionable to the viewer.

Note that, in principle, as $M \to \infty$, the speckle will vanish. However, suppression of speckle contrast is easier to achieve with a full-frame display than with a line-scan display because there is more possibility of achieving many diffuser realizations (large M) with a full-frame display.

6.5 Speckle in Projection Microlithography

A central task of microlithography[12] is the projection of an image of a mask onto a layer of photoresist overcoating a silicon wafer. Usually the optical system must also provide demagnification of the mask. The goal is, with the help of selective etching, to eventually imprint desired features onto the silicon, achieving the smallest feature sizes possible.

While there are many small contributing sources of error in microlithography in the semiconductor industry, speckle is generally not listed as one of those sources. Evidently the volume scattering of the light in the photoresist and random backscatter from the underlying silicon are too small to yield significant random interference effects in the photoresist.[13] However, it has recently been suggested ([140], [138]) that, for the excimer lasers currently in use as light sources, there is a speckle-like phenomenon that may be contributing to errors. As yet there is no definitive proof that this is the case, but the suggestion is sufficiently intriguing to be worthy of discussion here.

We begin with a discussion of the coherence properties of excimer lasers, and then turn to a discussion of the speckle-like phenomenon of concern. The effects of such speckle on line edge location errors are then analyzed.

6.5.1 Coherence Properties of Excimer Lasers

The excimer laser is currently the laser of choice for generating the deep-UV wavelengths currently used in microlithography. This laser oscillates in many different spatial modes and has a low degree of spatial coherence. In addition, it has a relatively broad spectrum, giving it low temporal coherence. These low degrees of coherence are desirable because they help avoid many of the undesired image artifacts that accompany imaging with coherent light. At the same time, the excimer laser does have a far higher brightness (watts per unit area and per unit solid angle) than non-laser sources in this same spectral region.

The low coherence of the excimer laser is in part due to the fact that, as a consequence of high loss in the laser cavity, a photon makes only a few bounces within the cavity before leaving. Typical parameters for a pulsed excimer laser for microlithography are as follows:

- Center wavelength: 193 nm;
- Gain spectrum bandwidth: 1 nm;
- Laser line width (gratings and/or prisms in the cavity): 0.15 pm;
- Pulse width: 35 nsec;
- Number of pulses integrated in one exposure: 40;
- Polarization state of the light: polarized.

---

12. For an excellent book covering the principles of lithography, see Ref. [101].
13. Nonrandom reflection from the silicon surface can cause standing waves in the photoresist and adversely affect the imprinted features, but this is not a speckle phenomenon.

Typically an exposure system would include a pulse stretcher to lengthen the pulses, a beam homogenizer to create a beam having as uniform an intensity as possible, and gratings and/or prisms in the cavity to narrow the linewidth. There are continuing efforts to further narrow the bandwidth of these lasers in order to minimize chromatic aberrations from the imaging optics.

6.5.2 Temporal Speckle

The speckle phenomenon of concern here is not the usual speckle caused by reflection of light from a rough surface, but rather the temporal intensity fluctuations associated with light from a source with coherence time that is much longer than the coherence time of typical incoherent sources. The authors of Ref. [140] have referred to this type of speckle as "dynamic" speckle. Here we prefer to use the term "temporal speckle", since "dynamic speckle" has a different meaning in the field of speckle metrology.

As was discussed in some detail on p. 224, the temporal fluctuations of intensity of a polarized nonlaser source obey the same statistics in time that conventional speckle obeys over space, namely the intensity obeys a negative-exponential distribution. Therefore it is reasonable to call such fluctuations "temporal speckle." Most of the spatial properties of conventional speckle also hold for the temporal properties of temporal speckle.

The energy to which the photoresist is subjected at any point consists of the integrated light intensity contributed by the multiple pulses used for a single exposure. There is a finite number of coherence times within that train of pulses, and as a consequence, there can remain residual fluctuations of intensity associated the integrated intensity. Consistent with the discussion in the subsection beginning on p. 109, for a single laser pulse with intensity pulse shape $P_T(t)$, the number of degrees of freedom M_1 is given by

$$M_1 = \frac{(\int_{-\infty}^{\infty} P_T(t)\,dt)^2}{\int_{-\infty}^{\infty} K_T(\tau)|\boldsymbol{\mu}_\mathbf{A}(\tau)|^2\,d\tau} \tag{6-76}$$

and the contrast of time-integrated speckle from one pulse is given by

$$C = \frac{[\int_{-\infty}^{\infty} K_T(\tau)|\boldsymbol{\mu}_\mathbf{A}(\tau)|^2\,d\tau]^{1/2}}{\int_{-\infty}^{\infty} P_T(t)\,dt}, \tag{6-77}$$

where $K_T(\tau)$ is the autocorrelation function of $P_T(t)$.

In the present case, we take the spectrum of the laser, normalized to have unity area, to be Gaussian, as in Eq. (6-24),

$$\hat{\mathcal{G}}(\nu) = \frac{2\sqrt{\ln 2}}{\sqrt{\pi}\Delta\nu} \exp\left[-\left(2\sqrt{\ln 2}\,\frac{\nu + \bar{\nu}}{\Delta\nu}\right)^2\right], \tag{6-78}$$

where $\Delta\nu$ is again the full width at half maximum of the spectrum. The squared magnitude of the complex coherence factor is

$$|\boldsymbol{\mu}_A(\tau)|^2 = \exp\left[-\frac{\pi^2 \Delta v^2 \tau^2}{2\ln 2}\right] \qquad (6\text{-}79)$$

In addition, we assume that the laser intensity pulses are Gaussian in shape, with full-width–half-maximum pulse width Δt and, for simplicity, unit area,

$$P_T(t) = \frac{2\sqrt{\ln 2}}{\sqrt{\pi}\Delta t}\exp\left[-\left(2\sqrt{\ln 2}\frac{t}{\Delta t}\right)^2\right]. \qquad (6\text{-}80)$$

The autocorrelation function of this pulse shape is given by

$$K_T(\tau) = \sqrt{\frac{2\ln 2}{\pi}}\frac{1}{\Delta t}\exp\left[-\left(\sqrt{2\ln 2}\frac{\tau}{\Delta t}\right)^2\right]. \qquad (6\text{-}81)$$

The number of degrees of freedom M_1 associated with a single pulse is then

$$\begin{aligned}M_1 &= \left[\int_{-\infty}^{\infty} K_T(\tau)|\boldsymbol{\mu}_A(\tau)|^2\, d\tau\right]^{-1}\\ &= \left[\sqrt{\frac{2\ln 2}{\pi}}\frac{1}{\Delta t}\int_{-\infty}^{\infty}\exp\left[-\left(\sqrt{2\ln 2}\frac{\tau}{\Delta t}\right)^2\right]\exp\left[-\frac{\pi^2\Delta v^2\tau^2}{2\ln 2}\right]d\tau\right]^{-1}\\ &= \left[1+\left(\frac{\pi\Delta v\Delta t}{2\ln 2}\right)^2\right]^{1/2}. \end{aligned} \qquad (6\text{-}82)$$

For a train of $\mathcal{K}$ pulses, the number of degrees of freedom is

$$M_\mathcal{K} = \mathcal{K}M_1, \qquad (6\text{-}83)$$

and with the substitution

$$\Delta v = \frac{c\Delta\lambda}{\bar{\lambda}^2} \qquad (6\text{-}84)$$

we obtain our final result

$$M_\mathcal{K} = \mathcal{K}\left[1+\frac{\pi^2\Delta t^2 c^2\Delta\lambda^2}{(2\ln 2)^2\bar{\lambda}^4}\right]^{1/2}. \qquad (6\text{-}85)$$

When the numbers assumed in Section 6.5.1 are substituted into this equation, the results are

$$M_\mathcal{K} = 3833 \qquad (6\text{-}86)$$

and

$$C = 0.016. \qquad (6\text{-}87)$$

Thus the standard deviation of the integrated speckle intensity is 0.016 times the mean integrated intensity.

It remains to determine how the predicted fluctuations of the integrated intensity would affect the line edge positions realized on the silicon wafer.

6.5.3 From Exposure Fluctuations to Line Position Fluctuations

In this section we use an extremely oversimplified model to show how variations of intensity cause variations in the positions of an edge. The example will be rather specialized, but it may be helpful in suggesting more comprehensive and accurate methods for calculating these effects.

Several assumptions are introduced at the start. First, assume that the goal is to produce an image of an edge between a bright region and a dark region. The photoresist is assumed to have a sharp threshold of integrated intensity, which is represented by W_T. Any value of integrated intensity larger than this level will cause the photoresist to be exposed, and any level smaller will leave the photoresist unexposed.

While the lithographic imaging system is actually partially coherent, we consider both an incoherent and a coherent system here, with the results for a partially coherent system expected to fall in between. Figure 6.18(a) shows spatial distributions of normalized integrated intensity, $W(x)/\overline{W}$, in the incoherent case for various levels of integrated intensity $W(0)$ at the edge relative to the threshold integrated intensity W_T. The form of the 1-D intensity spread function assumed here is

$$K(x) = \exp\left[-\left(\frac{2\sqrt{\ln 2}\, x}{\Delta x}\right)^2\right], \qquad (6\text{-}88)$$

where Δx represents the full-width–half-maximum (FWHM) width of the spread function. The curve for $W(0)/W_T = 1.0$ represents the ideal case, when the actual integrated intensity at the edge equals the threshold integrated intensity, and for this curve, the asymptote on the right should be twice the threshold in the incoherent case. In practice, due to the statistical properties of the integrated intensity, the actual value of $W(0)/W_T$ may be greater or less than this ideal value. The various other curves in the figure represent such cases.

Figure 6.18(b) shows similar plots for the case of fully coherent illumination. In this case the value of the average integrated intensity must be set so that the ideal curve for $W(0)/W_T = 1.0$ crosses the photoresist threshold at value 0.25 if the edge is to be at the correct position. Δx is again the FWHM width of the intensity distribution associated with the Gaussian amplitude blur.

Any value of integrated intensity larger than the photoresist threshold will cause the photoresist to be exposed, and any level smaller will leave the photoresist unexposed. A positive photoresist is assumed, that is, processing of the photoresist is assumed to remove exposed areas and not remove unexposed areas. Thus after processing it is possible to diffuse desired chemicals into the silicon in those areas where the photoresist is missing, thereby creating a slightly blurred edge between regions with different chemical composition.

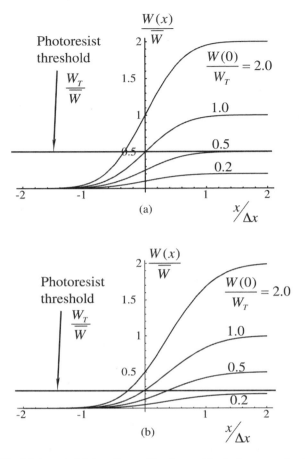

Figure 6.18 Normalized integrated intensity profiles for an ideal edge convolved with a Gaussian blur function, for various values of $W(0)/W_T$. (a) Incoherent case and (b) coherent case.

While the diffused edge would have an actual position at the position of the ideal step when the integrated intensity equals the mean integrated intensity, values of $W(0)$ other than $\overline{W}$ will result in displacements of the edge. Let s represent the slope of $x/\Delta x$ as a function of $W(0)/\overline{W}$. The magnitudes of this slope are found numerically to be $|s| = 0.64$ in the incoherent case and $|s| = 0.45$ in the coherent case.

Because the number of degrees of freedom $M_\mathcal{K}$ calculated above is large, the gamma distribution usually used to describe the distribution of integrated intensity can be replaced by a Gaussian approximation. Thus

$$p(W) \approx \frac{1}{\sqrt{2\pi}\sigma_W} \exp\left[-\frac{(W - \overline{W})^2}{2\sigma_W^2}\right], \qquad (6\text{-}89)$$

where

$$\sigma_W = \frac{\overline{W}}{\sqrt{M_\mathcal{K}}}. \qquad (6\text{-}90)$$

Using the slope information above, the normalized standard deviation σ_x of the edge position in response to fluctuations of $W(0)/\overline{W}$ can be described by the equation

$$\sigma_x/\Delta x = |s|\sigma_W/\overline{W} = |s|C, \qquad (6\text{-}91)$$

where C is the contrast of the speckle. For the particular example above with $C = .016$, we have

$$\begin{aligned}\sigma_x/\Delta x = 0.64 \times .016 = .010 & \quad \text{Incoherent case} \\ \sigma_x/\Delta x = 0.45 \times .016 = .007 & \quad \text{Coherent case.}\end{aligned} \qquad (6\text{-}92)$$

While the model and the calculation used here are very much oversimplified, nonetheless the result certainly suggests that, for the laser parameters assumed here, the classical model of temporal speckle predicts that fluctuations of edge position of the order of 1 percent of the resolution limit of the optical system may occur. In addition if changes to the laser characteristics are made, such as narrowing the spectral line width to minimize chromatic aberrations, or increasing the peak power of the pulses so that fewer pulses are utilized, the number of degrees of freedom may be smaller, and the effects would be predicted to be more severe.

As has been pointed out in Ref. [140], the classical model of speckle predicts that the situation can not be improved by superimposing spatially shifted or time delayed versions of the pulses on themselves, since the speckle cells then add on an amplitude basis, rather than an intensity basis, resulting in no improvement of contrast. This conclusion is consistent with the results of Appendix A, which proves that any linear transformation of a set of circular complex Gaussian random variates preserves the circular Gaussian character of the transformed random variables.

7 Speckle in Certain Nonimaging Applications

The speckle problems examined so far have arisen in imaging applications. In this chapter, attention is turned to certain applications in which obtaining an image is not the primary goal. First, speckle in multimode optical fibers is considered. Next, the effects of speckle on the performance of optical radars are analyzed.

7.1 Speckle in Multimode Fibers

Multimode fibers support a multitude of propagating modes with different phase velocities. From a geometrical-optics point of view, which has reasonable accuracy in highly multimoded fibers, various rays propagate down the fiber at different angles to the axis of the guide. The rays travel varying distances, depending on their angle of propagation, and as a consequence suffer different phase delays as they pass from the input to the output of the guide. A more accurate electromagnetic analysis of such a guide shows that there are many different possible propagating modes, and these modes propagate with a variety of different phase velocities. By either argument, the light at any point on the exit face of the guide consists of a sum of a multitude of individual field contributions. If the phase delays suffered by those contributions vary by more than 2π radians, and if the source is sufficiently coherent, then highly structured interference effects will be seen in the intensity distribution across the end of the guide. Figure 7.1 shows a photograph of the light emerging from a multi-mode fiber. This intensity pattern clearly resembles a speckle pattern. The number of modes supported by a multimode step-index fiber is given by the expression[1]

---

1. There are many books on the properties of optical fibers. One that has been particularly helpful in this writing is [144]. When results are presented in this section without proof and without an explicit reference, the derivations will be found in that book.

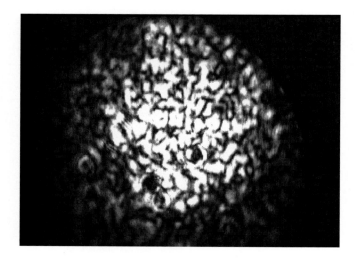

Figure 7.1 Intensity emerging from a multimode fiber. The fiber is a made from perfluorinated polymer glass, has a core diameter of 120 μm and a length of 35 cm. Photo courtesy of G. Durana and J. Zubia, University of Pais Vasco.

$$M \approx \frac{2\pi}{\lambda_0}(\mathrm{NA})a \qquad (7\text{-}1)$$

where a is the radius of the fiber core, λ_0 is the wavelength of light in air, and NA is the numerical aperture of the fiber. The numerical aperture is in turn related to the refractive indices of the core and the cladding of the fiber,

$$\mathrm{NA} \approx n_1\sqrt{2\Delta}, \qquad (7\text{-}2)$$

where n_1 is the refractive index of the core,

$$\Delta = \frac{n_1 - n_2}{n_1}, \qquad (7\text{-}3)$$

and n_2 is the refractive index of the cladding. Here the approximation entails an assumption that $\Delta \ll 1$. Thus,

$$M = (2\pi n_1 a/\lambda_0)\sqrt{2\Delta}. \qquad (7\text{-}4)$$

A graded-index fiber has an index profile given by

$$n(r) = \begin{cases} n_1[1 - 2\Delta(r/a)^\alpha]^{1/2} & r < a \\ n_2 & \text{otherwise,} \end{cases} \qquad (7\text{-}5)$$

where r is the radius from the center of the core, and α is an index profile parameter. When $\alpha = \infty$, the index profile becomes that of a step index fiber. When $\alpha = 2$, the fiber is referred to as a "parabolic index" fiber. For such a fiber, the gradually decreasing refractive index as the rays move away from the axis results in refraction, bringing the rays back to the center of the core periodically along the length of the fiber. The rays that travel farthest from the axis of the fiber have the longest physical path lengths, but since they encounter smaller refractive indices as they move away from the axis, the optical path length (the integral of the refractive index over the length of the path) is similar to that of the on-axis ray, and, with the right choice of α, modal dispersion is minimized. The number of modes supported by a graded index fiber can be shown to be ([144], Section 2.4.4)

$$M \approx \left(\frac{\alpha}{\alpha+2}\right)(2\pi n_1 a/\lambda_0)^2 \Delta. \tag{7-6}$$

Under changing environmental conditions, the phase velocities of the modes in such a fiber change with time and with respect to the phase velocities of other modes. Long lengths of such fiber are notoriously sensitive to changes of temperature and pressure, as well as to small vibrations and other types of fiber motion. In addition, when the wavelength of the source changes, the phase delays suffered by the various propagating modes change. When such changes occur, the detailed structure of the speckle across the end of the fiber changes, albeit rather slowly when the environmental changes are slow. Such changes can introduce what is known as modal noise, a phenomenon we now consider.

7.1.1 Modal Noise in Fibers

In 1978, R.E. Epworth [40] reported on a new source of noise in multimode fiber systems, which he named "modal noise." This noise arises from the speckle intensity fluctuations when the propagating light passes through a transformation that restricts the portion of the light that is transmitted. The restriction can be either a restriction of the light-carrying area of the core or a restriction of the modes that are propagating.

If coherent light is launched into a multimode fiber, and if a detector at the end of the fiber integrates all of the light emerging from the core, then neglecting radiation modes and light that escapes into the cladding, when environmental changes result in changes of the speckle pattern at the end of the fiber, the detector will sense no change in total power. To a good approximation for a low-loss fiber, all the power launched into the fiber is detected, regardless of the state of the environmental conditions.

However, if the detector integrates over only a portion of the core, part of the exiting power is not being detected, and as the speckle pattern on the end of the fiber changes under environmental changes, fluctuations of the detected power will occur. Such a case is illustrated in Fig. 7.2.

Let W represent the total optical power lying within a region D of the core over which the detector is integrating. Then

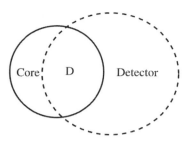

Figure 7.2 A fixed total power W_T is propagating within the core, represented by the solid circle. The detector photosensitive area is represented by the dashed circle. Power W is detected within the region D of overlap between the two circles.

$$W = \iint_D I(x, y)\, dx\, dy. \qquad (7\text{-}7)$$

The total power W_T exiting the fiber is constrained to be constant,

$$W_T = \iint_{\text{core}} I(x, y)\, dx\, dy. \qquad (7\text{-}8)$$

When the speckle pattern changes, W changes but W_T remains constant.

While we have illustrated the generation of modal noise with the case of a misaligned detector, there are many other ways that such noise can arise. A few examples of situations that can cause modal noise are as follows:

- A misaligned connector joining two sections of fiber;
- A splitter that divides the power two or more ways;
- A connection between a fiber with a large core and a fiber with a smaller core, or between a fiber with a large numerical aperture and a fiber with a smaller numerical aperture;
- Certain types of modal filtering devices.[2]

Note that changes in environmental conditions have been mentioned as a cause of modal noise, but frequency instability of the laser source can also cause such noise under the right conditions. For this to be the case, the frequency changes have to be larger than the frequency decorrelation interval $\Delta \nu$ discussed in Section 7.1.3.

---

2. If a device simply selects a certain subset of the propagating modes, the orthogonality of the modes guarantees that this subset would have constant power. If the detector used at the output of the fiber intercepts the full effective core area, the integrated intensity carried by that subset would not change under changing environmental conditions. However, if there is random mode coupling between modes before the subset is selected, then the modal coefficients of the selected modes will be random, and the subset of modes no longer carries constant power. As a consequence, fluctuations of power can be expected with changing environmental conditions.

7.1.2 Statistics of Constrained Speckle

Consider the speckle generated by a large set of CW modes propagating in the fiber. In a scalar formulation of the problem, the analytic signal representation of one polarization component of the field[3] observed at point (x, y) at the output of the core of the fiber can be modeled by

$$\mathbf{u}(x, y; t) = \mathbf{A}(x, y)e^{-j2\pi vt}, \quad (7\text{-}9)$$

where

$$\mathbf{A}(x, y) = \sum_{m=1}^{M_T} \mathbf{a}_m \boldsymbol{\psi}_m(x, y) e^{j\beta_m(v)L}. \quad (7\text{-}10)$$

The modal fields $\boldsymbol{\psi}_m(x, y)$ are assumed to be orthonormal over the finite extent of the fiber core (area $\mathcal{A}_T$), with the mth mode having an amplitude weighting $\mathbf{a}_m$. We assume further that the weights $\mathbf{a}_m$ of all the modes are constant across the entire core of area $\mathcal{A}_T$. The propagation constants of the modes are $\beta_m(v)$, and the quantity L represents the length of the fiber.

For fields represented by a set of such modes there is an important constraint that must be satisfied. Namely, if we neglect radiation modes and modes in the cladding, the total amount of instantaneous, spatially integrated power, W_T, across the any transverse section of the core of the fiber is a constant. Thus

$$W_T = \iint_{\mathcal{A}_T} A(x, y) A^*(x, y) \, dx \, dy$$

$$= \sum_{m=1}^{M_T} \sum_{n=1}^{M_T} \mathbf{a}_m \mathbf{a}_n^* \iint_{\mathcal{A}_T} \boldsymbol{\psi}_m(x, y) \boldsymbol{\psi}_n^*(x, y) \, dx \, dy \, e^{j(\beta_m(v)-\beta_n(v))L}$$

$$= \sum_{m=1}^{M_T} |\mathbf{a}_m|^2, \quad (7\text{-}11)$$

where the orthonormality of the $\boldsymbol{\psi}_m$ on $\mathcal{A}_T$ has been used. In other words, if we integrate the power across the finite area of the fiber core, that total power is constant and does not change if the phases of the interfering modes change. This conclusion remains valid even in the presence of random intermodal coupling, as long as there is negligible coupling to the radiation modes and the cladding modes. Such a property is not shared by the conventional speckle model, for which the total power obeys a gamma distribution and therefore is not constant.

Since the modes have different propagation constants β_m, their accumulated phase shifts after propagation over even a rather short length L become large and

---

3. Full vectorial solutions for the modes of dielectric waveguides can also be found. For an example of a numerical approach, see [26].

different on the scale of 2π radians, leading to a random walk of amplitudes in the complex plane that generates speckle. Changes in environmental conditions or changes in the frequency ν can change these relative phase shifts, thus changing the particular realization of speckle that is observed at the output, but always subject to the constant power constraint.

The statistical properties of the speckle appearing across the output of the fiber core in, for example, Fig. 7.1, are of considerable interest. On the one hand, if the detector were so badly misaligned that it covered only a single correlation area of that speckle, the fact that the field at a point is a superposition of a large number (equal to the number M_T of modes) of randomly phased phasors implies that the probability density function of the detected integrated intensity W would be very close to negative exponential, with a contrast C of unity. On the other hand, if the detector were perfectly aligned, that is, if it covered the entire core, the density function of the integrated intensity would become a delta function at integrated intensity W_T, and the contrast would be zero. It should be emphasized that, in this case the contrast would *not* be $1/\sqrt{M_T}$ as predicted by conventional speckle theory. Our goal in this section is to study the properties of integrated intensity that are pertinent to the current case. We refer to speckle for which the total integrated intensity is constant as "constrained speckle." For early work on this subject in the context of multimode fibers, see Refs. [164] and [72].

Several assumptions will be made at the start. First, we assume that the modal volume of the fiber is filled (for a discussion of the case of an under filled mode volume, see [123]), and that the number of modes is large. Second, we assume that the speckle is a spatially stationary random process within the core of the fiber. This is a reasonable assumption for a step-index fiber, but less so for a graded-index fiber, for which the modal contributions at various points on the exit face of the core may vary. Third, we assume that the number of speckle correlation cells across the core is equal to the number of modes propagating in the fiber, a good approximation if those modes are carrying nearly equal powers. In the event that the powers carried by the modes are not nearly equal, the number of correlation cells will be smaller than the number of modes. The equality of number of modes and number of speckles is a reasonable assumption for a step-index fiber, but less so for a graded-index fiber. A more complex analysis would be required if these assumptions are not valid.

The probability density function of interest is found in Appendix F. As shown in that appendix, if κ represents the ratio of the area covered by the detector to the total area of the core, the expression for the conditional probability density of the detected power for constrained speckle is

$$p(W|W_T) = \frac{1}{W_T} \left(\frac{W}{W_T}\right)^{\kappa M_T - 1} \left(1 - \frac{W}{W_T}\right)^{(1-\kappa)M_T - 1} \frac{\Gamma(M_T)}{\Gamma(\kappa M_T)\Gamma((1-\kappa)M_T)}, \quad (7\text{-}12)$$

valid for $0 \leq W \leq W_T$ and zero otherwise. In the statistics literature, this density function is known as the *beta* density function. The nth normalized moment of such a density is given by

$$\frac{\overline{W^n}}{\overline{W}_T^n} = \frac{\Gamma(M_T)\Gamma(\kappa M_T + n)}{\Gamma(\kappa M_T)\Gamma(M_T + n)}. \quad (7\text{-}13)$$

It is of interest to find expressions for the contrast C and the rms signal-to-noise ratio, $(S/N)_{\text{rms}}$ of constrained speckle. From the expression for the moments, we find

$$C = \frac{\sigma_W}{\overline{W}} = \sqrt{\frac{1-\kappa}{\kappa}} \frac{1}{\sqrt{M_T}} \quad (7\text{-}14)$$

and

$$\left(\frac{S}{N}\right)_{\text{rms}} = \frac{\overline{W}}{\sigma_W} = \sqrt{\frac{\kappa}{1-\kappa}} \sqrt{M_T}. \quad (7\text{-}15)$$

Note that the number of speckles influencing the contrast and rms signal-to-noise ratio can never be less than one (a single degree of freedom). Therefore, κ can never be smaller than $1/M_T$ in these expressions, in which case, for large M_T, $C \to 1$ and $(S/N)_{\text{rms}} \to 1$. At the other extreme, when κ approaches unity, $C \to 0$ and $(S/N)_{\text{rms}} \to \infty$.

It is also of interest to know how the beta density compares with the gamma density, that is, what is the effect of the constraint on the density function. Figure 7.3 illustrates two cases, one for $\kappa = 0.9$ and one for $\kappa = 0.1$. In the former case, for which most of the light emerging from the core is detected, the constraint plays a major role, while in the latter case, for which only a small fraction of the emerging light is detected, the constraint causes only minor changes. In addition, an unconstrained

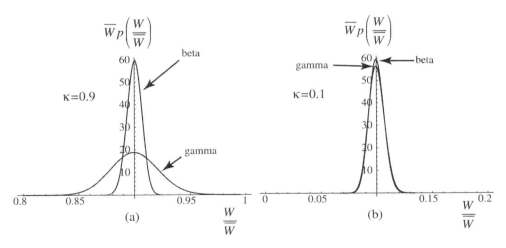

Figure 7.3 Probability density functions of integrated intensity for constrained (beta distribution) and unconstrained (gamma distribution) speckle for the case of $M_T = 2000$ modes. In the case of the beta distribution, $\overline{W}$ should be interpreted as W_T.

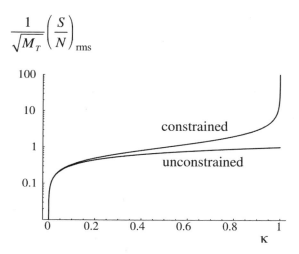

Figure 7.4 The rms signal-to-noise ratio, normalized by $\sqrt{M_T}$, plotted against the fractional area captured by the detector, κ. Note that κ can not be smaller than $1/M_T$ in these plots, for the smallest the signal-to-noise ratio can be is unity. Predictions of the unconstrained (conventional) and constrained speckle theories are shown.

assumption for W_T will be quite accurate when there is substantial modal loss into the cladding and radiation modes. (In that case W_T plays the role of W in the analysis above.)

A comparison of some interest is the rms signal-to-noise ratio of the fluctuations of integrated intensity as a function of the parameter κ for both constrained and unconstrained speckle. Figure 7.4 shows a plot of the signal-to-noise ratio, normalized by $\sqrt{M_T}$, as a function of the fractional area κ for the cases of both constrained and unconstrained speckle. Note that for the constrained theory, the signal-to-noise ratio approaches infinity as $\kappa \to 1$. For a misaligned connector, it is likely that the value of κ will be close to 1, and therefore the difference between the constrained and unconstrained predictions can be very significant.

In the preceding discussion, the number of degrees of freedom M_T has represented the total number of modes in both polarization components of the output of the fiber. If a polarizer is inserted between the end of the fiber core and the detector, the number of degrees of freedom is cut in half. The fraction of the modes participating in the measurement is therefore $\kappa/2$, and the two expressions for the signal-to-noise ratio become

$$\left(\frac{S}{N}\right)_{rms} = \begin{cases} \sqrt{\dfrac{\kappa M_T}{2-\kappa}} & \text{constrained case} \\ \sqrt{\dfrac{\kappa M_T}{2}} & \text{unconstrained case.} \end{cases} \quad (7\text{-}16)$$

7.1 Speckle in Multimode Fibers

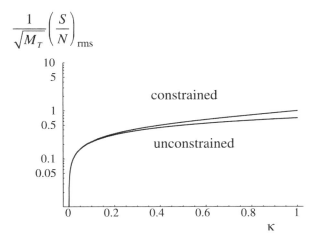

Figure 7.5 Normalized signal-to-noise ratio as a function of fractional area of the core covered by the detector. Again, κ can not be smaller than $1/M_T$ in these plots.

Figure 7.5 shows the normalized rms signal-to-noise ratio in this case. The difference between the two predictions is considerably less for values of κ near unity when a polarizer is present. However, for κ near unity, a significant price in terms of signal-to-noise ratio is paid when a polarizer is inserted.

This concludes our discussion of constrained speckle in multimode fibers with a CW monochromatic source. Attention is now turned to the frequency dependence of modal noise.

7.1.3 Frequency Dependence of Modal Noise

Time Delays in Propagation Through Multimode Fiber
It is important to understand the frequency decorrelation interval of speckle in fiber in order to ascertain whether frequency instabilities of the source will cause modal noise. We have seen in previous chapters (see in particular Eq. (5-113)) that the frequency dependence of speckle is determined by the spread of time delays associated with the various paths contributing to speckle at any given point. The frequency change required to de-correlate a speckle pattern at the end of a fiber is therefore approximately the reciprocal of the width of the impulse response of the fiber. Attention is therefore turned to estimating the width of such an impulse response. Throughout the discussion that follows, material dispersion will be ignored, and attention will be focused entirely on modal dispersion, which should be the more significant of the two effects. Detailed analyses of frequency decorrelation of speckle in waveguides can be found in Refs. [129] and [173].

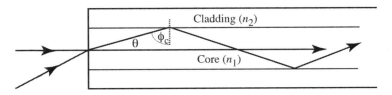

Figure 7.6 Shortest and longest paths traveled by rays in a step-index fiber. ϕ_c is the critical angle. The shortest path is down the axis of the fiber, while the longest path is traveled by the ray that strikes the core/cladding interface at the critical angle.

A geometrical picture of ray directions in a step-index multimode fiber[4] can aid in estimating the amount of modal time dispersion experienced as the light propagates from the fiber input to the fiber output. As illustrated in Fig. 7.6, the shortest propagation time is experienced by the ray that propagates down the axis of the fiber without ever reflecting off of the core/cladding interface. If the fiber length is L, then the shortest propagation time is

$$t_{\min} = Ln_1/c. \tag{7-17}$$

The longest propagation time is that suffered by the ray that reflects from the core/cladding interface at the critical angle ϕ_c with respect to the normal to the interface.[5] The critical angle is related to the propagation angle θ through

$$\sin \phi_c = \cos \theta = \frac{n_2}{n_1}, \tag{7-18}$$

and the longest propagation time is

$$t_{\max} = \frac{L/\cos\theta}{c/n_1} = \frac{Ln_1^2}{cn_2}. \tag{7-19}$$

The difference of propagation times for the fastest and the slowest rays is therefore[6]

$$\Delta t = \frac{Ln_1^2}{cn_2} - \frac{Ln_1}{c} = \frac{Ln_1}{cn_2}(n_1 - n_2) \approx \frac{Ln_1\Delta}{c} = \frac{L}{2cn_1}(\text{NA})^2, \tag{7-20}$$

---

4. Strictly speaking, the calculation to follow is valid for a step-index planar waveguide, but the same approach is often used in calculating the modal time spread of a step-index fiber (see, for example, [144], Sect. 2.4).
5. As is commonly done in analyses such as this, we have considered only so-called *meridional* rays, which are rays that pass through the fiber axis, and we have neglected *skew* rays that never pass through the axis of the fiber. Skew rays generally propagate near the core/cladding interface.
6. This expression assumes that there is no coupling between the modes. When mode coupling is present, this expression must be modified to $\frac{n_1\Delta}{c}\sqrt{LL_c}$ where L_c is a characteristic length of the fiber that is inversely proportional to coupling strength.

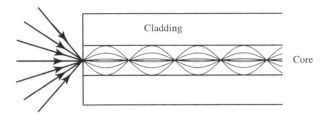

Figure 7.7 Meridional ray paths in a graded-index multimode fiber. With proper design of the index profile, the optical paths of the rays are nearly the same.

where in the next to last step we have replaced a factor n_2 in the denominator by n_1 because $n_2 \approx n_1$. We conclude that, for a step-index fiber, the time duration of the impulse response is on the order of the expression Δt given above.

The meridional ray paths in a typical graded index fiber are shown in Fig. 7.7. If the refractive index profile of the core is tailored to minimize modal dispersion, a geometrical optics analysis of pulse spreading in a graded-index fiber yields a time delay between the slowest mode and the fastest mode of (Ref. [59])

$$\Delta t \approx \frac{L n_1 \Delta^2}{2c} \approx \frac{L}{8 n_1^3 c} (\text{NA})^4 \qquad (7\text{-}21)$$

while a more rigorous electromagnetic analysis yields (Ref. [63])

$$\Delta t \approx \frac{L n_1 \Delta^2}{8c} \approx \frac{L}{32 n_1^3 c} (\text{NA})^4. \qquad (7\text{-}22)$$

However, 70% of the optical power appears in the first half of this time interval, so the effective time delay will be somewhat smaller than this expression. A reasonable approximation would be to assume a time delay that is half this result.

Frequency Covariance Functions

The frequency covariance function of two speckle intensity patterns, one a speckle intensity pattern $I(x, y; \nu)$ obtained when the source frequency is ν and the other a speckle intensity pattern $I(x, y; \nu + \Delta\nu)$ obtained when the source frequency is $\nu + \Delta\nu$, is defined by

$$\Psi(\Delta\nu) = \langle \overline{I(x, y; \nu) I(x, y; \nu + \Delta\nu)} \rangle - \langle \overline{I(x, y; \nu)} \rangle \langle \overline{I(x, y; \nu + \Delta\nu)} \rangle, \qquad (7\text{-}23)$$

where the overline indicates an ensemble average, and the angle brackets represent a spatial average applied after the ensemble average. The spatial average is necessary only when the speckle is spatially nonstationary over the spatial domain of interest. Often it is most convenient to deal with the normalized version of this covariance function,

$$\rho(\Delta v) = \frac{\Psi(\Delta v)}{\Psi(0)}. \tag{7-24}$$

The normalized frequency covariance functions of the intensities of two speckle patterns have been calculated in Ref. [129] using geometrical optics for the case of a step-index planar waveguide and in Ref. [173] for planar waveguides, step-index fibers and graded-index fibers using a modal approach. For the step-index planar waveguide, the geometrical-optics result is that the frequency correlation function is given by

$$\rho(\Delta v) = \left[\frac{C(y)}{y}\right]^2, \tag{7-25}$$

where $C(y)$ is the Fresnel cosine integral,

$$C(y) = \int_0^y \cos(\pi \eta^2/2)\, d\eta \tag{7-26}$$

and

$$y = \left(\frac{2L(\mathrm{NA})^2 \Delta v}{c n_1}\right)^{1/2} = \sqrt{\Delta v \Delta t}, \tag{7-27}$$

where Δt is given in Eq. (7-20). Figure 7.8 shows a plot of $\rho(\Delta v \delta t)$. This curve falls to value 1/2 when $\Delta v \Delta t = 1.18$, and therefore the frequency shift for which the correlation is reduced to 1/2 is

$$\Delta v_{1/2} = 0.59 \frac{n_1 c}{L(\mathrm{NA})^2}. \tag{7-28}$$

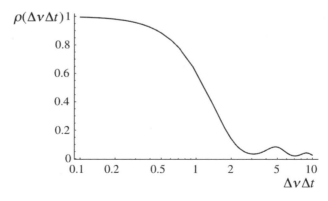

Figure 7.8 Normalized frequency covariance function for speckle intensities with an optical frequency shift Δv. The parameter Δt is given by Eq. (7-20).

Experimental results for step-index fibers show that this analytical result underestimates the correlation bandwidth by about 20% for the fibers investigated (Ref. [129], Fig. 7).

Finding the frequency covariance function for graded-index fibers requires a modal approach, which is beyond the scope of our discussion here, but can be found in Ref. [173]. Needless to say the values of $\Delta v_{1/2}$ are far larger for graded-index fibers than for step-index fibers. As a rough approximation, based on the nature of the result (7-28) and taking account of Eq. (7-21), a reasonable estimate for $\Delta v_{1/2}$ for a graded-index multimode fiber with a near-parabolic index profile would be

$$\Delta v_{1/2} \approx 1/\Delta t = \frac{8n_1^3 c}{L(\mathrm{NA})^4}, \qquad (7\text{-}29)$$

where we have used the geometrical-optics prediction of Δt rather than the wave-optics prediction because the calculation leading to the expression for $\rho(\Delta v)$ is strictly a geometrical calculation. Comparing the predictions of this equation with the wave-optics results of Ref. [173], we find that the predicted value of $\Delta v_{1/2}$ is about a factor of 2 too high.

Relationship of the Frequency Covariance Function to the Transfer Function of Multimode Fibers

The frequency covariance function and the transfer function of a fiber both depend on the time spread of the output intensity pulse when the input is excited by a very short impulse. The transfer function of the fiber predicts the response of the fiber to various intensity modulation frequencies at the input, while the frequency covariance function predicts the separation of two CW optical frequencies that results in decorrelation of the speckle intensities at the end of the fiber. It is natural to inquire about the relationship between these two related quantities. Such studies have been carried out in Refs. [115] using a geometrical optics approach and [173] using a modal analysis.

Fortunately, theory developed in earlier chapters can be directly applied to find the sought-after relationship. We begin with Eq. (5-38), plus modifications implied by Eq. (5-113), to write the normalized covariance function of the complex fields of the two speckle patterns as

$$\boldsymbol{\mu}_{\mathbf{A}}(\Delta q_z) = \mathbf{M}_l(\Delta q_z) = \int_0^\infty p_l(l) e^{j\Delta q_z l} \, dl, \qquad (7\text{-}30)$$

where

$$\Delta q_z \approx \frac{2\pi n_1 \Delta v}{c} \qquad (7\text{-}31)$$

and $p_l(l)$ is the probability density function of the path lengths encountered by the rays as they propagate to the end of the fiber. The factor $(n_1 - 1)$ appearing in Eq. (5-112) has been replaced by n_1 because the dispersion is all taking place in the fiber, rather than at a rough interface between a dielectric medium and air. As implied by Eq. (4-119), the probability density function of the path lengths is related to the average intensity impulse response (normalized to have unit area) of the fiber through

$$p_l(l)\,dl = \frac{\bar{I}(n_1 l/c)\,dl}{\int_0^\infty \bar{I}(n_1\xi/c)\,d\xi} = \frac{\bar{I}(t)\,dt}{\int_0^\infty \bar{I}(\eta)\,d\eta} = \hat{\bar{I}}(t)\,dt, \tag{7-32}$$

where the circumflex indicates that the function is normalized to unit area. It follows that

$$\boldsymbol{\mu}_\mathrm{A}(\Delta v) = \int_0^\infty \hat{\bar{I}}(t) e^{j2\pi\Delta v t}\,dt. \tag{7-33}$$

Noting that $\boldsymbol{\mu}_\mathrm{A}$ is the normalized covariance function of the *fields*, and using the fact that the fields obey circular Gaussian statistics, it follows that the normalized covariance function of the speckle *intensities* is

$$\rho(\Delta v) = |\boldsymbol{\mu}_\mathrm{A}(\Delta v)|^2 = \left| \int_0^\infty \hat{\bar{I}}(t) e^{j2\pi\Delta v t}\,dt \right|^2. \tag{7-34}$$

Consider now the transfer function of the optical fiber for sinusoidal intensity modulations. Again $\bar{I}(t)$ represents the average intensity impulse response of the fiber, normalized to have unit area. The transfer function is the Fourier transform of this quantity,

$$\mathbf{H}(f) = \int_0^\infty \hat{\bar{I}}(t) e^{-j2\pi f t}\,dt \tag{7-35}$$

where the normalization of $\hat{\bar{I}}(t)$ to have unit area causes $\mathbf{H}(f)$ to have unity value at the origin, and we have used the symbol f for intensity modulation frequency to distinguish it from the optical frequency v. From the previous two equations, it is clear that the relationship between the two quantities of interest is

$$\rho(\Delta v) = |\mathbf{H}(\Delta v)|^2. \tag{7-36}$$

Thus we have shown the important result that the normalized modulus of the fiber intensity-modulation transfer function can be found from measurements of the frequency covariance function of speckle intensities using a tunable CW optical source.

7.2 Effects of Speckle on Optical Radar Performance

The use of lasers as sources for optical radars has been an application of interest since the laser was first demonstrated. An oversimplified representation of such a system is shown in Fig. 7.9. A short pulse of optical energy is transmitted towards the distant target, and the receiver detects the returning energy.

7.2 Effects of Speckle on Optical Radar Performance

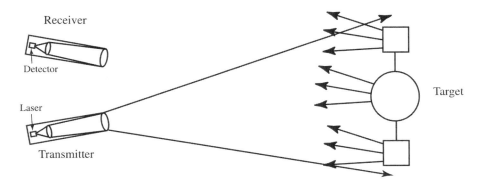

Figure 7.9 Diagrammatic representation of a laser radar. The transmitter contains a pulsed laser and the receiver contains a highly sensitive detector.

Two different types of detection are possible. For *incoherent* detection, the detector generates a photocurrent proportional to the power received as a function of time, possibly smoothed by the finite detector frequency response or by internal receiver circuitry. From the time delay of the pulse it is possible to estimate the distance z of the target through the simple relationship $z = \frac{1}{2}c\tau$, where c is the speed of light, τ is the time delay of the pulse and the factor of $1/2$ is due to the fact that the light is making a round trip over distance $2z$.

For *coherent* detection, the laser must have a coherence length that is long compared with $2z$. A portion of the laser light is split off in the transmitter and sent directly to the receiver, where it serves as a local oscillator. The detector measures the interference of the local oscillator and the return from the target. If the local oscillator is frequency shifted by a known amount before interfering with the returning signal, then the detector output contains a carrier frequency at the frequency difference between the target return and the local oscillator. Such detection is referred to as heterodyne detection. If no frequency shift of the local oscillator is introduced, then the detection is referred to as homodyne detection. Doppler shifts of the returned light can be detected with coherent detection, and therefore the component of target velocity towards or away from the transmitter/receiver can be measured.

Most targets are optically rough and therefore return diffusely reflected light, although they may also contain one or more specular glints. Therefore target returns generally exhibit speckle. We refer to a target that is purely a diffuse reflector (no specular glints) as a *diffuse* target. We refer to a target that returns a constant, nonfluctuating return as *specular* target. In what follows, we consider the detection statistics associated with both incoherent and coherent detection for both specular and diffuse targets. However, first we consider the spatial correlation properties of speckle returned from distant rough targets.

For rather comprehensive coverage of the subjects treated here, see [122]. For early work, see [66] and [67].

7.2.1 Spatial Correlation of the Speckle Returned from Distant Targets

The sizes of the spatial correlation areas of speckle amplitude and intensity will be important in determining the performance of optical radars, and for that reason we consider this issue in the current context. Two cases can be considered separately. First, and most common, is the case of a transmitted beam that is wider than the cross-sectional area of the target. Second, of importance in a limited class of problems, is the case of a focused transmitted beam that is largely contained within the target cross-sectional area.

In the first case, the size of the scattering spot is determined primarily by the size of the target. More specifically, let $I_t(u, v)$ represent the intensity of the light scattered back in the direction of the receiver as a function of transverse coordinates (u, v) on the target. Then by the van Cittert–Zernike theorem, as represented by Eq. (4-56), the normalized covariance function of the fields at the receiver is

$$\mu_A(\Delta x, \Delta y) = \frac{\iint\limits_{-\infty}^{\infty} I_t(u, v) e^{-j\frac{2\pi}{\lambda z}(\Delta x u + \Delta y v)} \, du \, dv}{\iint\limits_{-\infty}^{\infty} I_t(u, v) \, du \, dv}, \qquad (7\text{-}37)$$

where λ is the wavelength and z is the target–receiver distance. As an example, a uniformly bright circular object with diameter Δ would result in a field covariance function (cf. Eq. (4-67))

$$\mu_A(r) = 2 \frac{J_1\left(\frac{\pi \Delta}{\lambda z} r\right)}{\frac{\pi \Delta}{\lambda z} r}, \qquad (7\text{-}38)$$

where $r = \sqrt{\Delta x^2 + \Delta y^2}$.

In the second case, we assume that a diffraction-limited transmitting optical system focuses a small spot on a portion of the target. Assuming uniform target reflectivity within that spot, we see that the intensity distribution across that spot is proportional to the Fraunhofer diffraction pattern of the transmitting aperture. Let $\mathbf{P}(x, y)$ represent the (possibly complex-valued) aperture function of the transmitting optics.[7] The intensity distribution across the scattering spot is therefore (see [71], Eq. (4-25))

$$I(u, v) = \frac{I_0}{\lambda^2 z^2} \left| \iint\limits_{-\infty}^{\infty} \mathbf{P}(x, y) e^{-j\frac{2\pi}{\lambda z}(xu + yv)} \, dx \, dy \right|^2, \qquad (7\text{-}39)$$

where I_0 is the uniform intensity within the transmitting aperture. If we now apply the van Cittert–Zernike theorem to this intensity distribution and normalize the resulting field covariance function to unity at the origin, we obtain

---

7. If the transmitting optics are diffraction limited, then the aperture function is real-valued. Complex values are needed when the transmitting optics have phase aberrations.

$$\mu_{\mathbf{A}}(\Delta x, \Delta y) = \frac{\int\!\!\int_{-\infty}^{\infty} \mathbf{P}(x, y)\mathbf{P}^*(x - \Delta x, y - \Delta y)\, dx\, dy}{\int\!\!\int_{-\infty}^{\infty} |\mathbf{P}(x, y)|^2 \, dx\, dy}. \tag{7-40}$$

Thus the field covariance function is given by the normalized autocorrelation function of the aperture function of the transmitting optics. For the case of circular, diffraction-limited transmitting optics of diameter D, the result becomes (cf. Eq. (4-69))

$$\mu_{\mathbf{A}}(r) = \frac{2}{\pi} \left[\arccos\left(\frac{r}{D}\right) - \frac{r}{D}\sqrt{1 - \left(\frac{r}{D}\right)^2} \right], \tag{7-41}$$

where $r = \sqrt{\Delta x^2 + \Delta y^2}$. Note that the intensity autocovariance function is simply $|\mu_{\mathbf{A}}|^2$, a consequence of the circular Gaussian statistics of the speckle fields.

The number of speckle intensity correlation cells influencing the detector output is the same as the number of correlation cells captured by the receiving optics, and can be calculated by means of the usual formula (see Eq. (4-132)), which we repeat here for convenience,

$$M = \left[\frac{1}{\mathcal{A}_R^2} \int\!\!\int_{-\infty}^{\infty} K_D(\Delta x, \Delta y) |\mu_{\mathbf{A}}(\Delta x, \Delta y)|^2 \, d\Delta x\, d\Delta y \right]^{-1}. \tag{7-42}$$

The function $K_D(\Delta x, \Delta y)$ is the autocorrelation of the aperture of the receiving optics and $\mathcal{A}_R$ is the area of the receiving optics.

A case of special interest is one in which the same optics are used for both transmitting and receiving, and the entire transmitted beam is focused onto the target. Assuming the aperture of the optics is circular and of diameter D, we find

$$K_D(r) = \mathcal{A}_R \left(\frac{2}{\pi}\right) \left[\arccos\left(\frac{r}{D}\right) - \frac{r}{D}\left(\sqrt{1 - \left(\frac{r}{D}\right)^2}\right) \right], \tag{7-43}$$

where again $r = \sqrt{\Delta x^2 + \Delta y^2}$. It then follows that

$$M = \left[\frac{2\pi}{\mathcal{A}_R} \left(\frac{2}{\pi}\right)^3 \int_0^D r \left(\arccos\left(\frac{r}{D}\right) - \frac{r}{D}\sqrt{1 - \left(\frac{r}{D}\right)^2} \right)^3 dr \right]^{-1}$$

$$= \left[\frac{64}{\pi^3} \int_0^1 \xi(\arccos \xi - \xi\sqrt{1 - \xi^2})^3 \, d\xi \right]^{-1}, \tag{7-44}$$

where we have made the substitution $\xi = r/D$. Numerical integration can now be used to find that in this special case,

$$M = 3.77. \tag{7-45}$$

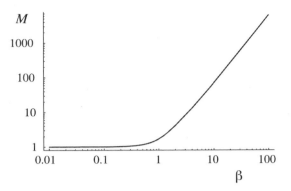

Figure 7.10 Number of degrees of freedom M vs. $\beta = D\Delta/\lambda z$.

Another case of interest occurs when a receiving aperture of diameter D is used and the angular divergence of the transmitted beam is much broader than the angular diameter of the target. Thus, to a good approximation, the transmitter uniformly illuminates the object. If the object is circular with diameter Δ, and uniformly bright when viewed from the receiver, we can use Eqs. (7-43) and (7-38) to write

$$M = \left[\frac{16}{\pi}\int_0^1 \xi(\arccos\xi - \xi\sqrt{1-\xi^2})\left(2\frac{J_1(\pi\beta\xi)}{\pi\beta\xi}\right)^2 d\xi\right]^{-1}, \quad (7\text{-}46)$$

where $\beta = D\Delta/\lambda z$. Using numerical integration, the values of M can be calculated for various values of β, with the results shown in Fig. 7.10. The critical parameter in these results is β, which is the ratio of the object diameter Δ to the diffraction limited resolution of the receiving optics at the target ($\approx \lambda z/D$). Thus if the receiving optics resolve the object well, $\beta \gg 1$ and $M \approx \beta^2$, while if the receiving optics are not capable of resolving the object, $\beta \ll 1$, in which case $M \approx 1$.

Knowing how to calculate the parameter M, we now turn to the subject of detection of target returns at very low light levels.

7.2.2 Speckle at Low Light Levels

In many optical radar detection problems, the energy returned from the object is very weak, and the current generated by the detector in the receiver consists of discrete "photoevents," each of which is the response of the detector to a single absorbed photon. In the case of direct detection, the discreteness of the photoevents changes the statistics of the measured signals, and any evaluation of radar performance must take account of these changes.

The theory used to describe the statistical properties of discrete detected signals is referred to as a "semiclassical" theory, and is based on certain assumptions [105] [70]:

- The probability of the occurrence of a single photoevent from a differential area dA at coordinates (x, y) on the sensitive photosurface of the detector and within a differential time dt is proportional to dA, dt, and to the intensity of the light incident at that point,

$$P(1; dt, dA) = \alpha\, dt\, dA\, I(x, y; t), \tag{7-47}$$

where α is the proportionality constant;
- The probability of more than 1 photoevent occurring in dA and dt is vanishingly small for differential area and time increments;
- The numbers of photoevents occurring in nonoverlapping time intervals are statistically independent.

These three assumptions can be shown to be sufficient to conclude that the number K of photoevents generated from an area A of the photodetector during a time interval T is given by a Poisson probability distribution,

$$P(K) = \frac{\bar{K}^K}{K!} e^{-\bar{K}}, \tag{7-48}$$

where the mean number of photoevents $\bar{K}$ is given by

$$\bar{K} = \alpha \iint_A \int_t^{t+T} I(x, y; \xi)\, d\xi\, dx\, dy. \tag{7-49}$$

It is convenient to express this result in terms of integrated intensity W, which has the dimensions of energy,

$$W = \iint_A \int_t^{t+T} I(x, y; \xi)\, d\xi\, dx\, dy, \tag{7-50}$$

in which case $\bar{K} = \alpha W$ and the probability of observing K photoevents is recognized as being a conditional probability

$$P(K \mid W) = \frac{(\alpha W)^K}{K!} e^{-\alpha W}, \tag{7-51}$$

which is conditioned by the incidence of energy W on the photodetector. The quantity α represents the efficiency with which incident energy is converted into photoevents, and can be expressed as

$$\alpha = \frac{\eta}{h\bar{\nu}} \tag{7-52}$$

where η is the quantum efficiency of the detector, h is Planck's constant, and $\bar{\nu}$ is the average frequency of the incident radiation. The unconditional probability of observing K photoevents would then be given by

$$P(K) = \int_0^\infty P(K|W)p(W)\,dW, \tag{7-53}$$

where $p(W)$ is the probability density function of the incident energy. It is straightforward to show that the mean and variance of K are given by

$$\bar{K} = \alpha\bar{W} \qquad \sigma_K^2 = \alpha\bar{W} + \alpha^2\sigma_W^2, \tag{7-54}$$

where $\bar{W}$ and σ_W^2 are the mean and variance of W, respectively.

Consider now the case when light from a single-mode, well-stabilized laser falls directly on the detector photosurface. The incident energy is then a known constant W_0, or equivalently, W obeys the probability density function

$$p(W) = \delta(W - W_0). \tag{7-55}$$

Substitution of this probability density function into Eq. (7-53) yields an unconditional probability density function for K that is Poisson,

$$P(K) = \frac{\bar{K}^K}{K!}e^{-\bar{K}}, \tag{7-56}$$

where $\bar{K} = \alpha W_0$.

Next, consider the case when the incident energy is speckle with a single (spatial or temporal) degree of freedom. In this case the energy W obeys negative exponential statistics and Eq. (7-53) becomes

$$p(W) = \int_0^\infty \frac{(\alpha W)^K}{K!}e^{-\alpha W} \times \frac{1}{\bar{W}}e^{-W/\bar{W}}\,dW = \frac{1}{1+\alpha\bar{W}}\left(\frac{\alpha\bar{W}}{1+\alpha\bar{W}}\right)^K. \tag{7-57}$$

Recognizing that $\alpha\bar{W} = \bar{K}$, we obtain

$$P(K) = \frac{1}{1+\bar{K}}\left(\frac{\bar{K}}{1+\bar{K}}\right)^K, \tag{7-58}$$

which is known in physics as the *Bose–Einstein* distribution, and in statistics as the *geometric* distribution.

Finally, consider the most general case when the incident energy is speckle with M degrees of freedom, in which case the density function of W is well approximated as a gamma density function with parameter M,

$$p(W) = \left(\frac{M}{\bar{W}}\right)^M \frac{W^{M-1}e^{-M\frac{W}{\bar{W}}}}{\Gamma(M)}, \tag{7-59}$$

valid for $W \geq 0$. Substitution of this expression in Eq. (7-53) yields

$$P(K) = \frac{\Gamma(K+M)}{\Gamma(K+1)\Gamma(M)}\left(\frac{\bar{K}}{\bar{K}+M}\right)^K\left(\frac{M}{\bar{K}+M}\right)^M, \tag{7-60}$$

which is known as the *negative binomial* distribution.

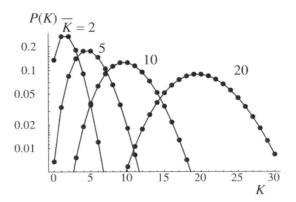

Figure 7.11 Poisson probability distributions for $\bar{K} = 2, 5, 10$ and 20.

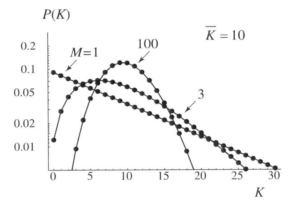

Figure 7.12 Negative-binomial probability distributions for $M = 1, 3$ and 100, with $\bar{K} = 10$ in all cases.

The above results are illustrated in Figs. 7.11 and 7.12. In the first of these figures, Poisson probabilities for various values of $\bar{K}$ are shown. The distributions are seen to shift to the right with increasing $\bar{K}$ and take on an increasingly symmetrical form. In the second figure, negative-binomial probabilities are shown for $M = 1, 3$ and 100, with $\bar{K} = 10$ for all curves. It is possible to show that, for a fixed $\bar{K}$, as $M \to \infty$, the negative binomial distribution with mean $\bar{K}$ approaches a Poisson distribution with the same mean. The critical parameter in determining this phenomenon is the degeneracy parameter, $\delta = \bar{K}/M$. When $\delta \ll 1$, there is little difference between the negative binomial distribution and the Poisson distribution (see, for example, [70], Section 9.3).

With this brief background on the statistical distributions of photoevents at low light levels, we now turn to considering the detection statistics of optical radars.

7.2.3 Detection Statistics—Direct Detection

In this section we consider direct detection under the assumption that the receiver counts photoevents and decides whether a target is present or not based on whether the count exceeds a predetermined threshold. Following the transmission of a pulse, the receiver counts photoevents occurring within a series of contiguous range intervals, those intervals having a length that matches the desired range resolution of the system.

Two types of decision error can occur. A *false alarm* error occurs if the decision is made that a target is present when in fact there is none. A *missed detection* error occurs if a target is actually present and the decision is made that there is none. Usually performance is specified by first setting an allowable false alarm probability, and then determining the *detection probability* (one minus the missed detection probability) as a function of the signal-to-noise ratio. The noise that leads to errors comes from two sources: the "dark" photoevents that occur due to thermal or steady incoherent background, which we assume is always present with or without a target, and the randomness of the photoevent count when a target is present.

In what follows we consider three different cases: (1) Poisson signal in Poisson noise, corresponding to the case of a specular target (no speckle) in the presence of dark counts, (2) Bose–Einstein signal in the presence of Poisson noise, corresponding to the case of a target return that is speckle with a single degree of freedom, and (3) negative-binomial signal in Poisson noise, corresponding to the case of a speckled target return with M degrees of freedom on the detector. We refer to targets that are rough and free from significant specular reflections as "diffuse" targets.

Poisson Signal in Poisson Noise

Let the photoevent distribution $P(K_n)$ from "dark" or noise events be Poisson with mean $\bar{K}_n$ and let the photoevent distribution $P(K_s)$ due to the target return be Poisson with mean $\bar{K}_s$. The total number of photoevents in a given range interval will be the sum of the numbers of photoevents generated by each of these independent Poisson variables, $K = K_s + K_n$.

The probability distribution of K can be found from a discrete convolution of the probability distributions of K_s and K_n, a fact that is true whenever two independent random variables are added. Thus

$$P(K) = \sum_{q=0}^{K} \frac{(\bar{N}_n)^q}{q!} e^{-\bar{N}_n} \frac{(\bar{N}_s)^{K-q}}{(K-q)!} e^{-\bar{N}_s}$$

$$= (\bar{N}_s)^K e^{-(\bar{N}_s + \bar{N}_n)} \sum_{q=0}^{K} \frac{1}{q!(K-q)!} \left(\frac{\bar{N}_n}{\bar{N}_s}\right)^q. \qquad (7\text{-}61)$$

The sum can be evaluated and is equal to $\frac{\left(1+\frac{\bar{N}_n}{\bar{N}_s}\right)^K}{K!}$, which when substituted into the above equation yields the following expression for $P(K)$:

$$P(K) = \frac{(\bar{N}_s + \bar{N}_n)^K}{K!} e^{-(\bar{N}_s + \bar{N}_n)}. \tag{7-62}$$

Thus the sum of two independent Poisson variates is itself Poisson, with a mean equal to the sum of the means of the constituents.

Throughout these illustrations, we shall assume an allowable false alarm probability P_{fa} of 10^{-6}, or more accurately, since we are dealing with discrete distributions, a false alarm probability that is as close as possible to 10^{-6} but not exceeding that number. To calculate the allowable count threshold in the absence of a target return, we determine the smallest N_T satisfying

$$\sum_{q=N_T}^{\infty} \frac{(\bar{N}_n)^q}{q!} e^{-\bar{N}_n} \leq 10^{-6}. \tag{7-63}$$

The following table shows the corresponding values of N_T for various values of $\bar{N}_n$:

$\bar{N}_n$	N_T
0	1
10^{-2}	3
10^{-1}	5
1	10
10	29
10^2	152
10^3	1155

(7-64)

The detection probability P_d is now found by summing

$$P_d = \sum_{K=N_T}^{\infty} \frac{(\bar{N}_s + \bar{N}_n)^K}{K!} e^{-(\bar{N}_s + \bar{N}_n)} = 1 - \sum_{K=0}^{N_T-1} \frac{(\bar{N}_s + \bar{N}_n)^K}{K!} e^{-(\bar{N}_s + \bar{N}_n)}. \tag{7-65}$$

Figure 7.13 shows plots of detection probabilities vs. $\bar{N}_s$ for a false alarm probability of 10^{-6} and for various average numbers of dark photoevents.

We now turn to investigating the effects of speckle on these detection statistics.

Bose–Einstein Signal in Poisson Noise

Let the photoevent distribution $P(K_n)$ from "dark" or noise events again be Poisson with mean $\bar{K}_n$, but now let the photoevent distribution $P(K_s)$ due to the target return be Bose–Einstein with mean $\alpha \bar{W} = \bar{K}_s$. The total number of photoevents in a given range interval will be the sum of the numbers of photoevents generated by each of these independent variables, $K = K_s + K_n$.

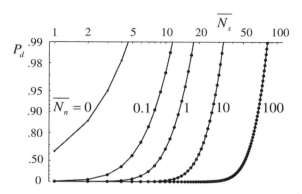

Figure 7.13 Detection probabilities P_d vs. average number of signal photoevents for a Poisson distributed signal in Poisson noise and for various average numbers of noise photoevents. The false alarm probability assumed is 10^{-6}, except in the case of $\bar{N}_n = 0$, for which the threshold is set at one count and there are no false alarms.

Again the probability distribution of K can be found from a discrete convolution of the probability distributions of K_s and K_n, which this time takes the form

$$P(K) = \sum_{q=0}^{K} \frac{(\bar{N}_n)^q}{q!} e^{-\bar{N}_n} \frac{1}{1+\bar{N}_s} \left(\frac{\bar{N}_s}{1+\bar{N}_s}\right)^{K-q}. \tag{7-66}$$

The sum can be performed (using *Mathematica* for example), yielding

$$P(K) = \frac{e^{\frac{\bar{N}_n}{\bar{N}_s}}(\bar{N}_s)^K \Gamma\left(1+K, \bar{N}_n\left(\frac{\bar{N}_s+1}{\bar{N}_s}\right)\right)}{(1+\bar{N}_s)^{1+K} K!}, \tag{7-67}$$

where $\Gamma(a,z)$ is an incomplete gamma function defined by

$$\Gamma(a,z) = \int_z^\infty t^{a-1} e^{-t}\, dt. \tag{7-68}$$

We again set the threshold so that the probability of a false alarm is 10^{-6}, and use the threshold settings shown in (7-64). The detection probability then becomes

$$P_d = 1 - \sum_{K=0}^{N_T-1} \frac{e^{\frac{\bar{N}_n}{\bar{N}_s}}(\bar{N}_s)^K \Gamma\left(1+K, \bar{N}_n\left(\frac{\bar{N}_s+1}{\bar{N}_s}\right)\right)}{(1+\bar{N}_s)^{1+K} K!}. \tag{7-69}$$

Figure 7.14 shows plots of detection probabilities vs. average number of signal photoevents for a false alarm probability of 10^{-6} and for various average numbers of noise photoevents.

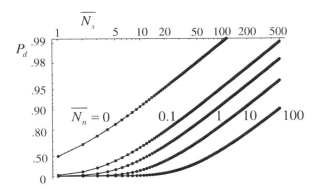

Figure 7.14 Detection probabilities for a Bose–Einstein signal in Poisson noise as a function of average number of signal photoevents, assuming a false alarm probability of 10^{-6} and various average numbers of noise photoevents, except for the case $\bar{N}_n = 0$, for which the threshold is 1 count and there are no false alarms.

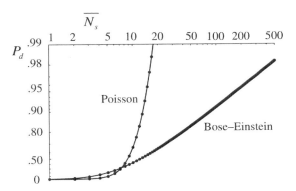

Figure 7.15 Detection probabilities as a function of $\bar{N}_s$ for a Poisson signal and a Bose–Einstein signal in Poisson noise for $\bar{N}_n = 1$ and a false alarm probability of 10^{-6}.

For comparison purposes, we show the Poisson and Bose–Einstein cases on the same plot for $\bar{N}_n = 1$ in Fig. 7.15. For detection probabilities higher than about 0.30, the signal level required to achieve a given detection probability is much greater for the Bose–Einstein signal than for the Poisson signal, a consequence of the significant probability that the integrated intensity W of the signal will be smaller than its mean, while for detection probabilities smaller than about 0.30 the Bose–Einstein signal outperforms the Poisson signal due to the smaller but still significant probability that the integrated intensity will be larger than its mean.

Negative-Binomial Signal in Poisson Noise

Let the noise again be Poisson distributed, but this time let the signal photoevents obey a negative-binomial distribution. Again we assume that the allowable false alarm probability is 10^{-6}. The probability of K photoevents again is a discrete convolution, this time most conveniently written[8]

$$P(K) = \sum_{q=0}^{K} \frac{(\bar{N}_n)^{K-q}}{(K-q)!} e^{-\bar{N}_n} \frac{\Gamma(q+M)}{q!\Gamma(M)} \left(\frac{\bar{N}_s}{\bar{N}_s + M}\right)^q \left(\frac{M}{\bar{N}_s + M}\right)^M. \quad (7\text{-}70)$$

Mathematica can perform this summation and finds the result to be

$$P(K) = \left(\frac{M}{M+\bar{N}_s}\right)^M \left(\frac{\bar{N}_s}{M+\bar{N}_s}\right)^k \frac{e^{-\bar{N}_n}}{k!} U\left(M, 1+K+M, \frac{\bar{N}_n(M+\bar{N}_s)}{\bar{N}_s}\right), \quad (7\text{-}71)$$

where $U(a,b,z)$ is the Tricomi confluent hypergeometric function[9] given by

$$U(a,b,z) = \frac{1}{\Gamma(a)} \int_0^\infty e^{-zt} t^{a-1} (t+1)^{-a+b-1} \, dt \quad (7\text{-}72)$$

with $\text{Re}(z) > 0$ and $\text{Re}(a) > 0$. The detection probability is then given by

$$P_d = 1 - \sum_{K=0}^{N_T - 1} P(K), \quad (7\text{-}73)$$

where $P(K)$ is given by Eq. (7-71).

Figure 7.16 shows plots of probability of detection vs. $\bar{N}_s$ for four values of M, for $\bar{N}_n = 1$, and for a false alarm probability of 10^{-6}. Note that for $M = 1$, the signal obeys Bose–Einstein statistics, and the curve coincides with the $\bar{N}_n = 1$ result from the previous subsection. When $M = \infty$, the detector integrates over an infinitely large set of speckles, with the result that the integrated signal intensity is a constant and the signal photoevent statistics are Poisson. This curve, too, was encountered earlier. The curves for $M = 2$ and $M = 4$ lie between these limits. The detection probability clearly improves as the number of degrees of freedom M increases.

7.2.4 Detection Statistics—Heterodyne Detection

We consider now an entirely different method of detection of target returns, namely heterodyne detection. A diagram of the structure of the simplest version of the transmitting and receiving system is shown in Fig. 7.17. A laser with a coherence length longer than the round-trip distance to the target is used. A portion of the light in the transmitter is split off, frequency shifted by $\Delta \nu$, and sent to the receiver, where it

---

8. Note that q is always an integer, while M need not be.
9. An equivalent expression in terms of Kummer's confluent hypergeometric function is found in [122], p. 247.

7.2 Effects of Speckle on Optical Radar Performance

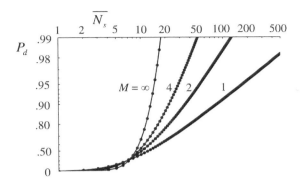

Figure 7.16 Detection probabilities as a function of $\bar{N}_s$ for a negative-binomial signal in Poisson noise, for various values of M. A false alarm probability of 10^{-6} and $\bar{N}_n = 1$ are assumed. The result for $M = 1$ coincides with the result for a Bose–Einstein signal in Poisson noise, while the result for $M = \infty$ coincides with the result for a Poisson signal in Poisson noise.

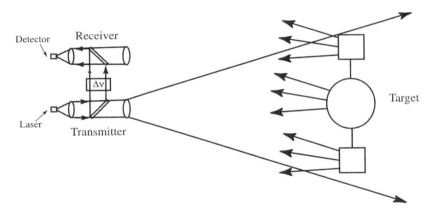

Figure 7.17 Diagram of an optical radar system using heterodyne detection. A highly coherent laser is used, and a portion of the light in the transmitter is split off, frequency shifted, and used as a local oscillator at the receiver. Not shown is an optical filter preceding the receiver to eliminate background noise.

serves as a local oscillator, interfering or beating with the light returning from the target. After amplification, the RF signal leaving the detector is passed through a bank of narrowband Doppler filters, each with a bandwidth of approximately the reciprocal of the transmitted pulse duration. If the target were completely stationary, then a filter centered on the difference frequency $\Delta \nu$ would pass the received pulse. However, a bank of filters is required because motion of the target towards or away from the transmitter and receiver causes the frequency of the light at the receiver to be Doppler shifted by an amount that is unknown, since the motion of the target is not known in advance. The frequency coverage of the bank of filters is broad enough to cover all anticipated Doppler shifts the returning light might experience. Following each Doppler filter is an envelope detector that detects the envelope of the RF signal transmitted by the filter. Decisions about the presence or absence of a target are made at the output of each envelope detector by comparing the detected envelope with a threshold, that threshold being determined by the allowable false alarm probability.

Note that the receiver shown in Fig. 7.17 has two important losses of light returned from the target. The first loss arises at the beam splitter in the signal path, which transmits only half the received power to the detector. If a nonpolarizing beam splitter is used, the splitting loss can be remedied by using balanced detectors, one in each beam leaving the beam splitter. The beat notes in the outputs of the two heterodyne receivers can be shown to be 180° out of phase and can therefore be subtracted electrically to obtain a single cosinusoidal electrical component with twice the amplitude of the outputs of the individual receivers. The noise outputs of the two receivers are independent and add electrically to double the noise power accompanying the beat note. The net result is a doubling of the output power signal-to-noise ratio. See Ref. [78], p. 814 for details.

A second loss arises because of the fact that the local oscillator is polarized, and the polarization component of the received light that is orthogonal to the local oscillator does not generate a beat note. Usually the returns from diffuse targets are depolarized. Polarization loss can be ameliorated by using a polarizing beam splitter and a pair of receivers ([78], p. 822). In the case of a depolarized signal return, the beat notes generated by the two heterodyne receivers will in general have random and independent phases and can not be combined electrically to double the signal amplitude. However, the detected signals can be combined electrically after each is envelope detected, with an increase in the power signal-to-noise ratio and a change in the probability density function of the detected signal. Both sources of loss can be eliminated by using four detectors in an appropriate receiver structure ([78], p. 826).

For simplicity, in what follows we consider only the simple receiver shown in Fig. 7.17. We represent the local oscillator and the target return as time-varying electric fields with the symbols $\vec{A}_{LO}(x, y; t)$ and $\vec{A}_S(x, y; t)$, respectively, in an (x, y) plane that is attached to the entrance pupil of the receiving optical system. To do so requires that the local oscillator field be conceptually back-propagated through the optical system to the entrance pupil of the collecting system, but in fact, if the target is in the far field of the receiving optics, the system is generally designed so that the local oscillator is equivalent to a plane wave in the entrance pupil. If the target is in the near field, the effective local oscillator should have the curvature of a spherical wave centered on the target. The two fields incident on the receiving aperture are written

$$\vec{A}_{LO}(x, y; t) = \vec{B}_0 e^{-j2\pi(v_0 + \Delta v)t}$$
$$\vec{A}_S(x, y; t) = \vec{B}_S(x, y) e^{-j2\pi(v_0 + v_d)t},$$

(7-74)

where v_0 is the frequency of the light transmitted by the radar, Δv is the offset frequency of the local oscillator, v_d is the Doppler shift of the light returning from the target, $\mathbf{B}_0$ is the (assumed uniform and constant) complex amplitude of the local oscillator, $\mathbf{B}_S(x, y)$ is the spatial distribution of complex amplitude of the target return, and the fields are represented as vectors to account for their polarizations. The finite duration of the pulse has been neglected here for simplicity.

Output of the Simple Coherent Receiver

The detector generates a current $i(t)$ that is proportional to the instantaneous integrated intensity, with a proportionality constant $\eta q/h v_0$, where η is the quantum efficiency of the detector, h is Planck's constant, q is the electronic charge, while $\Delta v \ll v_0$ and $v_d \ll v_0$ have been assumed.[10] Thus

$$i(t) = \frac{\eta q}{h v_0} \iint_{\mathcal{A}_R} [\vec{\mathbf{A}}_{LO}(x, y; t) + \vec{\mathbf{A}}_S(x, y; t)] \cdot [\vec{\mathbf{A}}_{LO}(x, y; t) + \vec{\mathbf{A}}_S(x, y; t)]^* \, dx \, dy, \quad (7\text{-}75)$$

where the "·" indicates an inner product of the vectors and $\mathcal{A}_R$ is the area of the entrance pupil of the receiver. Expanding this expression, we obtain

$$i(t) = \frac{\eta q}{h v_0} \iint_{\mathcal{A}_R} [\vec{\mathbf{A}}_{LO}(x, y; t) \cdot \vec{\mathbf{A}}^*_{LO}(x, y; t) + \vec{\mathbf{A}}_S(x, y; t) \cdot \vec{\mathbf{A}}^*_S(x, y; t)$$

$$+ \vec{\mathbf{A}}_{LO}(x, y; t) \cdot \vec{\mathbf{A}}^*_S(x, y; t) + \vec{\mathbf{A}}^*_{LO}(x, y; t) \cdot \vec{\mathbf{A}}_S(x, y; t)] \, dx \, dy$$

$$= \frac{\eta q}{h v_0} \Bigg[\iint_{\mathcal{A}_R} |\vec{\mathbf{B}}_0|^2 \, dx \, dy + \iint_{\mathcal{A}_R} |\vec{\mathbf{B}}_s(x, y)|^2 \, dx \, dy$$

$$+ e^{j 2\pi (\Delta v - v_d) t} \iint_{\mathcal{A}_R} \vec{\mathbf{B}}^*_0 \cdot \vec{\mathbf{B}}_S(x, y) \, dx \, dy$$

$$+ e^{-j 2\pi (\Delta v - v_d) t} \iint_{\mathcal{A}_R} \vec{\mathbf{B}}_0 \cdot \vec{\mathbf{B}}^*_S(x, y) \, dx \, dy \Bigg]. \quad (7\text{-}76)$$

The first term of this four-term expression yields a DC current generated by the local oscillator, which we represent by i_{LO} and can also be expressed as

$$i_{LO} = \frac{\eta q}{h v_0} \mathcal{A}_R I_{LO}, \quad (7\text{-}77)$$

---

10. We have neglected background optical noise in the interest of simplicity.

where I_{LO} is the intensity of the local oscillator, referred to the receiving aperture and assumed to be uniform there. In addition, since in general the local oscillator is much stronger than the signal, the second term can be ignored, for it is very small. This leaves the third and fourth terms, which represent the interference between the local oscillator and the signal.

From the above equation it is clear that only the component of the target return that is copolarized with the local oscillator contributes to the AC portion of the output. For simplicity, we assume from this point on that $\mathbf{B}_S(x, y)$ represents the complex amplitude of that copolarized component. The expression for the detector current can then be rewritten

$$i(t) = i_{LO} + \frac{2\eta q}{h\nu_0}|\mathbf{B}_0||\mathbf{B}_1|\cos[2\pi(\Delta\nu - \nu_d)t + \phi_1 - \phi_0], \tag{7-78}$$

where $\mathbf{B}_1 = \iint_{\mathcal{A}_R} \mathbf{B}_S(x, y)\, dx\, dy$, ϕ_0 is the constant and known phase associated with the local oscillator, and ϕ_1 is the unknown and random phase of $\mathbf{B}_1$.

A few words about the term $\mathbf{B}_1 = \iint_{\mathcal{A}_R} \mathbf{B}_S(x, y)\, dx\, dy$ are in order. If $\mathbf{B}_S$ were the signal field received from a specularly reflecting target, then this field would be constant across the receiving aperture and $\mathbf{B}_1 = \mathbf{B}_S \mathcal{A}_R$. However, if $\mathbf{B}_S(x, y)$ represents the complex field of a speckle pattern that is copolarized with the local oscillator, then as such, it is a circular complex Gaussian random process over space. The integral represents a linear transformation of that process, and as such, we know from Appendix A that it too must obey circular complex Gaussian statistics. It follows that $|\mathbf{B}_1|$ must obey Rayleigh statistics and ϕ_1 must be uniformly distributed on $(-\pi, \pi)$. We see the important fact that, *in a heterodyne system, integration over speckle lobes falling on the receiver does not reduce the contrast of the detected speckle signal*, due to the fact that the spatial averaging is over amplitude rather than intensity.

Signal-to-Noise Ratio in Coherent Detection—Specular Target

The problem of detecting a target return with a coherent receiver is precisely the same as the problem of measuring the resultant of the sum of a signal phasor and a noise phasor, with the length of the signal phasor being

$$s = 2\frac{\eta q}{h\nu_0}|\mathbf{B}_0||\mathbf{B}_1|. \tag{7-79}$$

As mentioned previously, if the signal is from a specular target, $\mathbf{B}_S$ is constant, $\mathbf{B}_1 = \mathbf{B}_S \mathcal{A}_R$, and the length of the signal phasor can be written

$$s = 2\frac{\eta q}{h\nu_0}\sqrt{P_0 P_S}, \tag{7-80}$$

where $P_0 = |\mathbf{B}_0|^2 \mathcal{A}_R$ is the equivalent power associated with the local oscillator at the detector, referred back to the receiving aperture, and $P_1 = |\mathbf{B}_S|^2 \mathcal{A}_R$ is the power from the specular target return falling on that aperture that actually reaches the detector. The time-averaged power associated with the cosinusoidal signal is

$$S = \frac{1}{2}s^2 = 2\left(\frac{\eta q}{h\nu_0}\right)^2 P_0 P_S. \tag{7-81}$$

The detection problem is then the problem of detecting the presence of a fixed phasor in the presence of circular complex Gaussian noise corresponding to shot noise generated by the local oscillator. The variance of the shot noise passed by a Doppler filter centered at the frequency of the target return and having bandwidth B is (see [111], Eq. (11-102a))

$$N = \sigma_{\text{shot}}^2 = 2qi_{LO}B = 2\frac{\eta q^2}{h\nu_0} P_0 B. \tag{7-82}$$

The variances σ_N^2 of the real part and the imaginary part of the circular complex Gaussian noise phasor are one half this amount, or

$$\sigma_N^2 = \frac{\eta q^2}{h\nu_0} P_0 B. \tag{7-83}$$

From the above results we see that, for a specular target, the postdetection signal-to-noise power ratio is

$$\frac{S}{N} = \frac{\eta}{h\nu_0} \frac{P_S}{B}. \tag{7-84}$$

Under the assumption that the bandwidth of the Doppler filter B is approximately the reciprocal of the transmitted pulse duration T, this expression becomes

$$\frac{S}{N} = \frac{\eta}{h\nu_0} P_S T = \overline{N}_s \tag{7-85}$$

where $\overline{N}_s$ represents the average number of target-return photoevents that would be observed at the detector (copolarized with the local oscillator) in time T in the absence of a local oscillator.

The problem of detecting such a signal in shot noise is the same as the problem of detecting the presence of a Rician phasor in circular complex Gaussian noise (cf. Sect. 2.3). In the absence of the signal, the phasor length is Rayleigh distributed. In the presence of the signal, the statistics of the phasor length are Rician. The false alarm probability can be found from

$$P_{\text{fa}} = \int_{V_T}^{\infty} \frac{A}{\sigma_N^2} \exp\left[-\frac{A^2}{2\sigma_N^2}\right] dA = \exp\left[-\frac{V_T^2}{2\sigma_N^2}\right], \tag{7-86}$$

where V_T is the threshold to be chosen to achieve the desired false alarm probability, and from Eq. (7-83),

$$\sigma_N^2 = \frac{\eta q^2}{h\nu_0} P_0 B. \qquad (7\text{-}87)$$

It follows that the proper choice for V_T is

$$\frac{V_T}{\sigma_N} = \sqrt{2 \ln \frac{1}{P_{\text{fa}}}}. \qquad (7\text{-}88)$$

The detection probability is given by the probability that the signal plus noise exceed the threshold when the signal is indeed present, which can equivalently be written

$$\begin{aligned} P_d &= \int_{V_T}^{\infty} \frac{A}{\sigma_N^2} \exp\left[-\frac{A^2+s^2}{2\sigma_N^2}\right] I_0\left(\frac{As}{\sigma_N^2}\right) dA \\ &= \int_{V_T/\sigma_N}^{\infty} q \exp\left[-\frac{q^2+s^2/\sigma_N^2}{2}\right] I_0(qs/\sigma_N)\, dq \end{aligned} \qquad (7\text{-}89)$$

This integral can be expressed in terms of tabulated functions known as the *Marcum Q-functions* [107], defined by

$$Q(a,b) = \int_b^{\infty} x \exp[-(x^2+a^2)/2] I_0(ax)\, dx. \qquad (7\text{-}90)$$

Figure 7.18 shows a plot of detection probability vs. $10 \log_{10}(S/N) = 10 \log_{10}(s^2/2\sigma_N^2)$ for three different false alarm probabilities.

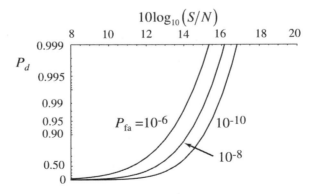

Figure 7.18 Detection probabilities P_d vs. $10 \log_{10}(S/N)$ for a specular target and for three different false alarm probabilities.

Signal-to-Noise Ratio in Coherent Detection—Diffuse Target

The problem of detecting the signal returned from a rough target in shot noise is equivalent to the problem of detecting a Rayleigh-distributed signal phasor in circular complex Gaussian noise. In fact, in this case, both the signal and the noise obey circular complex Gaussian statistics. When noise alone is present, the probability density function of the envelope detector output is

$$p_N(A) = \frac{A}{\sigma_N^2} \exp\left[-\frac{A^2}{2\sigma_N^2}\right], \tag{7-91}$$

where σ_N^2 is the variance of either the real or the imaginary part of the noise. When signal and noise are added together, the density function becomes

$$p_{S+N}(A) = \frac{A}{\sigma_S^2 + \sigma_N^2} \exp\left[-\frac{A^2}{2(\sigma_S^2 + \sigma_N^2)}\right], \tag{7-92}$$

where σ_S^2 is the variance of the real or imaginary part of the signal.

From Eq. (7-83) we have an explicit expression for the noise variance σ_N^2. Our goal now is to find an expression for the signal variance σ_S^2. To do so, we focus initially on the quantity $|\mathbf{B}_1|$ of Eq. (7-78). $|\mathbf{B}_1|$ obeys a Rayleigh probability density function of the form

$$p(|\mathbf{B}_1|) = \frac{|\mathbf{B}_1|}{\sigma_1^2} \exp\left(-\frac{|\mathbf{B}_1|^2}{2\sigma_1^2}\right), \tag{7-93}$$

where σ_1^2 is the variance of the real (or the imaginary) part of $\mathbf{B}_1$. To find σ_1^2 we use the fact, represented by Eq. (2-18), that

$$\overline{|\mathbf{B}_1|^2} = 2\sigma_1^2, \tag{7-94}$$

so if we can find an expression for $\overline{|\mathbf{B}_1|^2}$, one half of that expression will be σ_1^2. Towards this end, if $D(x, y)$ represents the aperture function of the receiving optics, with $D = 1$ in the collecting area and 0 otherwise, we find

$$\overline{|\mathbf{B}_1|^2} = \iint_{-\infty}^{\infty} \iint_{-\infty}^{\infty} D(x_1, y_1) D(x_2, y_2) \overline{\mathbf{B}_S(x_1, y_1) \mathbf{B}_S^*(x_2, y_2)} \, dx_1 \, dy_1 \, dx_2 \, dy_2$$

$$= \bar{I}_S A_R \iint_{-\infty}^{\infty} \mathcal{K}_D(\Delta x, \Delta y) \boldsymbol{\mu}_\mathbf{B}(\Delta x, \Delta y) \, d\Delta x \, d\Delta y, \tag{7-95}$$

where $\bar{I}_S = \overline{|\mathbf{B}_s|^2}$ is the average intensity of the signal at the receiving aperture, $\boldsymbol{\mu}_\mathbf{B}(\Delta x, \Delta y)$ is the normalized autocorrelation function of $\mathbf{B}_S(x, y)$, and

$$\mathcal{K}_D(\Delta x, \Delta y) = \frac{1}{\mathcal{A}_R} \iint\limits_{-\infty}^{\infty} D(x_1, y_1) D(x_1 + \Delta x, y_1 + \Delta y) \, dx_1 \, dy_1. \tag{7-96}$$

Note that $\mathcal{K}_D$ has been normalized to be unity at the origin. The integral in the second line of Eq. (7-95) has the dimensions of area. We refer to it as the "effective area" $\mathcal{A}_{\text{eff}}$ of the receiver, and write

$$\mathcal{A}_{\text{eff}} = \iint\limits_{-\infty}^{\infty} \mathcal{K}_D(\Delta x, \Delta y) \, \boldsymbol{\mu}_\mathbf{B}(\Delta x, \Delta y) \, d\Delta x \, d\Delta y. \tag{7-97}$$

When a single speckle lobe is very wide compared with the width of the receiving aperture, $\boldsymbol{\mu}_\mathbf{B} \approx 1$, and $\mathcal{A}_{\text{eff}} \approx \mathcal{A}_R$; that is, the effective area is approximately equal to the actual area of the receiving aperture. On the other hand, when there are many speckle lobes across the receiving aperture, $\mathcal{K}_D \approx 1$, and

$$\mathcal{A}_{\text{eff}} \approx \iint\limits_{-\infty}^{\infty} \boldsymbol{\mu}_\mathbf{B}(\Delta x, \Delta y) \, d\Delta x \, d\Delta y, \tag{7-98}$$

or in words, the effective area is approximately equal to the area of a single speckle amplitude lobe, which, under these assumptions, is much smaller than the actual receiver aperture area. Finally, we conclude that

$$\sigma_1^2 = \frac{1}{2} \bar{I}_S \mathcal{A}_R \mathcal{A}_{\text{eff}}. \tag{7-99}$$

Noting the terms that precede $|\mathbf{B}_1|$ in Eq. (7-79), we conclude that the length s of the signal phasor is Rayleigh distributed with parameter σ_S^2 given by

$$\sigma_S^2 = \left(\frac{\eta q}{h\nu_0}\right)^2 I_0 \overline{|\mathbf{B}_1|^2} = \frac{1}{2} \left(\frac{\eta q}{h\nu_0}\right)^2 (I_0 \mathcal{A}_R)(\bar{I}_S \mathcal{A}_{\text{eff}})$$

$$= \left(\frac{\eta q}{h\nu_0}\right)^2 P_0 (\bar{I}_S \mathcal{A}_{\text{eff}}). \tag{7-100}$$

Note that it is not the entire signal power falling on the receiving aperture that contributes to σ_S^2, only the power falling on the effective area $\mathcal{A}_{\text{eff}}$. Thus it is the power per speckle amplitude correlation area that determines the strength of the detected signal.

The expression for false alarm probability P_{fa} of Eq. (7-86) is still applicable in this case, since for a strong local oscillator, the shot noise properties are not affected by the nature of the target return. The detection probability is again found by integrating a Rayleigh density function, this time for both signal and noise present, with the result

7.2 Effects of Speckle on Optical Radar Performance

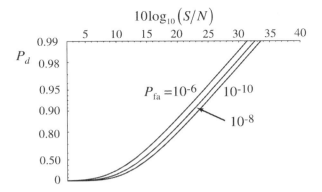

Figure 7.19 Detection probability for heterodyne detection with a rough target, as a function of $10 \log_{10}(S/N)$ for three different false alarm probabilities.

$$P_d = \exp\left[-\frac{V_T^2}{2(\sigma_N^2 + \sigma_S^2)}\right] = \exp\left[-\frac{V_T^2/2\sigma_N^2}{1 + \sigma_S^2/\sigma_N^2}\right] = \exp\left[\frac{\ln P_{\text{fa}}}{1 + S/N}\right], \quad (7\text{-}101)$$

where Eq. (7-88) has been used in the last step, while $S = 2\sigma_S^2$ and $N = 2\sigma_N^2$ are the powers associated with the signal and noise phasors, respectively. The ratio S/N is given by

$$\frac{S}{N} = \frac{\eta}{h\nu_0} \frac{\bar{I}_S \mathcal{A}_{\text{eff}}}{B} = \frac{\eta}{h\nu_0} (\bar{I}_S \mathcal{A}_{\text{eff}}) T, \quad (7\text{-}102)$$

where again the assumption that $B \approx 1/T$ has been made. Defining $\tilde{N}_s$ to be the number of photoevents that would result from a single pulse (in the absence of the local oscillator) due to the signal power copolarized with the local oscillator in a single effective area $\mathcal{A}_{\text{eff}}$, we see that

$$\frac{S}{N} = \tilde{N}_s = \frac{\mathcal{A}_{\text{eff}}}{\mathcal{A}_R} \bar{N}_s. \quad (7\text{-}103)$$

Figure 7.19 shows plots of probability of detection vs. $10 \log_{10} S/N$ for the same three false alarm probabilities assumed for a specular target. Note that the vertical and horizontal scales are different than those used in Fig. 7.18.

To facilitate a comparison of heterodyne system performances with specular and diffuse targets, Fig. 7.20 shows the curves from Figs. 7.18 and 7.19 on a common graph. As can be seen, the performance for a diffuse target is considerably worse than for a specular target. This situation will be exacerbated even further if the target return from the diffuse target has several speckle lobes across the receiving aperture, for then the signal power S is that associated with an area $\mathcal{A}_{\text{eff}}$, which may be smaller than that incident on full area of the detector.

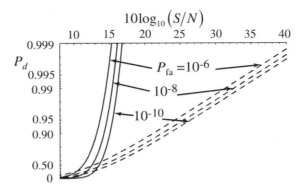

Figure 7.20 Detection probabilities for heterodyne detection with specular (solid lines) and diffuse targets (dashed lines) for various false alarm probabilities, plotted vs. $10\log_{10}(S/N)$. In effect, the diffuse target curves assume that $\mathcal{A}_{\text{eff}} = \mathcal{A}_R$, or that the target is not resolved by the receiving optics.

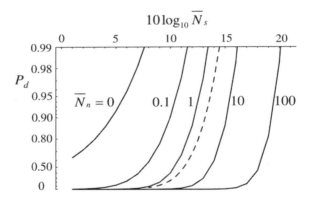

Figure 7.21 Comparison of direct detection and heterodyne detection for an unresolved specular target. Solid curves are for direct detection for various levels of dark counts. The dashed line is for heterodyne detection. A false alarm probability of 10^{-6} has been assumed, and for the heterodyne detection case the target return is assumed to be copolarized with the local oscillator.

7.2.5 Comparison of Direct Detection and Heterodyne Detection

A fair comparison of direct detection and heterodyne detection is difficult due to the fact that the two systems are likely to have different geometries and different system components. However, as a matter of curiosity, we pursue such a comparison, recognizing that it is of limited validity.

First consider the case of a specular target. Figure 7.21 shows plots of detection probability vs. $10\log_{10}\overline{N}_s$ for a specular target, for direct detection with several

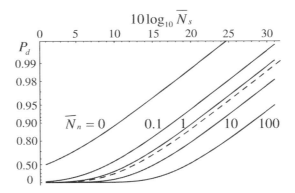

Figure 7.22 Comparison of direct detection and heterodyne detection for an unresolved diffuse target. Solid curves are for direct detection with various levels of dark counts. The dashed curve is for heterodyne detection. A false alarm probability of 10^{-6} is assumed, and only a single polarization component, copolarized with the local oscillator, is assumed.

different values of dark count $\bar{N}_n$, and for heterodyne detection. The target return is assumed to be copolarized with the local oscillator. For the case of heterodyne detection, calculation of $\bar{N}_s$ must take into account combining losses.

Next consider the case of a diffuse target. If the transmitted beam overfills the target and the receiver optics can not resolve the target, then a single speckle lobe contributes to the detected signal, with the result that $\mathcal{A}_{\text{eff}} = \mathcal{A}_R$, $\bar{I}_s \mathcal{A}_{\text{eff}} = P_S$, and $\bar{\bar{N}}_s = \bar{N}_s$. Figure 7.22 shows plots of detection probability vs. $10 \log_{10} \bar{N}_s$ for a diffuse object. The solid curves are for direct detection with various levels of dark counts. The dashed line is for heterodyne detection. A false alarm probability of 10^{-6} has been assumed, and for the heterodyne detection case only a single polarization component, copolarized with the local oscillator, is assumed. Again for the case of heterodyne detection, calculation of $\bar{N}_s$ must take into account combining losses.

When the receiver resolves the illuminated spot on a diffuse target, the comparison becomes more difficult. For direct detection, the value of the parameter M increases, thereby improving the detection probability. On the other hand, for heterodyne detection, the effective area $\mathcal{A}_{\text{eff}}$ shrinks as the target is resolved by the receiver, causing the detection probability to drop ($\bar{N}_s$ drops as the effective area shrinks, requiring more signal power to hold the detection probability constant). Unfortunately, since M depends on the correlation area of the speckle intensity while $\mathcal{A}_{\text{eff}}$ depends on the correlation area of the speckle amplitude, the respective changes in performance depend on the shape of the speckle amplitude correlation area. For example, if the illuminated spot on the target has approximately a circularly symmetric Gaussian intensity, the amplitude correlation function of the speckle at the receiver has an area that is twice the area of the intensity correlation function. Thus, for example, given a large value of M, the effective area of the receiving optics in the heterodyne case is to a good approximation $2\mathcal{A}_R/M$ provided the spot is Gaussian.

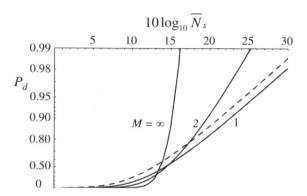

Figure 7.23 Comparison of direct detection and heterodyne detection, where the dotted curve represents the best possible performance for the heterodyne case (see the three assumptions mentioned in the text) and the solid curves represent the performance of direct detection under various degrees of resolving the target. A dark count of $\bar{N}_n = 10$ has been assumed for direct detection, and the probability of false alarm is 10^{-6}.

Different spot shapes result in different numerical multipliers[11] of A_R/M. It is difficult to generalize beyond the fact that the performance of direct detection improves as M increases, while the performance of the heterodyne detector gets worse.

Finally, in Fig. 7.23 we show detection probability vs. $10 \log_{10} \bar{N}_s$ for both direct detection and heterodyne detection. The solid curves show the results for direct detection for various numbers M of speckle correlation cells across the detector, all for a number of dark counts $\bar{N}_n$ per pulse of 10 and a probability of false alarm of 10^{-6}. The dashed line shows the best possible results for heterodyne detection, assuming the receiver can not resolve the target, the splitting losses have been eliminated by a balanced receiver, and the return is completely copolarized with the local oscillator. Violation of any of these three assumptions makes the performance of the heterodyne detector worse.

In closing this section, it should be emphasized that we have assumed for the direct detection receiver that the noise is in the form of discrete counts occurring within the measurement period for a returned pulse, which is a reasonable assumption for transmitted signals in the visible or even the near infrared portions of the spectrum, but probably not a reasonable assumption deeper in the infrared (for example at 10 μm wavelength), where photon counting receivers are much more difficult to realize. As we move further and further into the infrared, the receiver models must change and the advantages of heterodyne detection are found to increase.

---

11. For $M \gg 1$, the general form of the multiplier is $m = \dfrac{\iint_{-\infty}^{\infty} \mu_A(x,y)\,dx\,dy}{\iint_{-\infty}^{\infty} |\mu_A(x,y)|^2\,dx\,dy}$. A square spot on the target results in a multiplier of unity.

7.2.6 Reduction of the Effects of Speckle in Optical Radar Detection

We conclude this discussion of the effects of speckle on optical radar performance with a brief discussion of how these effects can be reduced. There are three main ways to reduce the effects of speckle. The first is by means of polarization diversity. In the case of direct detection, there should not be a polarizer in the optical train of the receiver, in which case, for many targets the return will be depolarized and the effective value of M will be doubled. For heterodyne detection, the use of a polarization diversity receiver will generate two independent Rician envelopes that can be added to reduce the speckle-induced fluctuations.

A second approach to speckle reduction is by means of multipulse integration. If the object is in motion, the speckle pattern incident on the detector may change from pulse to pulse, and if the detected pulses are added, the fluctuations of the received signal will be reduced.

A third approach to speckle reduction is by means of frequency diversity. The targets encountered are usually quite deep. As a consequence, the spread of depth from which the speckle originates makes the speckle quite sensitive to the optical frequency of the illumination. For example, for a target that is d meters deep, the speckle will de-correlate with a frequency change of approximately

$$\Delta \nu \approx \frac{c}{2d}, \qquad (7\text{-}104)$$

where c is the speed of light and the factor of 2 arises because it is the round-trip time delay to different parts of the target that is important. For a 1 meter deep target, the frequency decorrelation interval is $\Delta \nu \approx 150$ MHz. Transmitting consecutive pulses on different frequencies, separated by at least $\Delta \nu$, can thus assure independent speckle from pulse to pulse, even in the absence of target motion.

This concludes our discussion of the effects of speckle on optical radar performance and we now turn attention to the uses of speckle in optical metrology.

8 Speckle and Metrology

One could argue whether speckle metrology is an imaging or a nonimaging application of speckle. On the one hand, most speckle interferometry systems have an imaging system that is used in gathering information. On the other hand, it is not the image of the object that is really of interest; rather it is information about mechanical properties of the object, such as movement, vibration modes, or surface roughness that is desired. For this reason, plus the fact that speckle metrology is a well-developed field on its own, we have chosen to have a separate chapter devoted to this subject. While speckle has been a nuisance in almost all of the applications we have discussed previously, in the field of metrology, speckle is put to good use. Applications that use speckle for measurements of displacements arose in the late 1960s and early 1970s, often as an alternative to holographic interferometry. A number of survey articles and books cover the subject extremely well, including [41], [49], [39], [86], [146], and [128].

The field of speckle metrology has so many facets and so many applications that it is difficult to do it justice in a single chapter. At best we can only scratch the surface of such a rich and diverse field. We therefore restrict our goals to introducing some of the basic concepts while referring the reader to the more detailed treatments referenced above for a more in-depth study and more complete bibliographies.

8.1 Speckle Photography

The term "speckle photography" refers to a variety of techniques that use superposition of speckle intensity patterns, one from a rough object in an initial state and a second from the same object after it is subjected to some form of displacement. Particularly important early work on speckle photography includes that of Burch and Tokarski [22], Archibold, Burch and Ennos [4], and Groh [75]. Figure 8.1 shows

276   CHAPTER 8   Speckle and Metrology

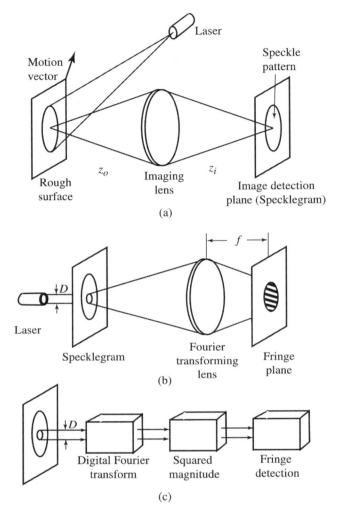

Figure 8.1 Recording and processing a doubly exposed "specklegram." (a) Recording the specklegram. (b) Optical Fourier transform for finding the fringe. (c) Digital Fourier transform for finding the fringe.

typical geometries for the measurement. Part (a) of the figure shows the recording geometry. A diffusely reflecting optically rough surface is illuminated by coherent light. An imaging lens forms an image of the surface, in some cases in focus and in other cases defocused, where a speckle pattern is recorded. The diffuse surface is then translated or deformed, and a second speckle pattern is recorded, typically on the same photographic plate or by the same detector array. This recording is often referred to as a "specklegram." Part (b) of the figure shows an optical method for fringe formation, in which the doubly exposed specklegram is optically Fourier transformed by illuminating a portion of the specklegram with coherent light and examining the intensity distribution in the rear focal plane of a positive lens, where a fringe pattern is observed. The period of the fringe pattern is determined by the distance the object moved between speckle recordings, and the fringes run normal to the direction of object motion. Part (c) of the figure is a block diagram of a more modern method of fringe analysis. In this case the specklegram is digitized, a portion of the digitized specklegram is Fourier transformed digitally, and the squared modulus of the result is taken to find the analog of intensity in the Fourier transform plane. Finally the resulting Fourier intensity pattern is analyzed to find the period and direction of the fringe pattern.

In the subsections that follow, we consider first the case of "in-plane" displacement of the object, meaning translation normal to the axis of the imaging system, and then the case of object rotation, which involves "out-of-plane" displacement, meaning movement in a direction parallel to the axis of the system. An excellent alternative description and analysis of speckle photography is found in [179].

8.1.1 In-Plane Displacement

Initially consider the case of in-plane displacement of the object. For simplicity, assume that the image plane and the object plane satisfy the lens law (the image of the diffuse surface is "in focus"), and also assume that the magnification is unity (the object and image distances are equal). The speckle pattern recorded in the first exposure can be represented by $I_1(x, y)$. A translation of the object in its own plane results in an overall translation of the speckle pattern and some degree of change of that pattern, due to the fact that some initial scatterers move out of the illumination beam and some new scatterers move in to replace them. Let the second speckle pattern intensity be represented by $I_2(x - x_0, y - y_0)$, where (due to image inversion) $(-x_0, -y_0)$ represents the in-plane vector motion of the object between exposures. For small motions, $I_2(x - x_0, y - y_0) \approx I_1(x - x_0, y - y_0)$, but for larger distances, I_1 and I_2 become partially de-correlated. The sum of the two intensity patterns is written $I_T(x, y) = I_1(x, y) + I_2(x - x_0, y - y_0)$.

As indicated in parts (b) and (c) of Fig. 8.1, in general a portion of the total specklegram may be analyzed at one time, with a real-valued weighting function $0 \leq W(x, y) \leq 1$ defining the portion of the specklegram that is processed. Often this weighting function may be a uniform circle with diameter D, but for the moment we retain more generality. Reasons for processing less than the full specklegram at one time will be discussed later. The Fourier transform of the weighted intensity region, which could be obtained optically[1] or digitally, is

---

1. If the Fourier transform is obtained optically, then $(v_X, v_Y) = \left(\frac{u}{\lambda f}, \frac{v}{\lambda f}\right)$, where λ is the optical wavelength used in the Fourier transform operation, f is the focal length of the Fourier transforming lens, and (u, v) are spatial coordinates in the optical Fourier transform plane.

$$\mathcal{F}\{W(x,y)I_T(x,y)\} = \mathcal{I}_1(\nu_X,\nu_Y) + \mathcal{I}_2(\nu_X,\nu_Y)e^{-j2\pi(x_0\nu_X+y_0\nu_Y)}, \qquad (8\text{-}1)$$

where $\mathcal{I}_1$ and $\mathcal{I}_2$ are the Fourier transforms of $W(x,y)I_1(x,y)$ and $W(x,y)I_2(x,y)$ respectively, (ν_X,ν_Y) are frequency variables, and the shift theorem of Fourier analysis has been used. The intensity or squared magnitude of this Fourier transform is

$$\begin{aligned}
|\mathcal{F}\{W(x,y)I_T(x,y)\}|^2 &= (|\mathcal{I}_1(\nu_X,\nu_Y)|^2 + |\mathcal{I}_2(\nu_X,\nu_Y)|^2) \\
&\quad + \mathcal{I}_1(\nu_X,\nu_Y)\mathcal{I}_2^*(\nu_X,\nu_Y)e^{j2\pi(x_0\nu_X+y_0\nu_Y)} \\
&\quad + \mathcal{I}_1^*(\nu_X,\nu_Y)\mathcal{I}_2(\nu_X,\nu_Y)e^{-j2\pi(x_0\nu_X+y_0\nu_Y)} \\
&= (|\mathcal{I}_1|^2 + |\mathcal{I}_2|^2) + 2|\mathcal{I}_1||\mathcal{I}_2| \\
&\quad \times \cos[2\pi(x_0\nu_X+y_0\nu_Y) + \phi_1 - \phi_2], \qquad (8\text{-}2)
\end{aligned}$$

where $\phi_1 = \arg\{\mathcal{I}_1\}$ and $\phi_2 = \arg\{\mathcal{I}_2\}$. Thus we see that the distribution of intensity in the Fourier plane consists of a varying bias and a fringe pattern with varying amplitude and phase modulations. The transverse displacement (x_0, y_0) of the surface can in principle be determined (up to a $\pm$ directional ambiguity) from the period $(1/x_0, 1/y_0)$ of the fringe pattern in the (ν_X, ν_Y) plane.[2]

A reasonable strategy to extract information about the fringe pattern period and orientation is to perform an additional Fourier transformation, this time on the power spectrum $|\mathcal{I}(\nu_X,\nu_Y)|^2$. According to the autocorrelation theorem of Fourier analysis, such a Fourier transformation yields the *autocorrelation function* of the specklegram $I_T(x,y)$. We would expect that autocorrelation function to have a central peak, arising from the components of intensity in the specklegram that do not represent interference, and two side peaks, located at coordinates (x_0, y_0) and $(-x_0, -y_0)$. The location of the side peaks yields the desired information about the surface displacement, up to a $\pm$ directional uncertainty.

As we shall see in the next subsection, the quantities $\mathcal{I}_1$ and $\mathcal{I}_2$ are in fact random quantities exhibiting many of the properties of speckle amplitudes. Often it may make more sense to discuss the *expected* or average fringe pattern that would be observed if the experiment were repeated over an ensemble of rough surfaces subjected to the same translation. Such an average expression is given by

$$\begin{aligned}
\overline{|\mathcal{F}\{W(x,y)I_T(x,y)\}|^2} &= (\overline{|\mathcal{I}_1(\nu_X,\nu_Y)|^2} + \overline{|\mathcal{I}_2(\nu_X,\nu_Y)|^2}) \\
&\quad + \overline{\mathcal{I}_1(\nu_X,\nu_Y)\mathcal{I}_2^*(\nu_X,\nu_Y)}e^{j2\pi(x_0\nu_X+y_0\nu_Y)} \\
&\quad + \overline{\mathcal{I}_1^*(\nu_X,\nu_Y)\mathcal{I}_2(\nu_X,\nu_Y)}e^{-j2\pi(x_0\nu_X+y_0\nu_Y)}. \qquad (8\text{-}3)
\end{aligned}$$

---

2. There is potential confusion for the reader if we speak of the spatial frequency of a fringe pattern in the (ν_X, ν_Y) plane, which is itself a spatial frequency plane. Think of the (ν_X, ν_Y) plane as just another spatial domain, as indeed it is if the Fourier transform is performed optically. In that domain there exists a fringe pattern with period $(1/x_0, 1/y_0)$.

8.1.2 Simulation

In this subsection we present the results of a simple simulation illustrating the results at various stages of the process described above. The simulation uses initial speckle patterns of dimension 1024×1024 pixels, with the size of a single speckle being approximately 8×8 pixels, and with one speckle pattern translated with respect to the other by various amounts ranging from 16 pixels to 128 pixels. The pupil function of the imaging system used to form these speckle patterns had a square exit pupil. The two speckle pattern intensities are added and a sub-window of size 512×512 is extracted for processing. Figure 8.2 illustrates the two speckle patterns and their sum for a small displacement of 16 pixels. The 512×512 specklegram is now subjected to a Fourier transformation, and the squared-modulus of the result is taken. This brings us to the fringe plane. A further Fourier transform of the fringe, followed by

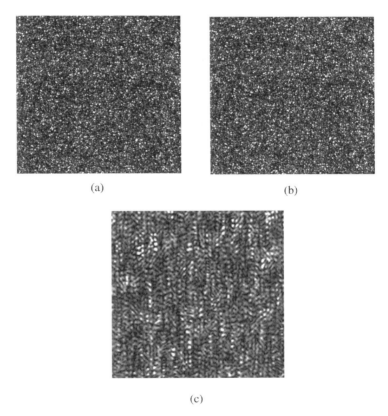

Figure 8.2 (a) Original 1024×1024 speckle pattern intensity. (b) The same speckle pattern shifted 16 pixels upwards. (c) A 512×512 window taken from the center of the sum of these two speckle patterns.

280 CHAPTER 8 Speckle and Metrology

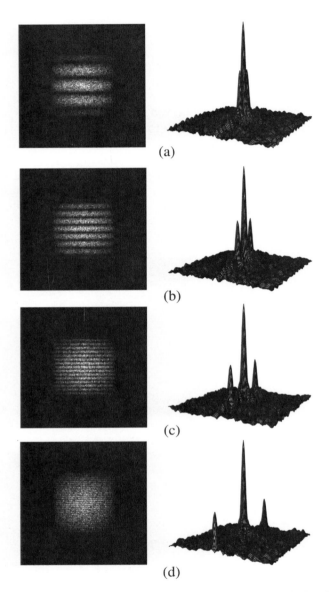

Figure 8.3 Spectral fringe patterns on the left and specklegram autocorrelation functions on the right, for object translations of (a) 16 pixels, (b) 32 pixels, (c) 64 pixels, and (d) 128 pixels.

a modulus operation, takes us to the final plane, where the autocorrelation function of the specklegram is obtained. Figure 8.3 shows on the left the fringe patterns obtained when the shift between speckle patterns is 16, 32, 64 and 128 pixels. On the right are the corresponding autocorrelation functions. In the case of perfect correlation between the spectra $\mathcal{I}_1(v_X, v_Y)$ and $\mathcal{I}_2(v_X, v_Y)$, the side peaks in the correlation function are expected to be half the height of the central peak. As the displacement increases, the height of the side peaks is reduced due to decorrelation of the two spectra, as will be explained in the subsection to follow.

We turn now to developing a deeper understanding of the properties of the spectra $\mathcal{I}_1(v_X, v_Y)$ and $\mathcal{I}_2(v_X, v_Y)$.

8.1.3 Properties of the Spectra $\mathcal{I}_k(v_X, v_Y)$

Extraction of the period of the above fringe pattern is complicated by the fact that $\mathcal{I}_1$ and $\mathcal{I}_2$ are random functions, and indeed can be shown to be the results of random walks in the complex plane. To see this, consider the explicit form of the Fourier transform (for $k = 1, 2$)

$$\mathcal{I}_k(v_X, v_Y) = \iint_{-\infty}^{\infty} W(x, y) I_k(x, y) e^{-j2\pi(v_X x + v_Y y)} \, dx \, dy. \tag{8-4}$$

At frequencies other than at or near the origin in the (v_X, v_Y) plane, the integral is the superposition of an infinite set complex exponentials $e^{-j2\pi(v_X x + v_Y y)}$ with random amplitudes $W(x, y) I_k(x, y)$ and phases that change with (x, y). As a consequence, to a good approximation, for frequencies not at or near the origin, the Fourier transform $\mathcal{I}_k(v_X, v_Y)$ is a complex circular Gaussian random variable, with magnitude $|\mathcal{I}_k|$ that is Rayleigh distributed and phase ϕ_k that is uniformly distributed. The requirements for this to be true are: (1) the frequency pair (v_X, v_Y) is far enough from the origin that, when the spatial coordinates (x, y) vary over the entire nonzero area of $W(x, y) I_k(x, y)$, the argument of the complex exponential will pass through at least 2π radians, and (2) the window defined by $W(x, y)$ contains many speckles.[3] If the average value of $W(x, y) I(x, y)$ is constant on a circle of diameter D, then the first requirement becomes approximately $\rho = \sqrt{v_X^2 + v_Y^2} > 1/D$.

Additional properties of the $\mathcal{I}_k$ require some discussion. First, it would be useful to know the average width and shape of the diffraction "halo" represented by $|\mathcal{I}_k(v_X, v_Y)|^2$. Second, since a particular realization of $|\mathcal{I}_k|^2$ has fluctuations, it would be useful to know the correlation width of those fluctuations in the frequency plane. Consider first the complex correlation of $\mathcal{I}_k$ at two different frequency pairs,

---

[3]. A proof of the assertion that $\mathcal{I}_k$ is a circular complex Gaussian variate for sufficiently large (v_X, v_Y) consists of proving that the means of its real and imaginary parts are zero, the variances of its real and imaginary parts are equal, and that the real and imaginary parts are uncorrelated. The Central Limit Theorem implies that the real and imaginary parts obey Gaussian statistics. We omit the details of such a proof here.

$$\overline{\mathcal{I}_k(v_{X_1},v_{Y_1})\mathcal{I}_k^*(v_{X_2},v_{Y_2})} = \iint_{-\infty}^{\infty} \iint_{-\infty}^{\infty} W(x_1,y_1)W(x_2,y_2)\overline{I_k(x_1,y_1)I_k(x_2,y_2)}$$

$$\times e^{-j2\pi(v_{X_1}x_1+v_{Y_1}y_1-v_{X_2}x_2-v_{Y_2}y_2)}\,dx_1\,dy_1\,dx_2\,dy_2. \qquad (8\text{-}5)$$

If the width of $W(x,y)$ is broad compared with its correlation width of the intensity $I(x,y)$, then we can make the following approximation:

$$W(x_1,y_1)W(x_2,y_2)\overline{I_k(x_1,y_1)I_k(x_2,y_2)} \approx \bar{I}^2 W(x_1,y_1)W(x_2,y_2)$$
$$+ \bar{I}^2 W^2(x_1,y_1)\mu_I(\Delta x, \Delta y), \qquad (8\text{-}6)$$

where μ_I is the normalized autocovariance of the intensity $I(x,y)$ as a function of difference coordinates $\Delta x = x_1 - x_2$ and $\Delta y = y_1 - y_2$. The complex correlation of interest then becomes

$$\overline{\mathcal{I}_k(v_{X_1},v_{Y_1})\mathcal{I}_k^*(v_{X_2},v_{Y_2})} \approx \bar{I}^2 \left| \iint_{-\infty}^{\infty} W(x,y)e^{-j2\pi(v_X x+v_Y y)}\,dx\,dy \right|^2$$

$$+ \bar{I}^2 \left[\iint_{-\infty}^{\infty} W^2(x_1,y_1)e^{-j2\pi(\Delta v_X x_1+\Delta v_Y y_1)}\,dx_1,dy_1 \right.$$

$$\left. \times \iint_{-\infty}^{\infty} \mu_I(\Delta x,\Delta y)e^{-j2\pi(v_X \Delta x+v_Y \Delta y)}\,d\Delta x\,d\Delta y \right], \qquad (8\text{-}7)$$

where $\Delta v_X = v_{X_1} - v_{X_2}$, $\Delta v_Y = v_{Y_1} - v_{Y_2}$, and it has been assumed that $\overline{I_k(x_1,y_1)} = \overline{I_k(x_2,y_2)} = \bar{I}$ over the window defined by $W(x_1,y_1)$. This result is entirely analogous to the generalized van Cittert–Zernike theorem of Eq. (4-86), with I playing the role of $\mathbf{a}$ and $\mathcal{I}_k$ playing the role of $\mathbf{A}$. If we make the further assumption that $W(x_1,y_1)$ is unity on a circle of diameter D and zero otherwise, we obtain the result

$$\overline{\mathcal{I}_k(v_{X_1},v_{Y_1})\mathcal{I}_k^*(v_{X_2},v_{Y_2})} \approx \bar{I}^2 \left[\pi\left(\frac{D}{2}\right)^2 2\frac{J_1(\pi D\rho)}{\pi D\rho} \right]^2$$

$$+ \bar{I}^2 \left[\pi(D/2)^2 2\frac{J_1(\pi D\Delta\rho)}{\pi D\Delta\rho} \iint_{-\infty}^{\infty} \mu_I(\Delta x,\Delta y)e^{-j2\pi(v_X\Delta x+v_Y\Delta y)}\,d\Delta x\,d\Delta y \right], \qquad (8\text{-}8)$$

where $\rho = \sqrt{v_X^2 + v_Y^2}$ and $\Delta\rho = \sqrt{\Delta v_X^2 + \Delta v_Y^2}$. The first term in this result is a diffraction-limited spike centered at the origin of the frequency plane, and will be neglected in what follows.

From this result we can calculate the average distribution $\overline{|\mathcal{I}_k|^2}$. In the above general result, we let $(\Delta v_X, \Delta v_Y) = (0,0)$, obtaining

8.1 Speckle Photography

$$\overline{|\mathcal{I}_k(\nu_X,\nu_Y)|^2} = \pi\left(\frac{D}{2}\right)^2 \bar{I}^2 \iint\limits_{-\infty}^{\infty} \mu_I(\Delta x, \Delta y) e^{-j2\pi(\nu_X\Delta x + \nu_Y\Delta y)}\, d\Delta x\, d\Delta y. \tag{8-9}$$

Thus the width of the envelope or "halo" in the frequency plane is determined by the Fourier transform of the correlation function of the intensity in the image plane where the specklegram is recorded. According to the van Cittert–Zernike theorem, the correlation function μ_I in the plane where the scattering spot is imaged is given by the properly scaled squared magnitude of the Fourier transform of the intensity distribution across the imaging lens pupil. Assuming that the pupil is circular with diameter L and uniformly bright, we can show that

$$\mu_I(\Delta x, \Delta y) = \left[2\frac{J_1\left(\frac{\pi L r}{\lambda z_i}\right)}{\frac{\pi L r}{\lambda z_i}}\right]^2, \tag{8-10}$$

where $r = \sqrt{\Delta x^2 + \Delta y^2}$ and z_i is the axial distance of the lens exit pupil from the image plane. Performing the additional Fourier transform required to obtain $\overline{|\mathcal{I}_k|^2}$, we find

$$\overline{|\mathcal{I}_k(\nu_X,\nu_Y)|^2} = \frac{2c}{\pi}[\arccos(\rho/\rho_0) - (\rho/\rho_0)\sqrt{1-(\rho/\rho_0)^2}] \tag{8-11}$$

(valid for $\rho \leq \rho_0$), where $\rho = \sqrt{\nu_X^2 + \nu_Y^2}$, $\rho_0 = L/\lambda z_i$, and the constant c is given by

$$c = \left(\lambda z_i \bar{I}\frac{D}{L}\right)^2. \tag{8-12}$$

When $\rho > \rho_0$, the result is zero. Thus we see that the average shape of the "halo," $\overline{|\mathcal{I}_k(\nu_X,\nu_Y)|^2}$, is given by Eq. (8-11), and the halo vanishes when the radius in the frequency plane exceeds $L/\lambda z_i$. This result is completely analogous to Eq. (4-69), which expresses the power spectral density of a speckle pattern in terms of the autocorrelation of the exit pupil function of an imaging system.

From the same general result we can calculate the correlation width of the fluctuations of $|\mathcal{I}_k|^2$ in the (ν_X, ν_Y) plane. For that purpose, we let (ν_{X_2}, ν_{Y_2}) approach a location $(\widetilde{\nu_{X_2}}, \widetilde{\nu_{Y_2}})$ in the frequency plane that it is far enough from the the origin to assure that circular complex Gaussian statistics hold for $\mathcal{I}_k$. In the vicinity of this location, the normalized covariance of $\mathcal{I}_k$ obeys

$$\mu_{\mathcal{I}}(\Delta\rho) = 2\frac{J_1(\pi D \Delta\rho)}{\pi D \Delta\rho}. \tag{8-13}$$

Since $\mathcal{I}_k$ is a circular complex Gaussian random variable, the normalized autocovariance of $|\mathcal{I}_k|^2$ is proportional to the squared magnitude of the right side of the previous equation,

$$\mu_{|\mathcal{I}|^2}(\Delta\rho) = \left|2\frac{J_1(\pi D\Delta\rho)}{\pi D\Delta\rho}\right|^2. \tag{8-14}$$

Since the function $2\frac{J_1(\pi x)}{\pi x}$ has its first zero when $x = 1.22$, we see that the correlation width in the (v_X, v_Y) plane is

$$\Delta\rho = \frac{1.22}{D} \approx \frac{1}{D}. \tag{8-15}$$

8.1.4 Limitations on the Size of the Motion (x_0, y_0)

The properties derived in the previous subsection imply certain limitations on the size of the in-plane motions (x_0, y_0) that are measurable by the speckle photography method. First, in order to measure the period of the fringe present in the frequency plane, it is normally necessary that there be at least one period present across the extent of the average diffraction halo $|\mathcal{I}_k(v_X, v_Y)|^2$ of Eq. (8-11). This will be true if

$$r_0 = \sqrt{x_0^2 + y_0^2} > \frac{\lambda z_i}{L}. \tag{8-16}$$

This is equivalent to requiring that the displacement be larger than the size of a single speckle in the (x, y) plane.

A second requirement limits the displacement from being too large. If the fringe period is so small that it is comparable with the size of a single speckle of $|\mathcal{I}(v_X, v_Y)|^2$ in the frequency plane, then it becomes impossible to measure the period of the fringe. This requirement can be stated mathematically as

$$r_0 < \frac{D}{1.22}. \tag{8-17}$$

Thus the displacement must be small compared with the full width of the speckle pattern window $W(x, y)$ in the (x, y) plane.

An additional factor limits the displacement from being too large. As the displacement of the scattering surface between exposures increases, not only does $I_2(x - x_0, y - y_0)$ move with respect to $I_1(x, y)$, but also the correlation of $W(x, y)I_1(x, y)$ and $W(x, y)I_2(x, y)$ decreases, due to the fact that the illuminated scatterers at that surface are displaced and the two sets of scatterers in the window are, to an extent that depends on the amount of motion, different. This decorrelation of $W(x, y)I_1(x, y)$ and $W(x, y)I_2(x - x_0, y - y_0)$ results in decorrelation of $\mathcal{I}_1(v_X, v_Y)$ and $\mathcal{I}_2(v_X, v_Y)$. In fact, due to the Fourier transform that relates them, the normalized cross-covariance between $W(x, y)I_1(x, y)$ and $W(x, y)I_2(x - x_0, y - y_0)$ must be the same as the normalized cross-covariance between $\mathcal{I}_1(v_X, v_Y)$ and $\mathcal{I}_2(v_X, v_Y)$. There is no mechanism that would increase or decrease the correlation in the frequency plane and make it different than the correlation in the space domain. The cross-covariance of $W(x, y)I_1(x, y)$ and

$W(x, y)I_2(x, y)$ must depend on the number of speckles they share in common within the window $W(x, y)$. This fraction of the speckles they share in common is in turn determined by the normalized overlap of two weighting functions $W(x, y)$ and $W(x - x_0, y - y_0)$. However, because $\mathcal{I}_1$ and $\mathcal{I}_2$ are the results of random walks, with each speckle in the (x, y) plane playing the role of a scatterer, the correlation $\overline{\mathcal{I}_1 \mathcal{I}_2}$ depends on the *square* of the number speckles they share in common. As a consequence, the normalized covariance of interest for a given displacement (x_0, y_0) is given by

$$\gamma = \frac{\overline{\mathcal{I}_1 \mathcal{I}_2^*}}{\sqrt{\overline{|\mathcal{I}_1|^2}\, \overline{|\mathcal{I}_2|^2}}} = \frac{\left| \int\int_{-\infty}^{\infty} W(x, y) W(x - x_0, y - y_0)\, dx\, dy \right|^2}{\left| \int\int_{-\infty}^{\infty} W^2(x, y)\, dx\, dy \right|^2}. \quad (8\text{-}18)$$

For a uniform circular weighting function $W(x, y)$ with diameter D, this result is

$$\gamma = \left[\frac{2}{\pi} \left(\arccos \frac{r_0}{D} - \frac{r_0}{D} \sqrt{1 - \left(\frac{r_0}{D}\right)^2} \right) \right]^2 \quad (8\text{-}19)$$

for $r_0 = \sqrt{x_0^2 + y_0^2} \leq D$ and zero otherwise. The quantity γ provides an estimate of the average contrast of the fringes observed in the (v_X, v_Y) plane. As the displacement radius r_0 approaches the diameter D of the window on the specklegram, the visibility of the fringes drops, making detection of the fringe harder and harder. For $r_0 > D$, the fringe vanishes entirely.

8.1.5 Analysis with Multiple Specklegram Windows

In the previous discussion it was assumed that the weighting function $W(x, y)$ limited the processed region of the specklegram to an area less than the full extent of the specklegram. It is always possible to process the entire specklegram at one time, but there are some circumstances under which processing a multitude of smaller areas can be desirable.

The primary reason to use a small region of the specklegram lies in what can be determined about the strain or displacement pattern in a nonrigid body, that is, one for which the amount of displacement changes in both magnitude and direction across a body under test. Each weighting function position analyzes a portion of the image of the object, and therefore if the displacement changes across the object, it is possible to determine the different vector displacements occurring on different parts of the object under test. Again the price paid is the limitation on how large the displacement can be and still be observed.

8.1.6 Object Rotation

Speckle is less sensitive to motions along the optical axis than it is to in-plane motions, due to the extended dimension of speckles along the axis of an optical system, as compared with the dimensions transverse to the axis. However, it has been demonstrated that speckle photography can be used to measure the out-of-plane tilts of rough surfaces quite accurately. Consider the geometry shown in Fig. 8.4, first proposed by Tiziani [163]. A rough surface present at distance z_o in front of the lens is illuminated by a collimated coherent beam, incident on the surface (before rotation) at angle ψ_i and the speckle pattern generated by that surface is recorded in the focal plane of a positive lens with focal length f. The angle of the central ray from the surface to the focal plane is assumed to be ψ_r with respect to the normal to the plane of the surface. The surface undergoes a tilt by a small angle $d\Phi$, with the result that the speckle pattern rotates in the direction of the tilt by angle $d\theta = \left(1 + \frac{\cos \psi_i}{\cos \psi_r}\right) d\Phi$. Assuming the tilt direction is parallel with the x-axis, the speckle pattern in the focal plane translates by distance $x_0 = f \tan d\theta$. For small angles ψ_i and ψ_r, this result becomes $x_0 = f \tan 2 d\Phi$.

A specklegram is obtained by recording the intensities $I_1(x, y)$ before rotation, and $I_2(x - x_0, y)$ after rotation. The specklegram is processed in the same manner as was done in the case of rigid body translation, with the result that x_0 can be determined after processing. From the measurement of x_0, coupled with knowledge of ψ_i, ψ_r and f, the angle of rotation $d\Phi$ can be calculated.

As in the case of rigid body motion, as the rotation angle $d\Phi$ and the amount of motion x_0 increase, the spectra $\mathcal{I}_1$ and $\mathcal{I}_2$ become less and less correlated. If the window function on the specklegram is again represented by $W(x, y)$, then as the body rotates, some speckles rotate out of the window and some new speckles rotate into the window. The correlation is again given by Eq. (8-18).

A modification of this geometry, in which the illumination beam does not have to be collimated, was demonstrated by Gregory [74].

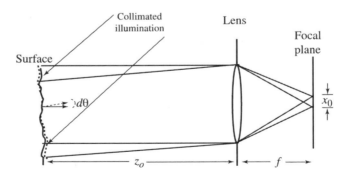

Figure 8.4 Geometry for tilt measurement using speckle photography. $d\theta$ is the angle of rotation of the speckle pattern.

8.2 Speckle Interferometry

While speckle photography generally involves the superposition of speckle pattern intensities, speckle interferometry generally involves the superposition of speckle pattern *amplitudes*. Systems for speckle interferometry have taken many forms since the early work of Leendertz [98]. In the earliest days of this field, the detector of choice for recording interferograms was photographic film. However, within a year of Leendertz's publication, two different groups reported different methods for performing speckle interferometry with electronic detectors [23], [104]. The primitive state of electronic detectors in the early 70s, as compared with their state today, limited the use of what became known as "electronic speckle pattern interferometry" (ESPI), but today the most capable speckle interferometers use electronic detection.

8.2.1 Systems That Use Photographic Detection

We begin our discussion of speckle interferometry with a description of the two-beam system introduced by Leendertz [98] for measuring in-plane displacements. We then turn attention other systems that use photographic detection.

Figure 8.5 shows the system of interest. Two mutually coherent collimated beams are simultaneously incident on a rough surface, inclined at angles $+\theta$ and $-\theta$ with respect to the average surface normal. A circular lens of diameter D images the illuminated region of the surface onto a photographic plate.

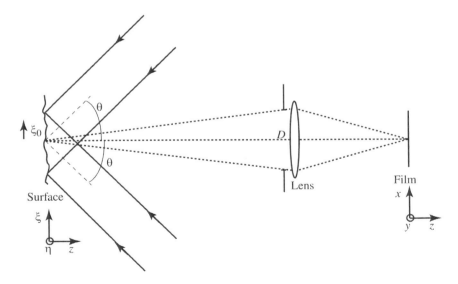

Figure 8.5 Rough surface illuminated by two coherent beams with equal but opposite inclinations. Two speckle patterns are added on an amplitude basis at the film plane, where they interfere.

If the the angular separation 2θ of the two illumination directions is greater than the angular subtense of the imaging lens, as viewed from the scattering surface, then the two superimposed speckle patterns are uncorrelated, a fact that is implied by Eq. (5-134). Nonetheless, unlike the case of two mutually incoherent illuminating beams, in this case the two speckle patterns add on an amplitude basis, yielding a single speckle pattern with full contrast. Note also that the two incident beams at the object interfere to form a fringe pattern, but the period of that fringe pattern is too small to be resolved by the optical system.

For simplicity, we assume that the magnification of the imaging system is unity and we neglect image inversion. If an element of the surface moves in the z direction there is no change of the interference pattern incident on the film plane, due to the fact that the two component speckle amplitudes are affected identically. The same is true of movement in the η direction. However, for small movement in the ξ direction by a distance $+\xi_0$ that is smaller than the size of a speckle, the illumination of a given point on the surface by the upper beam has travelled a shorter distance than previously by $\xi_0 \sin \theta$, while the illumination in the lower beam has travelled a longer distance by $\xi_0 \sin \theta$. At the image of this particular point, the phase of the interference pattern has changed by

$$\Delta\phi = \frac{4\pi}{\lambda} \xi_0 \sin \theta. \qquad (8\text{-}20)$$

There will then be a resulting change in the recorded speckle pattern in the region where this motion occurred.

Let one interference pattern be recorded on a photographic plate with the object in its unperturbed state. After the plate is developed, place the resulting negative back into the apparatus at the exact same location where it was recorded. Now observe the object through this photographic mask. If the object has not been deformed, bright speckles in the second pattern will lie at positions where the absorption of the plate is maximum, and therefore very little light will be transmitted. However, if the object is translated or distorted in the ξ direction, the location of the speckle maxima in the distorted region will move, with the result that more light will be transmitted by the mask in those areas. By observing regions of the image where there is significant light, the regions on the object where motion has taken place can be deduced.

A more quantitative understanding of this system can be obtained by considering the two intensity distributions associated with the speckled images of the object when it is in its undeformed state and the corresponding image for the deformed state. In the undeformed state the intensity is

$$I_1(x, y) = |\mathbf{A}_u(x, y) + \mathbf{A}_l(x, y)|^2 = |\mathbf{A}_u|^2 + |\mathbf{A}_l|^2 + 2|\mathbf{A}_u||\mathbf{A}_l| \cos(\phi_u - \phi_l), \qquad (8\text{-}21)$$

where $\mathbf{A}_u$ is the complex amplitude of the speckle incident on the film plane due to the upper illumination beam, $\mathbf{A}_l$ is the complex amplitude of the incident speckle due to the lower illumination beam, while ϕ_u and ϕ_l are the respective phases of these contributions. Note that all of these quantities are functions of (x, y). In the deformed state, the intensity is

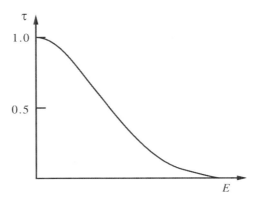

Figure 8.6 Typical intensity transmission-vs-exposure curve for a photographic plate.

$$I_2(x, y) = |\mathbf{A}_u(x, y) + \mathbf{A}_l(x, y)|^2$$
$$= |\mathbf{A}_u|^2 + |\mathbf{A}_l|^2 + 2|\mathbf{A}_u||\mathbf{A}_l|\cos(\phi_u - \phi_l - \Delta\phi). \quad (8\text{-}22)$$

The photographic plate that records the intensity distribution I_1 can be characterized by its intensity-transmission-vs-exposure curve, $\tau(E)$, where τ represents intensity transmission and E represents the exposure the plate was subjected to, with exposure being the product of incident intensity and exposure time. A typical $\tau(E)$ curve is shown in Fig. 8.6. If the exposure time T is chosen so that the *variations* of the intensity lie predominantly in a linear region near the center of this characteristic, then to a good approximation, when the intensity I_2 illuminates the developed transparency, the transmitted intensity will be

$$I_T(x, y) = I_2(x, y)\tau(I_1 T) \approx I_2(x, y)\tau_0 - KI_2(x, y)I_1(x, y), \quad (8\text{-}23)$$

where τ_0 is a bias point on the $\tau(E)$ curve and K is a constant that incorporates both the magnitude of the slope of the curve and the exposure time. The average transmitted intensity is seen to depend on the correlation of I_1 and I_2,

$$\overline{I_T(x, y)} = \overline{I_2(x, y)}\tau_0 - K\overline{I_1(x, y)I_2(x, y)}. \quad (8\text{-}24)$$

Thus the correlation of I_1 and I_2 is of interest. To find the locations of the maxima and minima of this correlation, we calculate the average of the squared difference $|I_1 - I_2|^2$. Before averaging we have

$$|I_2(x, y) - I_1(x, y)|^2 = 4I_u I_l |\cos(\phi_u - \phi_l - \Delta\phi) - \cos(\phi_u - \phi_l)|^2$$
$$= 16 I_u I_l \sin^2 \frac{\Delta\phi}{2} \sin^2\left(\phi_u - \phi_l - \frac{\Delta\phi}{2}\right), \quad (8\text{-}25)$$

where $I_u = |\mathbf{A}_u|^2$ and $I_l = |\mathbf{A}_l|^2$.

Now consider a small spatial region on the object within which the phase difference $\phi_u - \phi_l$ changes by several times 2π radians (a region that extends over several of the unresolved fringes on the object). We approximate this finite spatial average by an ensemble average assuming that I_u, I_l and $\phi_u - \phi_l$ are statistically independent and $\phi_u - \phi_l$ is uniformly distributed over $(-\pi, \pi)$. Thus

$$\overline{(I_2 - I_1)^2} = 16\overline{I_u I_l} \sin^2\left(\frac{\Delta\phi}{2}\right) \overline{\sin^2\left(\phi_u - \phi_l - \frac{\Delta\phi}{2}\right)}$$
$$= 8\bar{I}^2 \sin^2\left(\frac{\Delta\phi}{2}\right), \qquad (8\text{-}26)$$

where $\overline{I_u} = \overline{I_l} = \bar{I}$, and the average of the last $\sin^2$ term is $1/2$. As can be seen, the two patterns are most highly correlated when $\Delta\phi = n2\pi$, where n is an integer. Indeed, referring back to the original expressions for I_1 and I_2, we see that they are identical under this condition. Thus we will see a minimum transmitted intensity when the object displacement is

$$\xi_0 = \frac{n\lambda}{2\sin\theta}. \qquad (8\text{-}27)$$

Maximum transmitted intensity will occur when $\Delta\phi = (2n+1)\pi$, (n an integer) or when

$$\xi_0 = \frac{(2n+1)\lambda}{4\sin\theta}. \qquad (8\text{-}28)$$

Unfortunately the visibility of the fringes obtained by this photographic method is quite low, the maximum being a visibility of 1/3 (see [86], p. 149). The geometry using two beams equally inclined to the optical axis does have some advantages:

- The measurement is insensitive to out-of-plane motions. This allows the measurement of purely in-plane motion, without inaccuracies caused by out-of-plane motion.
- The sensitivity to in-plane motion is adjustable within bounds by symmetrically changing the angles $\pm\theta$ of the incident beams.

An alternative approach to obtaining in-plane displacement data with the same basic system described and analyzed above records I_1 and I_2 sequentially on the same photographic plate, with a small translation of the plate between exposures. The developed plate is then illuminated by a collimated beam, and the intensity of the Fourier spectrum is examined in the focal plane of a positive lens. The translation of the plate must be small, in order to assure that correlated speckles still overlap to a large degree. Typically the translation of the plate should be enough to cause about two fringes to be seen in the diffraction halo when the two speckle patterns are highly correlated. When transverse strain of the object has caused the speckle to be decorrelated in a region of the image, no fringes will be present in the contribution of that region to the spectrum. In general, both types of regions will be present at various locations on the object, and the Fourier plane intensity will consist of a superposition of an unmodulated diffraction halo and a modulated one. If a small aperture is placed at a minimum of the fringes, and an image is formed from the portion of the spectrum that is passed, the observer will see only those regions where the speckles are uncorrelated. Maximum brightness will occur where Eq. (8-28) holds.

In closing this subsection, we note that sensitivity to out-of-plane displacements, or combinations of in-plane and out-of-plane displacements, can be obtained if the two illuminating beams come from the same side of the optical axis and are properly spaced in angle. See [86], pp. 153–156, for further details.

8.2.2 Electronic Speckle Pattern Interferometry (ESPI)

The emergence of electronic detection in speckle interferometry began in 1971 [23] [104], and has accelerated since that time as detector technology has improved by leaps and bounds.[4] A review of the history of this subject, as well as an introduction to a wide variety of systems and applications, is found in a chapter by O.J. Løkberg in [146]. An additional authoritative reference is found in the contribution by Butters to Ref. [41]. Two different approaches are distinguished in this field. In one, the speckle field from a diffuse object under test is interfered with a uniform reference field, namely a plane or spherical reference wave. In the other, the reference field is itself a speckle field generated by a reference surface that also is a diffuse reflector. The reference surface can be the object under test, as in the case shown in Fig. 8.5, or a separate surface. Figure 8.7 shows examples of the two different types of speckle interferometers. In part (a) of the figure, the reference at the detector is a spherical wave that is coherent with the light imaged from the surface under test. In part (b) of the figure, the reference arises as diffuse reflection from a rough surface (in this case a different surface than the surface under test). In both cases, because the object is illuminated and observed normally, the phase changes induced by loading of the object are $\Delta\phi(x, y) = 4\pi \Delta z(x, y)/\lambda$, where $\Delta z(x, y)$ is the axial displacement of the object at each point (x, y). The two cases can be analyzed by a common mathematical framework, as will now be seen.

The two different intensity distributions, I_1 and I_2, are detected sequentially as separate frames of data, one with the object in its rest state and one with the object under stress. We represent the two intensities as

$$I_1(x, y) = I_r + I_o + 2\sqrt{I_r I_o} \cos(\phi_r - \phi_o)$$
$$I_2(x, y) = I_r + I_o + 2\sqrt{I_r I_o} \cos(\phi_r - \phi_o - \Delta\phi),$$
(8-29)

where I_r and I_o are the intensities of the reference and object waves at the detector, ϕ_r and ϕ_o are their respective phases, and $\Delta\phi$ (the (x, y) dependence of this phase

---

4. Early ESPI systems used analog video techniques, but most modern methods digitize the collected data. Sometimes the term "Digital Speckle Pattern Interferometry (DSPI)" is used to describe the latter techniques, but with the present use of computers being so ubiquitous, it seems unnecessary to distinguish between the two methods today.

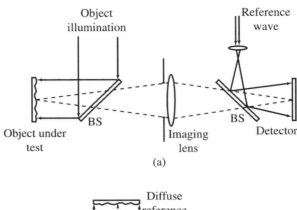
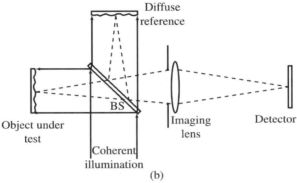

Figure 8.7 Two types of electronic speckle interferometry systems: (a) use of a spherical reference wave, and (b) use of a speckled reference wave. The elements labeled BS are beam splitters.

difference will be left implicit for simplicity) is the phase change induced in the object wave after the object has been subjected to a load. The two frames are typically subtracted electronically, and the difference image is then subjected to full-wave rectification, yielding (cf. Eq. (8-25))

$$|I_1 - I_2| = 4\sqrt{I_r I_o}\left|\sin\frac{\Delta\phi}{2}\right|\left|\sin\left(\phi_r - \phi_o - \frac{\Delta\phi}{2}\right)\right|. \tag{8-30}$$

In the case of a spherical reference wave, the difference is averaged with respect to I_o and ϕ_o, assuming they are statistically independent, that the intensity I_o obeys negative-exponential statistics, and that the phase ϕ_o is uniformly distributed on $(-\pi, \pi)$. Thus

$$\overline{|I_1 - I_2|} = 4\sqrt{I_r}\overline{\sqrt{I_o}}\left|\sin\frac{\Delta\phi}{2}\right|\overline{\left|\sin\left(\phi_r - \phi_o - \frac{\Delta\phi}{2}\right)\right|}. \tag{8-31}$$

Simple calculations show that $\overline{\sqrt{I_o}} = \sqrt{\pi \overline{I_o}}/2$ and
$$\overline{|\sin(\phi_r - \phi_o - \Delta\phi/2)|} = 2/\pi,$$
giving the result
$$\overline{|I_1 - I_2|} = \frac{4}{\sqrt{\pi}} \sqrt{I_r \overline{I_o}} \left|\sin \frac{\Delta\phi}{2}\right|. \quad (8\text{-}32)$$

Before interpreting this result, we consider the case of the diffuse reference, for which we must average with respect to I_r, I_o, ϕ_r, and ϕ_o, assuming all are statistically independent, I_r and I_o obey negative-exponential statistics, while ϕ_r and ϕ_o are uniformly distributed on $(-\pi, \pi)$. Thus the average to be calculated is

$$\overline{|I_1 - I_2|} = 4\overline{\sqrt{I_r I_o}} \left|\sin \frac{\Delta\phi}{2}\right| \overline{\left|\sin\left(\phi_r - \phi_o - \frac{\Delta\phi}{2}\right)\right|}. \quad (8\text{-}33)$$

Again with simple calculations we find $\overline{\sqrt{I_r I_o}} = (\pi/4)\sqrt{\overline{I_r} \overline{I_o}}$, and
$$\overline{|\sin(\phi_r - \phi_o - \Delta\phi/2)|} = 2/\pi,$$
yielding the result
$$\overline{|I_1 - I_2|} = 2\sqrt{\overline{I_r} \overline{I_o}} \left|\sin \frac{\Delta\phi}{2}\right|. \quad (8\text{-}34)$$

Thus we see that for both a spherical (or plane) reference wave and a speckled reference wave, the average of the magnitude of the difference of the two image frames is proportional to $|\sin(\Delta\phi/2)|$. The minimum value of this result (perfect correlation of I_1 and I_2) occurs again at $\Delta\phi = n2\pi$, while the maximum value (zero correlation) occurs again at $\Delta\phi = (2n+1)\pi$. Fringes of constant amounts of object displacement are thus seen on the display, modulating an image $\sqrt{\overline{I_o(x,y)}}$ of the object (assuming $\sqrt{I_r}$ or $\sqrt{\overline{I_r}}$ is constant). While we have calculated the average magnitude of the image difference and the resulting fringe pattern, it should be remembered that in any one result obtained by this method, the average will not have taken place, and the fringes will be speckled, making the determination of their exact peaks and nulls difficult. One of the beauties of this approach to interferometry is that the fringe data can be digitized, making it possible to apply various kinds of image processing algorithms to smooth the speckle, thus helping in the determination of fringe peaks and nulls somewhat.

The great advantage of electronic speckle pattern interferometry is the speed with which results can be obtained. Typically individual images can be obtained in 1/30 sec (a single TV frame time) and processed results can be displayed very shortly after two frames have been gathered. The availability of multi-megapixel detector arrays makes the complexity of the images that can be obtained competitive with photographic approaches, but with the added advantage of much greater speed.

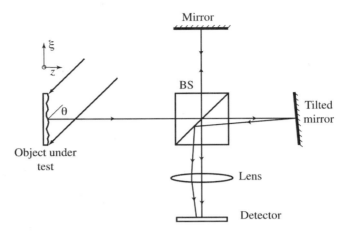

Figure 8.8 Speckle shearing interferometer using a Michelson interferometer with a tilted mirror. BS represents a cube beam splitter.

8.2.3 Speckle Shearing Interferometry

A common modification of the speckle interferometers discussed above is the *speckle shearing* interferometer. The number of different configurations used by various workers in the field is almost equal to the number of those workers, but we will illustrate with one simple example. For a more complete discussion of this subject, see the chapter by P.K. Rastogi in [128], in particular Section 3.5, and the book on the subject by W. Steinchen and L. Yang [152]. Figure 8.8 shows one such system. Light from a laser illuminates the rough surface under test. Light scattered from that surface enters the beam splitter BS, where half the light is transmitted and half is reflected. The reflected light passes to the top mirror, where it is again reflected. Half of this reflected light is transmitted downward through the beam splitter, through the imaging lens, to the detector, where an image of the object is formed. For the second path, the light transmitted through the beam splitter on the first pass travels to a *tilted* mirror, where it is reflected with an angular offset. Half of this returning light is reflected at the beam splitter and passes to the detector, where it arrives with a slight spatial offset due to the tilt of the mirror on the right.

The wave amplitude $\mathbf{A}_u(x, y)$ arriving at the detector due to the untilted mirror and and the amplitude $\mathbf{A}_t(x, y)$ arriving due to the tilted mirror interfere on the detector. The amplitude $\mathbf{A}_t(x, y)$ can be expressed in terms of the amplitude $\mathbf{A}_u(x, y)$ through

$$\mathbf{A}_t(x, y) = \mathbf{A}_u(x - \delta x) e^{-j\frac{2\pi}{\lambda}\sin(\delta\alpha)x} \tag{8-35}$$

where the shift or shear of the wave has been assumed to be in the x direction by amount δx and $\delta\alpha$ is the accompanying small angular shift. In what follows, we assume that $\delta\alpha$ is small enough to be ignored.

The intensity incident on the detector when the object is not deformed is given by

$$I_1(x, y) = |\mathbf{A}_u + \mathbf{A}_t|^2 = |\mathbf{A}_u(x, y) + \mathbf{A}_u(x - \delta x, y)|^2 = I_u(x, y) + I_u(x - \delta x, y)$$
$$+ 2\sqrt{I_u(x, y)I_u(x - \delta x, y)} \cos[\phi_u(x, y) - \phi_u(x - \delta x, y)], \qquad (8\text{-}36)$$

where $\phi_u(x, y)$ is the phase distribution of $\mathbf{A}_u(x, y)$. We now make the assumption that the shift δx is small compared with the average width of a single speckle, allowing the following approximations to be made:

$$I_u(x, y) \approx I_u(x - \delta x, y)$$
$$\phi_u(x, y) - \phi_u(x - \delta x, y) \approx \frac{\partial \phi_u}{\partial x} \delta x, \qquad (8\text{-}37)$$

yielding

$$I_1(x, y) \approx 2I_u(x, y)\left[1 + \cos\left(\frac{\partial \phi_u}{\partial x} \delta x\right)\right]. \qquad (8\text{-}38)$$

Suppose now that the object is subjected to a mechanical load, and a deformation occurs. The phase distribution associated with the image of the object is changed by amount $\Delta\phi(x, y)$, and the second intensity incident on the detector is

$$I_2 \approx 2I_u(x, y)[1 + \cos[\phi_u(x, y) - \Delta\phi(x, y) - \phi_u(x - \delta x, y) + \Delta\phi(x - \delta x, y)]]$$
$$\approx 2I_u(x, y)\left[1 + \cos\left(\frac{\partial \phi_u}{\partial x} \delta x - \frac{\partial \Delta\phi}{\partial x} \delta x\right)\right]. \qquad (8\text{-}39)$$

Again two frames are detected, one recorded when the object is not under load and one when the object is under load. The average of the magnitude of this difference is

$$\overline{|I_1(x, y) - I_2(x, y)|} = 2\overline{I}_u(x, y)\left|\sin\left(\frac{1}{2}\frac{\partial \Delta\phi(x, y)}{\partial x}\delta x\right)\right|, \qquad (8\text{-}40)$$

where it has been assumed that I_u and ϕ_u are statistically independent random variables and that $\frac{\partial \phi_u}{\partial x}\delta x$ is uniformly distributed on $(-\pi, \pi)$. Thus we see that the average of the magnitude of the frame difference depends on the *slope* of $\Delta\phi$ in the x direction. By tilting the mirror in a direction different than the x-direction, the slope in that direction will replace the slope with respect to x. A zero of Eq. (8-40) occurs, and we see a fringe null when

$$\frac{\partial \Delta\phi}{\partial x} = \frac{n2\pi}{\delta x}, \qquad (8\text{-}41)$$

and a fringe peak is seen when

$$\frac{\partial \Delta\phi}{\partial x} = \frac{(2n + 1)\pi}{\delta x}. \qquad (8\text{-}42)$$

Finally we ask how the change $\Delta\phi$ is related to changes of the surface position in the ξ and z directions on the object. If we assume unit magnification for simplicity, and ignore image inversion, consideration of the geometry shows that

$$\frac{\partial \Delta\phi}{\partial x}\delta x = \frac{2\pi\delta x}{\lambda}\left[\sin\theta\frac{\partial x_0}{\partial x} + (1+\cos\theta)\frac{\partial z_0}{\partial x}\right], \qquad (8\text{-}43)$$

where x_0 and z_0 are the amounts by which the object moved at the image plane in the x direction and the z direction, respectively, and θ is the angle of illumination of the surface.[5] As can be seen, sensitivity only to out-of-plane movements can be assured if $\theta = 0$, that is, if the object is illuminated normally. In this case, fringe nulls are seen wherever

$$\frac{\partial z_0}{\partial x} = \frac{n\lambda}{2\delta x}, \qquad (8\text{-}44)$$

and a fringe peak wherever

$$\frac{\partial z_0}{\partial x} = \frac{(n+1/2)\lambda}{2\delta x}. \qquad (8\text{-}45)$$

An enormous variety of speckle shearing systems are known, some specialized to particular applications, others more general. It is impossible to do justice to this subject in this short overview, so the reader is referred to the references cited at the beginning of this section for a fuller treatment of the subject.

8.3 From Fringe Patterns to Phase Maps

In previous sections of our treatment of speckle metrology, we have seen how, through detection of two interferometrically generated speckle patterns, one for the object in its undisturbed state and one for the object in its changed state, and subtraction of those detected patterns, a fringe pattern can be obtained that contains information on how the object has changed between detected frames. To determine quantitative information about the changes of the object, it is necessary to form contour plots of the resulting fringe patterns and to locate the peaks and the nulls. Such detection is hindered by the fact that the fringes are usually modulated by speckle, making the exact location of the peaks and nulls difficult to determine.

In this section we consider other methods for determining a phase map implied by the fringes, either by suitable postdetection processing, or by a combination of a modified interferometric approach and post-detection processing. We then turn attention briefly to the problem of phase unwrapping.

---

5. This result assumed that the angle subtended by the lens, as viewed from the object, is not too large.

8.3.1 The Fourier Transform Method

In 1982, Takeda, Ina and Kobayashi [158] published a Fourier transform–based method for determining a phase map from recorded fringe patterns. The method is quite useful for interferometry in general, but in particular it can be applied to speckle interferometry. For this approach to be applicable, it is necessary that there be a carrier frequency term in the fringe pattern, that is, the detected fringe pattern of interest must be of the form

$$f(x, y) = a(x, y) + b(x, y)\cos[2\pi f_0 x + \phi(x, y)]$$
$$= a(x, y) + \mathbf{c}(x, y)e^{j2\pi f_0 x} + \mathbf{c}^*(x, y)e^{-j2\pi f_0 x}, \quad (8\text{-}46)$$

where

$$\mathbf{c}(x, y) = \frac{b(x, y)}{2}e^{j\phi(x,y)}.$$

A fringe of this form can be obtained in speckle interferometry if, for example, the reference wave in the first exposure that records I_1 is tilted by a sufficiently large angle, and not tilted in the second exposure that records I_2. If the difference between I_1 and I_2 is squared and the average taken, the resulting fringe can be shown to be

$$\overline{(I_1 - I_2)^2} = 4\overline{I_r I_o}[1 + \cos(2\pi f_0 x + \Delta\phi)], \quad (8\text{-}47)$$

where f_0 is the carrier frequency, given by $f_0 = \sin\alpha/\lambda$ where α is the angular shift imparted to the reference wave. It is also possible to generate such fringe patterns from a single interferogram [52].

A one-dimensional Fourier transform of Eq. (8-46) with respect to x yields

$$\mathbf{F}(v_X, y) = \mathbf{A}(v_X, y) + \mathbf{C}(v_X - f_0, y) + \mathbf{C}^*(-(v_X + f_0), y), \quad (8\text{-}48)$$

where capital letters represent the one-dimensional Fourier transform of the lower-case quantities. Figure 8.9(a) shows a typical plot of the magnitude of this spectrum for a fixed y. The first processing step is to select the upper sideband centered at frequency f_0 and translate it down to be centered at the origin, eliminating both the central part of the spectrum of $\mathbf{F}$ and the lower sideband in the process. The resulting spectrum is shown in Fig. 8.9(b).

Having isolated the Fourier spectrum $\mathbf{C}(v_X, y)$, it is now possible to inverse Fourier transform it with respect to v_X to yield the complex function $\mathbf{c}(x, y)$. The next step in the process is to numerically calculate the complex logarithm of $\mathbf{c}$,

$$\ln[\mathbf{c}(x, y)] = \ln[b(x, y)/2] + j\phi(x, y), \quad (8\text{-}49)$$

from which we see that the imaginary part of the complex logarithm is the phase distribution $\phi(x, y)$. The phase obtained in this way is "wrapped" into the interval $(-\pi, \pi)$. In a later subsection we will discuss methods for unwrapping the phase. For the moment it is clear that by following the prescription outlined above, it is possible to arrive at a wrapped version of the phase distribution ϕ. Once the phase is

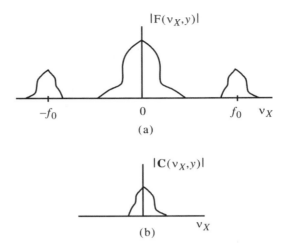

Figure 8.9 One-dimensional Fourier spectra of (a) the fringe pattern, and (b) a translated single sideband. The figure is for a fixed value of y.

unwrapped, it is then usually possible to arrive at a map of the surface deformation that gave rise to the phase shift.

There are distinct advantages to finding the phase in this manner, rather than trying to find fringe contours: (1) as we shall see when we consider phase unwrapping, it is possible to determine the sign of the phase change, as well as its magnitude, and (2) the variations of $a(x, y)$ and $b(x, y)$, which can limit the accuracy of finding of fringe contours, have been eliminated by this process. This approach to phase extraction is generally considerably more accurate than methods based on finding phase contours.

8.3.2 Phase-Shifting Speckle Interferometry

A method for achieving extremely high accuracy in electronic speckle pattern interferometry was introduced by K. Creath in 1985 [28]. This method, termed *phase-shifting speckle interferometry*, records several interferograms with a reference wave that is stepped in phase between recordings. From this set of interferograms the phase distribution associated with the changes of an object can be calculated modulo 2π radians, following which phase unwrapping can again be used.

In this method of speckle interferometry, the light from the diffuse object under test is reflected off of a mirror before it interferes with the reference, and the mirror is capable of being moved with great precision by means of a piezoelectric transducer (PZT), introducing a component of phase term change that depends on the phase change introduced by the mirror, in addition to the phase change $\Delta\phi$ caused by the deformation of the object under test. Again we represent the two different interferometric exposures gathered in conventional speckle interferometry by I_1 and I_2, and consider the average of the square of their difference, finding (cf. Eq. (8-26))

$$\overline{(I_1 - I_2)^2} = 8\overline{I_1}\overline{I_2}\sin^2(\Delta\phi/2) = 4\overline{I_1}\overline{I_2}(1 - \cos\Delta\phi). \tag{8-50}$$

Now consider the collection of four pairs of different interferograms of the type written above, but with the PZT moving through four different phase shifts, $-3\pi/4$, $-\pi/4$, $+\pi/4$ and $+3\pi/4$. The four interferograms are

$$\begin{aligned} A(x, y) &= 4\overline{I_1}\overline{I_2}(1 - \cos(\Delta\phi - 3\pi/4)) \\ B(x, y) &= 4\overline{I_1}\overline{I_2}(1 - \cos(\Delta\phi - \pi/4)) \\ C(x, y) &= 4\overline{I_1}\overline{I_2}(1 - \cos(\Delta\phi + \pi/4)) \\ D(x, y) &= 4\overline{I_1}\overline{I_2}(1 - \cos(\Delta\phi + 3\pi/4)). \end{aligned} \tag{8-51}$$

There are several ways these quantities can be combined to recover $\Delta\phi$. We choose to let these four quantities be combined according to

$$\psi = \arctan\left(\frac{C + D - A - B}{A + D - B - C}\right). \tag{8-52}$$

Mathematica finds the value of ψ to be

$$\psi = \arctan\left(\frac{\sin\Delta\phi}{\cos\Delta\phi}\right) = \Delta\phi \tag{8-53}$$

on the interval $(-\pi, \pi)$, and the phase is recovered modulo 2π. Other choices of phase increments for the reference can be used as well. See [152], Chapter 4, and [155] for more details. There are a number of practical issues associated with these methods that we have not covered here. See [28], as well as the references just cited, for a more detailed discussion. Very accurate measurements of $\Delta\phi$ are possible with phase-shifting techniques.

A method has recently been reported [114] for collecting all four phase-shifted interferograms in parallel in less than 10 nsec using a pulsed laser and a phase plate that introduces all of the necessary phase shifts simultaneously.

8.3.3 Phase Unwrapping

The phase unwrapping problem is an important one in optics in general, and especially important in optical interferometry. For references on this general subject, see the book on this subject by Ghiglia and Pritt [60], a review paper by Takeda [157], as well as review articles by Robinson [134] and by Judge and Bryanston-Cross [88]. We assume that we are given a two-dimensional wrapped or folded phase map $\mathcal{W}\{\phi(x, y)\}$, where $\mathcal{W}\{\ \}$ is the wrapping operator that wrapped all values of phase into the interval $(-\pi, \pi)$. Our goal is to unwrap the phase values to obtain $\phi(x, y)$, which should be proportional to the surface deformations that took place in the interferometric measurement.

We begin by describing the simplest of phase unwrapping methods, and then explain why it often will not work. We then turn to a brief discussion of a method to remove the problems encountered with this simple method. The simplest approach is to take the sampled values of $\phi(x, y)$, which we assume exist on a two-dimensional array (y-samples in the vertical dimension and x-samples in the horizontal dimension), and to examine adjacent samples, initially in the y-direction for a fixed value of x. Each phase sample (except the first) is subtracted from the phase sample that preceded it and if the magnitude of the difference exceeds some predetermined large fraction of 2π, say $0.9 \times 2\pi$, then 2π is added to or subtracted from the current phase value, depending on whether the previous difference of phase values is positive or negative (positive slope or negative slope). Having performed the unwrap along one vertical column, say the column furthest to the left, the unwrapped phases so obtained are used as starting points to unwrap the phases along all the rows of the phase matrix, one-by-one. A complete array of unwrapped phase values would be obtained in the ideal case.

The procedure described is of course limited by noise in the data, but this is not the most serious limitation. The chief difficulty comes from the fact that the phase map that results can be dependent on the particular paths taken in the (x, y) plane during the unwrapping procedure. For example, to find the unwrapped phase at some location (x_n, y_m) in the array, the procedure described started at the upper left corner of the array, moved down the first column of data to the particular row of interest, and then moved across the row to the location where the unwrapped phase is desired. Such a path is shown by the solid line in Fig. 8.10. Would we have obtained the same unwrapped phase if we had proceeded along the dotted path, rather than the solid path? If the object is a smooth and gradually changing reflective surface, then the answer is "yes," and the procedure can work very well, as it did in the example used by Takeda et al. [158]. On the other hand, if the wavefront under test results

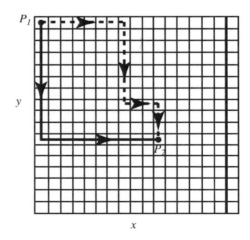

Figure 8.10 Two different paths for unwrapping phase between points P_1 and P_2.

from a field distribution that has zeros of intensity, the phase vortices at these zeros can cause serious problems (see Section 4.8 for a discussion of phase vortices). The two paths illustrated in Fig. 8.10, if taken together, constitute one closed path in the (x, y) plane. If that closed path encloses a single phase vortex, then the integral (or sum) of the phases around that path will be $\pm 2\pi$ radians, with the sign depending on whether the vortex was positive or negative, as well as the direction taken around the closed path. If the path encircles both a positive and a negative vortex, the effects of the two vortices cancel, and the integral around the closed path yields zero, meaning that the two paths yield the same result when used for phase unwrapping. The requirement for completely unambiguous phase unwrapping is that the sum of the phase differences around any closed path is zero. The continuous equivalent of this condition is that the integral of the phase gradient $\nabla\phi(\vec{r})$ around any closed path in the (x, y) plane is zero:

$$\oint_C \nabla\phi(\vec{r})\, d\vec{r} = 0, \tag{8-54}$$

where $\vec{r} = (x, y)$. Any field distribution that is a speckle pattern is known to have approximately the same number of phase vortices as there are speckles, and unwrapping the phase of such a field can be extremely difficult.

Detecting the presence and locations of vortices is a central part of any phase-unwrapping strategy. In the discrete case, a vortex map can be found by summing phase differences around every local 2×2 pixel square in the pixel array.[6] When any one of these sums yields $\pm 2\pi$ (or a multiple of $\pm 2\pi$ in the case of very rare higher-order vortices), a vortex is present within that square.

A multitude of approaches for avoiding the effects of vortices are known, but we will not review them all here. Rather, the reader is referred to the references above for a more complete discussion of this rather fascinating mathematical problem. We note in closing that the problem of unwrapping phase is essentially the same as the problem of determining a phase distribution from measurements of an array of phase differences. This problem has been studied extensively in the field of adaptive optics. See for example Ref. [156].

8.4 Vibration Measurement Using Speckle

A variety of speckle-related techniques exist for measuring vibration amplitudes and/or modes for rough surfaces that reflect diffusely. Early works by Archibold, Ennos and Taylor [5] and by Tiziani [162] are notable. Many approaches depend on reduction of speckle contrast in regions where the vibration is most severe, while regions that are relatively stationary retain high speckle contrast.

---

6. M. Takeda, private communication.

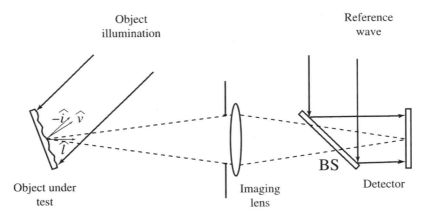

Figure 8.11 Imaging geometry for vibration analysis.

We consider one illustrative approach here [3]. An imaging geometry is used, as shown in Fig. 8.11. Light from a laser illuminates the rough reflecting object under test. A plane reference wave is added to the field incident on the detector. An image of the object is formed, where it can be detected. The unit vector $\hat{i}$ represents the direction of the incident object illumination, the unit vector $\hat{l}$ represents the direction normal to the lens aperture, and the unit vector $\hat{v}$ represents the direction of vibration of the object (up to a sign). The total intensity at the detector is

$$I_D(x, y; t) = I_r + I_o(x, y) + 2\sqrt{I_r I_o(x, y)} \cos \phi(x, y; t), \tag{8-55}$$

where I_r is the constant intensity of the reference wave, $I_o(x, y)$ is the time invariant, speckled intensity of the object wave, and $\phi(x, y; t)$ is the phase difference between the reference and object waves at each point (x, y) on the detector. Let the object undergo vibration in the direction $\pm \hat{v}$ of the form $a(x, y) \sin \omega t$, where $a(x, y)$ is the peak amplitude of the vibration at (x, y), while ω is its angular frequency. Let α represent the angle between $-\hat{i}$ and $\hat{v}$, while β represents the angle between $\hat{v}$ and $\hat{l}$. Then

$$\phi(x, y; t) = \phi_0(x, y) + \frac{2\pi}{\lambda}(\cos \alpha + \cos \beta) a(x, y) \sin \omega t \tag{8-56}$$

where $\phi_0(x, y)$ is a spatially varying but time-independent phase difference between the reference wave and the object wave.

Let the detector integrate the incident intensity for a time T that is long compared with the vibration period $2\pi/\omega$. Then

$$E(x, y) = \int_0^T I_D(x, y; t)\, dt$$

$$= TI_r + TI_o(x, y) + 2\sqrt{I_r I_o(x, y)} \int_0^T \cos \phi(x, y; t)\, dt. \tag{8-57}$$

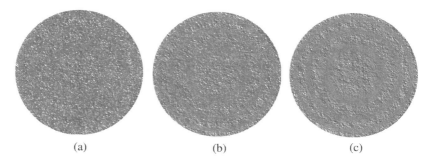

Figure 8.12 Simulated fringe patterns for an object with a parabolic vibration mode, with a maximum of approximately 3 waves peak-to-peak motion at the center, tapering to zero waves motion at the edges. Reference to mean object intensity ratios are (a) 32, (b) 128 and (c) 640. Considerable bias has been suppressed to enhance the contrast. The fringes are least visible in (a) and most visible in (c).

For $T \gg 2\pi/\omega$, the last integral yields, to a good approximation,

$$\int_0^T \cos \phi(x, y; t)\, dt \approx T \cos \phi_0(x, y) J_0\left(\frac{2\pi}{\lambda}(\cos \alpha + \cos \beta) a(x, y)\right), \quad (8\text{-}58)$$

where $J_0(x)$ is a Bessel function of the first kind, order zero. The result is exact if the integration is over an integer number of periods, and a very good approximation otherwise; then

$$E(x, y) \approx T\left[I_r + I_o(x, y) + 2\sqrt{I_r I_o(x, y)} \cos \phi_0(x, y) J_0\left(\frac{2\pi}{\lambda}(\cos \alpha + \cos \beta) a(x, y)\right)\right]. \quad (8\text{-}59)$$

Equation (8-59) is a fringe pattern containing speckle. The fringes are not sinusoidal, but rather follow the envelope of $J_0\left(\frac{2\pi}{\lambda}(\cos \alpha + \cos \beta)a(x, y)\right)$, thus allowing the vibration amplitude $a(x, y)$ to be determined at each point (x, y) on the object under study. The fringe pattern contains speckle by virtue of the term $\sqrt{I_o(x, y)} \cos \phi_0(x, y)$, which is the real part of the complex speckle amplitude, but it is the speckle associated with the interference of a strong constant phasor (the reference wave) with a weaker circular complex Gaussian phasor (the object wave), rather than fully developed speckle with a contrast of unity.

Figure 8.12 shows the results of a simulation of the above method for three different ratios of reference intensity to average object intensity, $I_r/\overline{I_o}$, and for a peak-to-peak vibration amplitude that varies quadratically from three waves of path length difference at the center to zero waves at the edge. The illumination direction, the observation direction and the vibration direction have all been chosen to be normal to the object. The contrast of the fringes is actually much lower than is shown in the figures, considerable bias being suppressed to show the structure of the fringes. The fringes are seen to contain speckle, and their contrast is very low, especially for the smaller beam ratios. When the beam ratio is too small, the speckle from the second term in the expression for $E(x, y)$ begins to dominate and hide the fringe, while when the beam ratio is too high, the contrast of the fringes becomes too low to be easily seen.

The contrast of the fringes can be much enhanced if two such images are taken, with the phase of the reference wave changed by 180 degrees between exposures while the object wave is unchanged (except for vibration), and those images are subtracted. The second exposure is then

$$\tilde{E}(x, y) \approx T \left[I_r + I_o(x, y) \right.$$
$$\left. - 2\sqrt{I_r I_o(x, y)} \cos \phi_0(x, y) J_0 \left(\frac{2\pi}{\lambda} (\cos \alpha + \cos \beta) a(x, y) \right) \right], \quad (8\text{-}60)$$

and the squared difference of the two exposures is then

$$(E(x, y) - \tilde{E}(x, y))^2 \approx 16 T^2 I_r I_o(x, y) \cos^2 \phi_0(x, y)$$
$$\times J_0^2 \left(\frac{2\pi}{\lambda} (\cos \alpha + \cos \beta) a(x, y) \right). \quad (8\text{-}61)$$

Figure 8.13 shows the results corresponding to parts (a), (b) and (c) of Fig. 8.12 when the square of the fringe difference is calculated. The contrast of the fringes is good in all cases. However, the fact that the fringe patterns appear independent of the beam ratio in this simulation masks an important effect that occurs in a real experiment. Each of the detected fringe patterns will contain measurement noise, as well as small changes in I_r and $\bar{I}_o$. For very large contrast ratios, the subtraction of the background will be imperfect, and for sufficiently large beam ratio, the fringes will be-

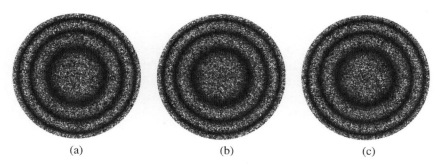

Figure 8.13 Fringes obtained from the square of the exposure difference, under the same conditions as Fig. 8.12. The fringe patterns are the same, independent of the beam ratio.

come masked by noise and residual errors that have not been cancelled by subtraction. Nonetheless, with careful experimentation and a proper choice of beam ratio, not too large and not too small, good results can be obtained.

Many other geometries and techniques exist for measurement of vibrations using speckle. For example, rather than subtracting a phase-shifted frame from the original frame, it is possible to high-pass filter and full-wave rectify the first frame to obtain more visible fringes. Alternatively, a more complex phase-shifting method can combine more than two frames to obtain high-contrast fringes. An excellent review article has been written by Tiziani ([41], Chapter 5).

8.5 Speckle and Surface Roughness Measurements

A final application of speckle in metrology to be treated here is to the measurement of surface roughness. For excellent treatments of this subject, see the chapter by Asakura in [41] and the chapter by Briers in [146].

Two different aspects of surface roughness are generally of special interest:

1. The rms value of the surface-height variations, and
2. The covariance function of surface height, which reveals the correlation length of the surface-height variations.

Various techniques for measuring both of these properties will be treated. However, before embarking on a discussion of such techniques, a cautionary note should be sounded. Since the speckle of interest here is formed by the interference of optical waves, *any attempt to measure surface roughness through speckle is really measuring properties of the optical wave reflected by the surface under test, rather than the properties of the surface itself.* If the correlation length of the surface is shorter than the wavelength of the light being used, the optical phase variations of the reflected wave can not follow the undulations of the surface, since the wave can not change significantly in distances shorter than an optical wavelength. Thus a major problem in obtaining meaningful results is relating the reflected wave fluctuations to the surface fluctuations. An excellent reference for this subject is the book by Beckmann and Spizzichino [16]. A further problem arises in choosing a model for the statistical fluctuations of surface height, for which most commonly a Gaussian assumption is made, purely for analytical convenience.

With these cautions in mind, we assume that the surface-height variations are gradual enough that the phase of the reflected wave is accurately proportional to the surface-height at each point on the surface.

In what follows we review several different ways to measure surface roughness and correlation length.

8.5.1 RMS Surface Height and Surface Covariance Area from Speckle Contrast

We first consider a measurement technique pioneered by Asakura and his group (see, for example, [120] and [54]) for deducing surface roughness and surface-height correlation area from speckle contrast when the scattering surface is relatively smooth. In the cases to be considered, the measurement geometry is as illustrated previously in Fig. 4.20.

This imaging case has already been analyzed in Section 4.5.5, and the dependence of speckle contrast C on the surface-height standard deviation σ_h and the number N_0 of surface correlation areas covered by the imaging system weighting function in the object plane has been plotted in Fig. 4.21(a). This plot assumes that the surface height fluctuations obey Gaussian statistics, and that the covariance function of surface height has a circularly symmetric Gaussian shape. From this result we can see that if N_0 is known or can be accurately estimated, then a measurement of contrast yields a unique value for the normalized surface-height standard deviation σ_h/λ, provided $\sigma_h \ll \lambda$. When σ_h becomes comparable with λ, the curves saturate at a value of unity, as they should for fully developed speckle, and it is no longer possible to deduce a value of σ_h from the contrast.

By the same token, for small surface-height fluctuations, Fig. 4.21(b) suggests a way to measure surface-height correlation radius r_c if σ_h is known or can be estimated. The value of N_0 can be changed by opening or closing the pupil in the imaging system. Opening the pupil reduces the number of surface correlation areas covered by the point-spread function and closing the pupil increases that number. The number N_0 is given explicitly by

$$N_0 = \frac{\lambda^2 f^2}{\pi r_c^2 A_p}, \qquad (8\text{-}62)$$

where A_p is the area of the pupil in the imaging system. Each curve of contrast vs. N_0 has a maximum, and the value of N_0 that achieves the maximum contrast can be determined from Eqs. (4-118) and (4-117). In this way, a known value of σ_h can lead to an estimated value of the surface correlation radius r_c.

In this imaging geometry, the diffusely scattered component of light does not obey circular statistics. Rather, the phasor component of light normal to the phasor direction of the specular component is stronger than the component parallel to the direction of the specular phasor. If the measurement geometry is changed so that the plane where contrast is measured is out of the image plane, then the diffusely scattered light becomes statistically circular, and the contrast of the speckle is somewhat larger. See [126] and [68] for more details.

The techniques described above are accurate for rms surface roughness in the range 0.02 to 0.3 wavelengths ([146], Chapter 8, p. 393).

8.5.2 RMS Surface Height from Two-Wavelength Decorrelation

Consider now a free-space geometry, in which the rough surface is illuminated normally and speckle intensity is measured in a plane that is parallel to the scattering surface. We have seen already in Section 5.3.1 that the normalized correlation of two amplitude speckle patterns recorded at wavelengths λ_1 and λ_2 is (cf. Eq. (5-42))

$$\mu_A(\vec{q}_1,\vec{q}_2) \approx \mathbf{M}_h\left(4\pi \frac{|\lambda_2-\lambda_1|}{\overline{\lambda_1\lambda_2}}\right), \tag{8-63}$$

where $\mathbf{M}_h(\omega)$ is the characteristic function of the surface-height fluctuations. Since the speckle amplitudes are circular complex Gaussian random variates, it follows that the intensity correlation function is

$$|\mu_A|^2 = \left|\mathbf{M}_h\left(4\pi \frac{|\lambda_2-\lambda_1|}{\overline{\lambda_1\lambda_2}}\right)\right|^2 \approx \left|\mathbf{M}_h\left(4\pi \frac{|\Delta\lambda|}{\bar{\lambda}^2}\right)\right|^2. \tag{8-64}$$

If the surface roughness obeys Gaussian statistics, then this result takes the form of Eq. (5-46),

$$|\mu_A|^2 = \exp(-\sigma_h^2 |q_z|^2) = \exp\left[-\left(2\pi \frac{\sigma_h}{\bar{\lambda}} \frac{|\Delta\lambda|}{\bar{\lambda}}\right)^2\right]. \tag{8-65}$$

Thus the measurement system would detect one speckle intensity pattern with illumination wavelength λ_1 and a second with wavelength λ_2, and the two speckle intensity patterns would be correlated. The value of that correlation establishes a curve on Fig. 5.10. Knowing the wavelength change that leads to this correlation, it is then possible from the cited figure to find the value of $\sigma_h/\bar{\lambda}$ that must have caused this amount of decorrelation. In this way the surface roughness is obtained.

Surface roughness standard deviations in the range from a few tenths of a micron to a few microns can be measured this way, depending on the wavelength separations that are possible. Note that for rough surfaces, the wavelength spacing should be small while for smooth surfaces, larger wavelength separations are needed, since the correlation depends on the product of $\sigma_h/\bar{\lambda}$ and $|\Delta\lambda|/\bar{\lambda}$.

Note that the correlation operation referred to above can be performed using a double exposure technique such as described in earlier sections of this chapter, or digitally.

An extension of these ideas is to illuminate light with a finite bandwidth [150], [37] and measure the contrast of the speckle, again in a free-space geometry. When light with a Gaussian spectrum is used and the surface-height fluctuations obey Gaussian statistics, a measurement of the contrast of the resulting speckle pattern, together with the results shown in Fig. 5.18, can be used to determine surface roughness.

8.5.3 RMS Surface Height from Two-Angle Decorrelation

Just as the correlation of two speckle patterns taken with different wavelengths can yield information about surface roughness, so too can the correlation of two speckle patterns taken with the same wavelength but with different angles of illumination. The sensitivity to angle changes has been evaluated in the subsection starting on page 162. If the translational effects of changing angle are removed by shifting one pattern with respect to the other by the proper amount, the intensity correlation becomes the result shown in Eq. (5-60), which we repeat here for the reader's convenience and which has been plotted in Fig. 5.13:

$$|\mu_A|^2 \approx \exp\left[-\left(2\pi\frac{\sigma_h}{\lambda}\Delta\theta_i \sin\theta_i\right)^2\right], \tag{8-66}$$

where σ_h is the standard deviation of the surface-height fluctuations, θ_i is the initial angle of illumination, $\Delta\theta_i$ is the change in angle of illumination, λ is the wavelength, and it has been assumed that the observation direction is normal to the surface. The shift of the second pattern required is $\Delta\theta_i z$ in the direction mirror opposite to the direction of the illumination shift, where z is the distance from the surface to the observation plane. Thus, given a measured intensity correlation between the properly shifted patterns, and knowing the change in angle that was made between measurements, an estimate of the normalized surface roughness can be made.

Generalizations from this result can be made when a single exposure with an extended source is used. The reduction of the contrast of the speckle can be related to the surface roughness [125]. See Section 5.4.4 for relevant results.

8.5.4 Surface-Height Standard Deviation and Covariance Function from Measurement of Angular Power Spectrum

The generalized van Cittert–Zernike theorem of Section 4.5.2 provides us with a connection between the normalized covariance function μ_a of the fields leaving the scattering surface and the broad, average intensity distribution $\bar{I}$ observed in the focal plane of a positive lens. In particular,

$$\overline{I(x,y)} \propto \iint\limits_{-\infty}^{\infty} \mu_a(\Delta\alpha, \Delta\beta) e^{-j\frac{2\pi}{\lambda f}(x\Delta\alpha + y\Delta\beta)} d\Delta\alpha\, d\Delta\beta, \tag{8-67}$$

where f is the focal length of the positive lens being used. Thus if $\overline{I(x,y)}$ is carefully measured, using a detector that is large enough to average over many individual speckles, but at the same time small enough so that the structure of $\overline{I(x,y)}$ is not smoothed, then a properly normalized Fourier transform of the result should yield the field autocovariance function μ_a at the scattering surface.

The problem remains to relate μ_a to the covariance function μ_h of the scattering surface. To make headway on this problem, two assumptions also made in previous subsections are necessary:

1. The field leaving the surface is of the form $\mathbf{a}(\alpha,\beta) = e^{j\frac{4\pi}{\lambda}h(\alpha,\beta)}$, where h is the surface height (assumes normal illumination and observation directions), and
2. The surface height fluctuations $h(\alpha,\beta)$ obey Gaussian statistics and are a spatially stationary random process.

With these assumptions it follows (see Eq. (4-103)) that

$$\mu_a(\Delta\alpha, \Delta\beta) = \exp\left[-\left(\frac{4\pi}{\lambda}\sigma_h\right)^2(1 - \mu_h(\Delta\alpha, \Delta\beta))\right], \tag{8-68}$$

where σ_h is the standard deviation of the surface height fluctuations.

The normalized field correlation μ_a has an asymptotic value for large arguments,

$$\sqrt{\Delta\alpha^2 + \Delta\beta^2} \to \infty \Rightarrow \mu_a \to \exp\left[-\left(\frac{4\pi}{\lambda}\sigma_h\right)^2\right]. \tag{8-69}$$

This asymptotic value is caused by specular reflection and has significant value only for relatively smooth surfaces. For such surfaces, since we know μ_a, by determining its value at large arguments we can deduce the standard deviation of surface height fluctuations through

$$\sigma_h = \frac{\lambda}{4\pi}\sqrt{\ln\left(\frac{1}{\mu_a}\right)}, \tag{8-70}$$

where in this equation μ_a is understood to be the real-valued asymptotic value at large arguments. Now, knowing the value of σ_h, we can return to Eq. (8-68) and solve for μ_h,

$$\mu_h(\Delta\alpha, \Delta\beta) = 1 - \left(\frac{\lambda}{4\pi\sigma_h}\right)^2 \ln\left(\frac{1}{\mu_a}\right). \tag{8-71}$$

Thus in principle, for relatively smooth surfaces and from measurements such as we have described, both the standard deviation and the covariance function of surface-height fluctuations can be found. In practice, such an approach is limited both by the possible lack of accuracy of the assumption of Gaussian statistics for the surface-height fluctuations and by measurement noise encountered in detecting $\overline{I(x, y)}$. In addition, for surfaces that are rough on the scale of a wavelength, it is impossible to measure the asymptotic value of μ_a. As a consequence, under such conditions the standard deviation σ_h and the covariance function μ_h can not be determined separately.

9 Speckle in Imaging Through the Atmosphere

9.1 Background

Emphasis in the previous chapters has been on speckle that arises either from scattering from rough surfaces or scattering from diffuse volumes. In this chapter, attention is devoted to a different kind of speckle, namely intensity fluctuations in the image of a distant incoherent object when the light from that object has propagated through random inhomogeneities associated with atmospheric turbulence. When the exposure or detection time is short compared with the fluctuation time of those inhomogeneities, atmospherically induced speckle is observed in the image. An explanation of the origin of this phenomenon will be given in the following subsections.

Unlike discussions of speckle in previous chapters, which assumed reasonably monochromatic light, the speckle of interest here can occur in relatively broadband light, although if the bandwidth is too large the speckle can begin to wash out. In addition, since propagation through atmospheric turbulence does not change the polarization of the light [154], the speckle of interest here will be independent of polarization.

The most widely quoted and influential text on propagation through the atmosphere is that of Tatarski [159]. For additional references, see [80] (Volume 2, Part IV), [137], [135], and [70], Chapter 8.

9.1.1 Refractive Index Fluctuations in the Atmosphere

Figure 9.1 illustrates a general case of image formation in the presence of atmospheric turbulence. The object on the left is assumed to be spatially incoherent, as are most naturally occurring objects, whether self-luminous or illuminated by sunlight. The gray "blobs" between the object and the lens are an idealized representation of regions of refractive index inhomogeneities or departures from the refractive

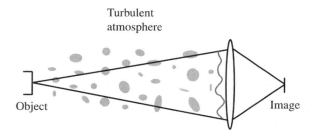

Figure 9.1 Imaging through an inhomogeneous atmosphere.

index of the undisturbed air. The origin of these inhomogeneities is a complex physical process driven by uneven heating of the surface of the Earth by the Sun, and a cascading of temperature-induced refractive index inhomogeneities to smaller and smaller scale sizes through turbulent flow and convection. In general, propagation through a distributed set of random inhomogeneities of varying scale size results in both amplitude and phase fluctuations of the light entering the lens. The irregularly distorted line to the left of the lens is meant to represent the distorted wavefront arriving at the imaging aperture—the amplitude fluctuations are difficult to represent in such a picture. The straight-line rays passing from the center of the object to the center of the image represent ray-paths that would be taken if there were no inhomogeneities present. The lens focuses the incident complex amplitude distribution to an image, which is not an ideal diffraction-limited image, due to the wavefront errors introduced by the atmosphere.

In astronomy, which will be the main application of interest in this chapter, the geometry is somewhat different. In this case the imaging system most commonly uses a reflecting mirror for image formation, rather than a refractive lens, and is looking up from the surface of the Earth, rather than parallel to that surface. In addition, the turbulence is present only in the atmosphere that lies directly above the telescope, and is absent through most of the path to the distant astronomical object.

It is customary to describe the spatial properties of the refractive-index fluctuations through a *structure function*, defined by

$$D_n(\vec{r}_1, \vec{r}_2) = \overline{[n(\vec{r}_1) - n(\vec{r}_2)]^2}. \tag{9-1}$$

From the classic work of Kolmogorov [92] on the theory of turbulence, it is known that the refractive-index structure function takes the form (under isotropic conditions)

$$D_n(r) = C_n^2 r^{2/3}, \tag{9-2}$$

where C_n^2 is known as the *structure constant* (dimensions meters$^{-2/3}$) and is a measure of the strength of the turbulence. Turbulence with such a structure function is referred to as *Kolmogorov turbulence*. C_n^2 generally diminishes with height, except in the vicinity of the tropopause, where the strength may locally increase, although generally not to the level found near the ground. Generally speaking, the turbulence strength becomes extremely small beyond the tropopause[1]—beyond a height of perhaps 20 km. The best astronomical images gathered by ground-based telescopes are obtained from mountain-top observatories, which, because of their height, experience the least effects of turbulence, although those effects are far from negligible.

9.2 Short-Exposure and Long-Exposure Point-Spread Functions

The inhomogeneities in the atmosphere are constantly in motion, and as a consequence, the deformations of the wavefront entering the telescope are changing with time. Suppose that the telescope is focused on a star, and that a simple plane wave is arriving from that source. For the moment, assume that the detector is fast and in a short exposure captures an image that arises from what are essentially stationary and fixed atmospherically induced wavefront deformations. If we consider the light captured at a single point on the detector, it will consist of contributions from all points on the telescope mirror; since different parts of that aperture have suffered different amounts of atmospherically induced phase delay, the complex field at that point on the detector will undergo a random walk.[2] The number of steps in the random walk depends on the number of independent wavefront correlation cells embraced by the telescope aperture. If the standard deviation of the phase delays is on the order of 2π radians or larger, then a random walk with uniform statistics in angle in the complex plane will result. This walk is complicated by the fact that under some circumstances there may be significant intensity fluctuations accompanying the wavefront distortions, in which case the random walk has random step lengths. However, if the amplitude fluctuations in a wavefront correlation cell are independent of the phase fluctuations in that cell, and if there are many wavefront correlation cells contributing to the intensity at each image point, then the statistics of the intensity in the short-exposure intensity spread function will, to a good approximation, be negative-exponential, but with a mean that varies spatially over the point-spread function.

Figure 9.2 shows a simulation of a short-exposure point-spread function and a long-exposure point-spread function, for which the detector integrates over many realizations (in this particular case, 1000 realizations) of the wavefront fluctuations. The "speckled" nature of the short-exposure spread function is readily seen. The dimensions of the individual speckles are approximately those of a diffraction-limited spread function, while the width of the total spread function is determined by the transverse correlation function of the wave amplitude entering the aperture of the imaging system. In the long-exposure case, the speckles have largely averaged out, leaving a relatively smooth and broad spread function.

---

1. The height of the tropopause varies with both latitude and season.
2. We assume here, and throughout this chapter, that the light is narrowband.

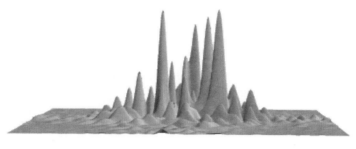

(a) Short exposure PSF

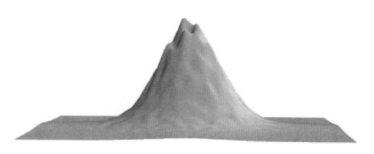

(b) Long exposure PSF

Figure 9.2 Short- and long-exposure intensity point-spread functions (PSFs). The phase fluctuations have been taken to have a uniform probability distribution on $(-\pi, \pi)$, and an isotropic Gaussian-shaped correlation function, which is a simplification of the true situation, but serves to illustrate our points. For the short exposure case, one realization of such a phase distribution, limited by the finite extent of a circular aperture, has been Fourier transformed and the resulting intensity has been found. For the long-exposure case, 1000 independent realizations of the short-exposure spread function have been averaged on an intensity basis.

As mentioned above, the speckle observed in the short-exposure point-spread function has a mean intensity distribution that varies over space, with a maximum generally falling at the "center" of the spread function and the average intensity then dropping off as we move away from the center. It is tempting to say that the distribution of the average intensity in this spread function is equal to the long-exposure spread function, but this is not quite true. The distribution of wavefront distortions across the aperture of the imaging system has a component that is a linear phase tilt, plus higher order distortions. The effect of the phase tilt component is to shift the point-spread function but not to broaden it. The shift of the spread function shifts the resulting image but does not blur it, and therefore in most applications it can be neglected in calculating the average short-exposure spread function. On the other hand, for a long exposure, the changing linear component of phase does broaden the spread function and therefore does affect the average spread function.

How short is a "short" exposure? The full answer is complex (see, for example, [137], pp. 76–78). The temporal changes in the point-spread function are often modeled as arising from flow of turbulent layers of the atmosphere transverse to the light paths followed to the collecting aperture of the imaging system; as a consequence the fluctuation times often depend on the wind speed. Generally such flow must be modeled by a spectrum of time fluctuations, that spectrum containing both low-frequency and high-frequency fluctuations. We will not delve into the details here. As a rule of thumb, at an astronomical observatory, a time exposure that is a few milliseconds or less is generally regarded as a "short" exposure, while an exposure much longer than that number is regarded as a "long" exposure. However, like most rules of thumb, under extreme conditions this number may under- or over-estimate the critical time.

9.3 Long-Exposure and Short-Exposure Average Optical Transfer Functions

It is not our goal here to derive some of the key mathematical expressions that make quantitative analysis of the effects of atmospheric distortions possible, but rather, to present a few of the results that are relevant to speckle. Detailed derivations can be found in, for example, Chapter 8 of [70] and the references therein, as well as [137]. Throughout what follows, we assume that the turbulence is isotropic.

For our purposes, we begin with a quantity that can be called the "atmospheric coherence diameter," but is also known as the "Fried parameter" after D.L. Fried who first derived it [53]. The symbol used for this parameter is r_0 (even though it is a diameter) and it is given by

$$r_0 = 0.185 \left[\frac{\bar{\lambda}^2}{\int_0^z C_n^2(\xi)\, d\xi} \right]^{3/5}, \tag{9-3}$$

where $C_n^2(\xi)$ is the structure constant as a function of distance ξ, z is the maximum path length in significant turbulence, and $\bar{\lambda}$ is the mean wavelength. As can be seen, the larger the integrated value of C_n^2, the smaller the value of r_0.

The overall average optical transfer function of an imaging system operating within the atmosphere can be written as a product of the diffraction-limited OTF,[3] $\mathcal{H}_0(\vec{v})$, times an average atmospheric OTF; thus

$$\overline{\mathcal{H}}_L(\vec{v}) = \mathcal{H}_0(\vec{v}) \times \overline{\mathcal{H}}_{LE}(\vec{v}) \quad \text{(long exposure)} \tag{9-4}$$

$$\overline{\mathcal{H}}_S(\vec{v}) = \mathcal{H}_0(\vec{v}) \times \overline{\mathcal{H}}_{SE}(\vec{v}) \quad \text{(short exposure)}, \tag{9-5}$$

---

3. It will be assumed here and throughout that the diffraction-limited OTF is real and positive, and hence its symbol is not written in boldface.

where $\overline{\mathcal{H}}_{LE}(\vec{v})$ represents the average long-exposure OTF of the atmosphere, $\overline{\mathcal{H}}_{SE}(\vec{v})$ is the average short-exposure OTF of the atmosphere, $\overline{\mathcal{H}}_L$ and $\overline{\mathcal{H}}_S$ are the overall long- and short-exposure average OTFs (including the effect of the finite collection aperture), and $\vec{v} = v_X \hat{x} + v_Y \hat{y}$. For a diffraction-limited circular aperture of diameter D,

$$\mathcal{H}_0(\vec{v}) = \frac{2}{\pi}\left[\arccos\left(\frac{\bar{\lambda} z_i v}{D}\right) - \frac{\bar{\lambda} z_i v}{D}\sqrt{1 - \left(\frac{\bar{\lambda} z_i v}{D}\right)^2}\right], \qquad (9\text{-}6)$$

where $v = |\vec{v}|$, $\bar{\lambda}$ is the average wavelength, and z_i is the distance from the exit pupil to the image plane.

The significance of r_0 becomes evident when we consider the average long-exposure optical transfer function of the atmosphere (which is the Fourier transform of the average intensity point-spread function). Representing this OTF by $\overline{\mathcal{H}}_{LE}(\vec{v})$, for isotropic turbulence we have (see the references cited above)

$$\overline{\mathcal{H}}_{LE}(\vec{v}) = \exp\left[-3.44 \left(\frac{\bar{\lambda} z_i v}{r_0}\right)^{5/3}\right]. \qquad (9\text{-}7)$$

Comparing the roles of $D/(\bar{\lambda} z_i)$ in Eq. (9-6) and $r_0/(\bar{\lambda} z_i)$ in Eq. (9-7), we see that to a good approximation, when $r_0 \ll D$, the resolution of the system approaches that of a system with aperture diameter r_0.

In many respects, the short-exposure case is of more interest in our context than the long-exposure case, due to the fact that the speckle-based imaging techniques to be discussed assume short-exposure images. In the short-exposure case, the average OTF can be expressed as

$$\overline{\mathcal{H}}_{SE}(\vec{v}) = \exp\left\{-3.44 \left(\frac{\bar{\lambda} z_i v}{r_0}\right)^{5/3}\left[1 - \alpha\left(\frac{\bar{\lambda} z_i v}{D}\right)^{1/3}\right]\right\}, \qquad (9\text{-}8)$$

where α is a parameter that varies between $1/2$ when there are both intensity and phase variations across the collecting aperture and 1 when there are only phase distortions across the collecting aperture. In deriving this expression, a least-squares estimate of the wavefront tilt has been removed from the wavefront, as discussed above.

Figure 9.3 shows the plots of the four OTFs involved, the diffraction-limited OTF, the average long-exposure OTF, and the two short-exposure OTFs. The plots are specifically for the case $D = 5r_0$.

9.4 Statistical Properties of the Short-Exposure OTF and MTF

The above results provide descriptions of the *average* OTFs, but the actual OTF achieved on any single short exposure generally will not look much like the average. It will in fact be highly irregular with speckle-like behavior, particularly in the mid

9.4 Statistical Properties of the Short-Exposure OTF and MTF

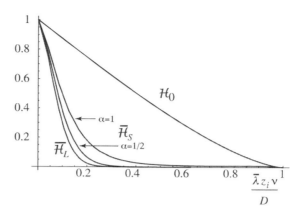

Figure 9.3 Plots of $\mathcal{H}_0$, $\overline{\mathcal{H}}_L$, and $\overline{\mathcal{H}}_S$ (for $\alpha = 1$ and $1/2$) vs. normalized radial frequency. r_0 is $1/5$ of D.

frequencies. Figure 9.4 shows a simulated short-exposure point-spread function and the corresponding modulation transfer function (MTF, the modulus of the OTF). The value of the MTF at zero frequency is unity, but the value quickly falls off outside the central lobe to a speckle-like structure that is clearly visible in the plot. The simulation was similar to that used in Fig. 9.2, except that the diameter of the circular aperture is somewhat larger in this case. In order to make the speckle structure in the MTF more obvious, only values of the MTF between 0 and 0.12 are shown (recall that by convention, the MTF is normalized to have value unity at the origin). Associated with the speckle-like structure of the MTF is a complicated speckle phase function that is uniformly distributed on $(-\pi, \pi)$, as well as phase vortices wherever the MTF goes to zero.

A better understanding of the statistical nature of the short-exposure OTF and MTF of an optical imaging system can be gained by remembering how the OTF of an optical imaging system can be calculated. The OTF is the (properly normalized) deterministic autocorrelation function of the complex pupil function of the system. Thus if $\mathbf{P}(u,v)$ represents the distribution of amplitude and phase across the exit pupil of the system (coordinates (u,v)), then the short-exposure OTF $\mathcal{H}_S(\vec{v})$ is given by

$$\mathcal{H}_S(\vec{v}) = \frac{\iint\limits_{-\infty}^{\infty} \mathbf{P}\left(\xi + \frac{\bar{\lambda} z_i v_X}{2}, \eta + \frac{\bar{\lambda} z_i v_Y}{2}\right) \mathbf{P}^*\left(\xi - \frac{\bar{\lambda} z_i v_X}{2}, \eta - \frac{\bar{\lambda} z_i v_Y}{2}\right) d\xi\, d\eta}{\iint\limits_{-\infty}^{\infty} |\mathbf{P}(\xi, \eta)|^2 \, d\xi\, d\eta}. \quad (9\text{-}9)$$

In the case of interest, the phase aberration is introduced as the wave propagates to the imaging system through the atmosphere. In what follows, we shall assume that the atmospherically induced intensity variations within the pupil are sufficiently small to be ignored, or equivalently that the pupil function is of the form

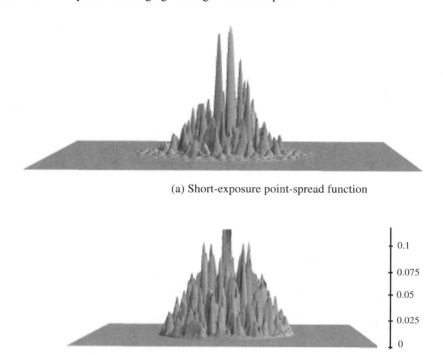

(a) Short-exposure point-spread function

(b) Short-exposure MTF

Figure 9.4 Short-exposure case (a) PSF, (b) MTF, showing only the region from 0 to 0.12.

$$\mathbf{P}(\xi,\eta) = \begin{cases} \exp[j\psi(\xi,\eta)] & \text{in the pupil} \\ 0 & \text{outside the pupil} \end{cases} \quad (9\text{-}10)$$

where $\psi(\xi,\eta)$ is the *phase aberration function*. Note that the denominator of the OTF for the pupil function above is simply the area of the pupil, and has no statistical fluctuation, again assuming that there are no atmospherically induced intensity fluctuations across the pupil.

A somewhat intuitive explanation of the character of the OTF can be found by considering Fig. 9.5. The pupil of the system is represented by a large circle (diameter D). Smaller circles (diameter r_0) are used to represent the atmospheric coherence diameter. As a first approximation, the phase aberration within each of the small circles is constant during the short exposure, but random with Gaussian statistics. There are

$$N_0 = \frac{D^2}{r_0^2} \quad (9\text{-}11)$$

different coherence cells within the aperture.

9.5 Astronomical Speckle Interferometry 319

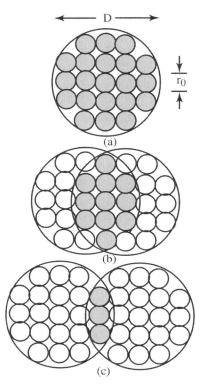

Figure 9.5 OTF calculation at three different frequencies: (a) zero frequency, (b) a mid frequency, and (c) a high frequency. The shaded small circles represent coherence areas that lie within the area of overlap.

As the autocorrelation function is performed, we displace one version of the pupil from a second identical version by amount $(\bar{\lambda} z_i \vec{v})$ and calculate the area under the resulting product of the two displaced functions, one of which has been subjected to complex conjugation. The shaded coherence areas are meant to indicate those that lie in the area of overlap. When there is no displacement of the two circles, we are evaluating the OTF at zero spatial frequency. Each coherence area then perfectly overlaps itself, and the normalized product of $\exp[j\psi(\xi, \eta)]$ with its complex conjugate yields unity everywhere. Once we displace by distance r_0, then each coherence area overlaps a different, independent coherence area, and the product of the two exponentials yields a complex number with a random phase and length proportional to r_0^2. To this complex phasor must be added $N - 1$ additional phasors from other coherence areas in the area of overlap, all with the same length but with independent random phases. For any particular frequency $\bar{\lambda} z_i \vec{v}$, the number N of random phasors added is given by

$$N = N_0 \mathcal{H}_0(\vec{v}), \tag{9-12}$$

provided $|\vec{v}| > r_0/(\bar{\lambda}z_i)$. Here, $\mathcal{H}_0$ is again the diffraction-limited OTF of the imaging system. In the case of a circular aperture,

$$\frac{N}{N_0} \approx \begin{cases} 1 & v < r_0/(\bar{\lambda}z_i) \\ \frac{2}{\pi}\left[\arccos\left(\frac{\bar{\lambda}z_i v}{D}\right) - \left(\frac{\bar{\lambda}z_i v}{D}\right)\sqrt{1 - \left(\frac{\bar{\lambda}z_i v}{D}\right)^2}\right] & r_0/(\bar{\lambda}z_i) < v < D/(\bar{\lambda}z_i) \\ 0 & v > D/(\bar{\lambda}z_i), \end{cases} \quad (9\text{-}13)$$

where again $v = |\vec{v}|$.

Thus at each spatial frequency greater than $r_0/(\bar{\lambda}z_i)$, the numerator of the OTF can be regarded to be the resultant of an amplitude random walk composed of N equal-length, randomly phased phasors, with the length of the individual phasors being proportional to r_0^2. If the intensity of the wavefront at the exit pupil is represented by unity (we are interested in normalized quantities here, so the actual intensity will not matter), then the (identical) amplitudes of the individual phasors are

$$a = \pi\left(\frac{r_0}{2}\right)^2. \quad (9\text{-}14)$$

If $D \gg r_0$, then for the mid frequencies, the number of independent phasors contributing to $\mathcal{H}_S(\vec{v})$ is large, and the statistics of $|\mathcal{H}_S(\vec{v})|$ can be expected to be well approximated by a Rayleigh distribution in this frequency range. We can approximate the value of $\overline{|\mathcal{H}_S|}$ in the mid-frequency range from the results in the next paragraph regarding $\overline{|\mathcal{H}_S|^2}$.

A quantity that will prove to be important when the subject of speckle imaging is discussed is the average of the squared modulus of the short-exposure OTF. The average of the numerator of the squared magnitude of the OTF is analogous to the average of the intensity of the resultant of a random walk and is given by $Na^2 = N\pi^2\left(\frac{r_0}{2}\right)^4$. The numerator must be normalized by $\pi^2\left(\frac{D}{2}\right)^4$ to obtain the square of an OTF, with the result

$$\overline{|\mathcal{H}_S(\vec{v})|^2} = \frac{\pi^2\left(\frac{r_0}{2}\right)^4}{\pi^2\left(\frac{D}{2}\right)^4}N = \left(\frac{r_0}{D}\right)^4 N_0 \mathcal{H}_0(\vec{v}) = \left(\frac{r_0}{D}\right)^2 \mathcal{H}_0(\vec{v}). \quad (9\text{-}15)$$

Thus we see that $\overline{|\mathcal{H}_S(\vec{v})|^2}$ behaves approximately as follows:

$$\overline{|\mathcal{H}_S(\vec{v})|^2} \approx \begin{cases} \left(\frac{r_0}{D}\right)^2 \mathcal{H}_0(\vec{v}) & r_0/(\bar{\lambda}z_i) < v < D/(\bar{\lambda}z_i) \\ 0 & v > D/(\bar{\lambda}z_i). \end{cases} \quad (9\text{-}16)$$

This approximate argument does not lead to an expression that is valid when $v < r_0/(\bar{\lambda}z_i)$, but $\overline{|\mathcal{H}_S|^2}$ must start at value 1 at the origin and then smoothly drop to the vicinity of $\left(\frac{r_0}{D}\right)^2 \mathcal{H}_0(r_0/(\bar{\lambda}z_i))$.

The results presented in the previous paragraph are quite general, regardless of whether the number of independent phasors is large or small. If $D \gg r_0$, then for the mid frequencies, the number of independent phasors contributing to $\mathcal{H}_S(\vec{v})$ is large, and the statistics of $|\mathcal{H}_S(\vec{v})|^2$ can be expected to be well approximated by a negative exponential distribution. From Eq. (3-13), the mean value of a negative exponential variate is equal to twice the variance σ^2 of the real or the imaginary part of the complex amplitude underlying the negative exponential variate. Thus for the case at hand, at mid frequencies,

$$\sigma^2 = \frac{1}{2}\overline{|\mathcal{H}_S(\vec{v})|^2} \approx \frac{1}{2}\left(\frac{r_0}{D}\right)^2 \mathcal{H}_0(\vec{v}). \qquad (9\text{-}17)$$

Drawing on the result of Eq. (2-18), the mean of the Rayleigh distribution obeyed by $|\mathcal{H}_S(\vec{v})|$ is given by

$$\overline{|\mathcal{H}_S(\vec{v})|} = \sqrt{\frac{\pi}{2}}\sigma = \frac{\sqrt{\pi}}{2}\left(\frac{r_0}{D}\right)\sqrt{\mathcal{H}_0(\vec{v})}. \qquad (9\text{-}18)$$

Note that this is not the same quantity as $\overline{\mathcal{H}}_S(\vec{v})$ presented in Eqs. (9-5) and (9-8); the two quantities are different, since changes in the phase of the OTF reduce the former, but not $\overline{|\mathcal{H}_S(\vec{v})|}$. In addition, Eq. (9-18) applies only in the mid-frequency range, whereas Eqs. (9-5) and (9-8) apply for all frequencies.

Finally we note that, in the mid frequencies, the variance of the MTF is given by

$$\sigma^2_{|\mathcal{H}_S|} = \overline{|\mathcal{H}_S|^2} - (\overline{|\mathcal{H}_S|})^2 = \left(1 - \frac{\pi}{4}\right)\left(\frac{r_o}{D}\right)^2 \mathcal{H}_0(\vec{v}). \qquad (9\text{-}19)$$

The most important conclusion of this section is that, while the average short-exposure OTF falls to nearly zero when $v \approx r_0/(\bar{\lambda}z_i)$, the quantities $\overline{|\mathcal{H}_S|^2}$ and $\overline{|\mathcal{H}_S|}$ reach intermediate levels above this frequency and fall much more slowly as frequency increases.

To test some of these results, a simulation can be performed, constructing samples of $|\mathcal{H}_S(\vec{v})|^2$ and averaging over many independent realizations of the atmospherically induced phase errors.[4] Figure 9.6 shows the result of such a simulation for 5000 independent short-exposure images. Also shown is a plot of the average of 5000 values of the short-exposure OTF (Fourier transform of the short-exposure point-spread function), without tilt removal from the pupil wavefront, from the same simulation. The average of $|\mathcal{H}_S(\vec{v})|^2$ can be seen to track $\overline{\mathcal{H}}_L^2(\vec{v})$ for frequencies smaller than $r_0/(\bar{\lambda}z_i)$, then level off and follow what can reasonably be assumed to be $(r_0/D)^2 \mathcal{H}_0$ through the mid- and high-frequency ranges.

---

4. The simulation began by generating a square array of 256×256 independent Gaussian random phases, each with zero mean and standard deviation 1.8138. The array of random phases was then convolved with a 16×16 truncated Gaussian-shaped kernel, the $1/e$ radius of which was approximately 5. The resulting smoothed phase was then used as the atmospherically induced phase error. The resulting complex wavefront array was truncated by a circular pupil of radius 32, and the resulting wave was Fourier transformed. The intensity of the result was found, and then Fourier transformed and normalized to find the short-exposure OTF. The squared modulus of this result was found, and averaged over 5000 trials.

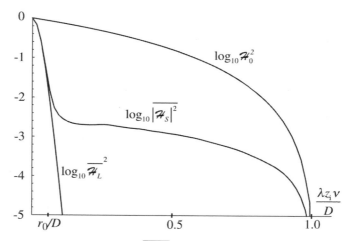

Figure 9.6 Plots of the logarithms of $\mathcal{H}_0^2$, $\overline{|\mathcal{H}_S|^2}$, and $\overline{\mathcal{H}_L}^2$ vs. normalized radial frequency.

We turn now to considering ways to exploit this behavior of $\overline{|\mathcal{H}_S(\vec{v})|^2}$ in the mid-frequency range.

9.5 Astronomical Speckle Interferometry

In a seminal paper in 1970, A. Labeyrie [97] proposed a novel technique for extracting certain kinds of object information from a sequence of short-exposure astronomical images. This technique held the promise that limited kinds of information could be retrieved at frequencies far beyond the usual long-exposure limit imposed by atmospheric turbulence. The technique was first demonstrated in astronomical imaging in 1972 by Gezari, Labeyrie and Stachnik [58].

The basis of the method lies in the fact already discussed, that the average of the squared MTF falls as $(r_0/D)^2 \mathcal{H}_0$ at mid and high frequencies. The method for taking advantage of this fact is outlined in the following subsection.

9.5.1 Object Information that Is Retrievable

Let the object intensity of interest have a Fourier spectrum $\mathcal{O}(\vec{v})$ and let the image detected with a short exposure have a Fourier spectrum $\mathcal{I}(\vec{v})$. Then, under the assumption that the imaging system is space-invariant, these two spectra are related by

$$\mathcal{I}(\vec{v}) = \mathcal{H}_S(\vec{v}) \mathcal{O}(\vec{v}). \tag{9-20}$$

Now suppose that, rather than using the image Fourier spectrum directly, we take its squared modulus and average this quantity over many short-exposure images. Then the average in question becomes

$$\overline{|\mathcal{I}(\vec{v})|^2} = \overline{|\mathcal{H}_S(\vec{v})|^2}|\mathcal{O}(\vec{v})|^2, \qquad (9\text{-}21)$$

where the object spectrum has been assumed to be the same for all images. However, when $v > r_0/(\bar{\lambda}z_i)$,

$$\overline{|\mathcal{H}_S(\vec{v})|^2} \approx \left(\frac{r_0}{D}\right)^2 \mathcal{H}_0(\vec{v}), \qquad (9\text{-}22)$$

and therefore

$$\overline{|\mathcal{I}(\vec{v})|^2} \approx \left(\frac{r_0}{D}\right)^2 \mathcal{H}_0(\vec{v})|\mathcal{O}(\vec{v})|^2, \qquad (9\text{-}23)$$

valid from $v = r_0/(\bar{\lambda}z_i)$ to the diffraction-limited cutoff. Thus, provided the value of $(r_0/D)^2 \mathcal{H}_0(\vec{v})$ can be estimated in advance by use of a reference star (small enough to assure that $|\mathcal{O}(\vec{v})|$ is constant for all frequencies), then the squared magnitude of the object spectrum can be determined from

$$|\mathcal{O}(\vec{v})|^2 \approx \frac{\overline{|\mathcal{I}(\vec{v})|^2}}{(r_0/D)^2 \mathcal{H}_0(\vec{v})}. \qquad (9\text{-}24)$$

The "approximate" sign here is because of the approximate nature of our analysis.

There are, of course, several sources of inaccuracy in this procedure. First, the average over a finite number of short-exposure images is never exactly equal to the corresponding ensemble average. Second, there is inevitably noise in the detection process, and that noise must be taken into account. Third, the averaging process assumes that we add many spectral distributions with *independent* speckle patterns, implying that the time duration from which the short exposures are drawn spans many atmospheric fluctuation times. However, putting these issues aside, it is reasonable to ask what information about the object can be obtained from knowledge of the magnitude (or squared magnitude) of its spectrum. Since the phase information in the object spectrum is not available from this measurement, it is not possible in general to directly find the object itself. While methods do exist for finding the full complex spectrum from spectral magnitude information when certain other a priori information is available (see, for example, [45]), such a subject is beyond our scope here.

A useful way of looking at the operation performed by speckle interferometry is to note that Eq. (9-21) resembles the frequency-domain representation of a linear filtering operation, in that a spectral distribution associated with the image is given by a spectral distribution associated with the object times a spectral distribution associated with the imaging system. However, unlike the usual case of an OTF, the object and image spectral distributions are the squared magnitudes of the object and image spectra. Recall from the autocorrelation theorem of Fourier analysis (see [71], p. 8) that the inverse Fourier transform of the squared magnitude of a spectrum is the autocorrelation function of the object (or image) that gave rise to that spectrum, that is,

$$\mathcal{F}^{-1}\{|\boldsymbol{\mathcal{O}}(\nu_X,\nu_Y)|^2\} = \iint_{-\infty}^{\infty} O(\xi,\eta)O^*(\xi-x,\eta-y)\,d\xi\,d\eta$$
$$\mathcal{F}^{-1}\{|\boldsymbol{\mathcal{I}}(\nu_X,\nu_Y)|^2\} = \iint_{-\infty}^{\infty} I(\xi,\eta)I^*(\xi-x,\eta-y)\,d\xi\,d\eta,$$
(9-25)

where $\mathcal{F}^{-1}$ is an inverse Fourier transform operator, and since $O(x,y)$ and $I(x,y)$ are real-valued intensity distributions, the asterisk representing complex conjugate has no effect in this case. We conclude that the quantity $|\mathcal{H}_S(\vec{\nu})|^2$ can be regarded to be the transfer function that relates the autocorrelation functions of the object and image intensity distributions. For this reason, this quantity is often referred to as the "speckle transfer function". This discussion also points out that the information directly obtainable from $|\boldsymbol{\mathcal{O}}|^2$ in Eq. (9-24) is in fact the autocorrelation function of the object intensity distribution.

One case in which this type of speckle interferometry has been put to extensive use is in measuring the separation of binary stars. To explore this application analytically, suppose that the binary star is modeled as having an intensity distribution

$$O(x,y) = I_1\delta\left(x-\frac{\Delta}{2},y\right) + I_2\delta\left(x+\frac{\Delta}{2},y\right) \quad (9\text{-}26)$$

where the unresolved individual stars are represented by the delta functions, and the separation of the stars is Δ. The squared magnitude of the Fourier spectrum of this object intensity distribution is easily shown to be

$$|\boldsymbol{\mathcal{O}}(\nu_X,\nu_Y)|^2 = (I_1^2 + I_2^2)\left[1 + \frac{2I_1 I_2}{I_1^2 + I_2^2}\cos(2\pi\nu_X\Delta)\right]. \quad (9\text{-}27)$$

Note in particular the cosinusoidal fringe that appears in the spectral domain, a fringe with period $1/\Delta$. The farther apart the components of the binary star, the finer the period. If the signal-to-noise ratio in the measurement of $|\boldsymbol{\mathcal{O}}(\vec{\nu})|^2$ is high enough, the fringe period can be measured, and from that measurement the value of the separation Δ can be determined. Note also that it would be possible to inverse Fourier transform $|\boldsymbol{\mathcal{O}}(\nu_X,\nu_Y)|^2$ in this case and measure the separation of the stars from the autocorrelation function of the object intensity distribution. Finally, we mention that it is also possible to find the ratio of the intensities of the two sources from the fringe pattern of Eq. (9-27).

Figure 9.7 shows the results of a simulation[5] producing fringes for the case of two equally bright point-sources when K different measurements of $|\mathbf{I}(\vec{\nu})|^2$ are obtained from K independent short-exposure images. The fringe pattern is obvious

---

5. The simulation used the same parameters as discussed in the previous footnote.

9.5 Astronomical Speckle Interferometry

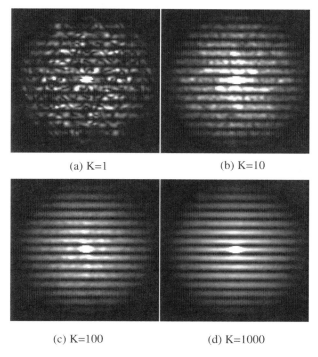

(a) K=1   (b) K=10

(c) K=100   (d) K=1000

Figure 9.7 Fringe patterns in an average over K independent measurements of $|I(\vec{v})|^2$ for (a) $K = 1$, (b) $K = 10$, (c) $K = 100$ and (d) $K = 1000$ short-exposure images for the case of an object consisting of two equally bright point sources. The sharp peak in the center of the fringe pattern is about 450 times stronger than the maximum of the fringe pattern just beyond the sharp peak.

even for $K = 1$, but the ability to measure the exact value of its period improves as the value of K increases.

9.5.2 Results of a More Complete Analysis of the Form of the Speckle Transfer Function

A more exact calculation of the form of the speckle transfer function was performed by Korff [93]. Alternative discussions of his analysis can be found in [70] and [137]. Here, only the results of the analysis will be presented and discussed.

Assume that the atmospheric effects consist primarily of phase distortions and that amplitude fluctuations are small enough to be neglected. Assume in addition that the atmospheric turbulence is statistically isotropic and obeys Kolmogorov statistics, and that the exit pupil of the imaging system is circularly symmetric, in which case the speckle transfer function depends only on $v = |\vec{v}|$. Then (see [70], Section 8.8.3)

$$|\mathcal{H}_S(v)|^2 = \frac{\int\!\!\int_{-\infty}^{\infty} Q(\xi,\eta,s)L(\xi,\eta,s)\,d\xi\,d\eta}{\left[\int\!\!\int_{-\infty}^{\infty}|P(\xi,\eta)|^2\,d\xi\,d\eta\right]^2}, \qquad (9\text{-}28)$$

where $s = \bar{\lambda}z_i v$ (without loss of generality, we can take s to be a displacement in the ξ direction). Here

$$Q(\xi,\eta,s) = \exp\left\{-6.88\left[\left(\frac{s}{r_0}\right)^{5/3} + \left(\frac{\sqrt{\xi^2+\eta^2}}{r_0}\right)^{5/3}\right]\right\}$$

$$\times \exp\left\{-6.88\left[-\frac{1}{2}\left(\frac{\sqrt{(\xi+s)^2+\eta^2}}{r_0}\right)^{5/3}\right.\right.$$

$$\left.\left. -\frac{1}{2}\left(\frac{\sqrt{(\xi-s)^2+\eta^2}}{r_0}\right)^{5/3}\right]\right\}, \qquad (9\text{-}29)$$

and

$$L(\xi,\eta,s) = \int\!\!\int_{-\infty}^{\infty} \mathbf{P}\!\left(\frac{p_x+\xi-2s}{2},\frac{p_y+\eta}{2}\right)\mathbf{P}^*\!\left(\frac{p_x+\xi}{2},\frac{p_y+\eta}{2}\right)$$

$$\times \mathbf{P}^*\!\left(\frac{p_x-\xi-2s}{2},\frac{p_y-\eta}{2}\right)\mathbf{P}\!\left(\frac{p_x-\xi}{2},\frac{p_y-\eta}{2}\right) dp_x\,dp_y \qquad (9\text{-}30)$$

with $\mathbf{P}$ being the pupil function of the system without atmospheric phase distortion. For a clear pupil ($\mathbf{P} = 1$ or $\mathbf{P} = 0$), the function L simply determines the region over which the function Q is integrated, and the weight given to that region.

Evaluation of this expression for the speckle transfer function must be done numerically, and is a formidable computational task. In Fig. 9.8 we show plots of the speckle transfer function and the square of the diffraction-limited OTF for the particular value $D/r_0 = 14.3$. (see also [137], Fig. 4.2). The results are generally consistent with the results obtained earlier by simulation.

It is important to note that the recovery of object information by means of speckle interferometry is a *noise-limited* process. That is, while in a noise-free environment, the autocorrelation function of the object intensity distribution can be recovered in great detail, in practice, the finite number of photoevents involved in the measurement, and the associated noise that results, is the most significant limiting factor for this procedure.

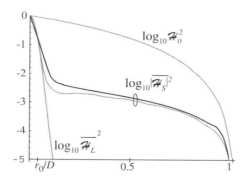

Figure 9.8 The dark curve represents the result of an exact calculation of the speckle transfer function for $D/r_0 = 14.3$, which is approximately the same value as for the simulation shown in Fig. 9.6. The curves from that figure are shown in gray for comparison purposes.

9.6 The Cross-Spectrum or Knox–Thompson Technique

The cross-spectrum method is a modification Labeyrie's speckle interferometry that allows full images to be recovered from a sequence of atmospherically degraded, short-exposure images. The technique is also known as the Knox–Thompson technique, after the authors who first proposed it [91]. Our explanation here follows that of Roggemann [137]. We first consider properties of the cross-spectrum that preserve object information. We then turn to consideration of how full object information can be recovered from the image cross-spectrum.

9.6.1 The Cross-Spectrum Transfer Function

The *cross-spectrum* $\mathbf{C_I}(\vec{v}, \vec{\Delta v})$ of a single image $i(\vec{x})$ is defined in terms of the image's Fourier transform $\mathbf{I}(\vec{v})$ through the relation

$$\mathbf{C_I}(\vec{v}, \vec{\Delta v}) = \mathbf{I}(\vec{v})\mathbf{I}^*(\vec{v} + \vec{\Delta v}). \tag{9-31}$$

The cross-spectrum of a single image is in general complex-valued. In addition, it has the following symmetry property:

$$\mathbf{C_I}(-\vec{v}, -\vec{\Delta v}) = \mathbf{C_I^*}(\vec{v}, \vec{\Delta v}), \tag{9-32}$$

a consequence of the Hermitian property of the Fourier transform of a real-valued image.

If $\mathbf{O}(\vec{v})$ is the spectrum of the object and $\mathcal{H}_S(\vec{v})$ is the short-exposure OTF of the combined optical system and atmosphere, then

$$\mathbf{I}(\vec{v}) = \mathcal{H}_S(\vec{v})\mathbf{O}(\vec{v}) \tag{9-33}$$

and
$$\mathbf{C_I}(\vec{v}, \vec{\Delta v}) = \mathbf{O}(\vec{v})\mathbf{O}^*(\vec{v}+\vec{\Delta v})\,\mathcal{H}_S(\vec{v})\,\mathcal{H}_S^*(\vec{v}+\vec{\Delta v}). \tag{9-34}$$

Now suppose that an entire sequence of short-exposure images is captured, with the atmosphere changing between images but the object not changing. Consider calculating the *average* of the image cross-spectrum over that set of images. Since the object is not changing, the average applies only to the terms involving $\mathcal{H}_S$. Approximating that finite average by a statistical average, we obtain an *average* cross-spectrum given by

$$\overline{\mathbf{C}_I}(\vec{v}, \vec{\Delta v}) = \mathbf{O}(\vec{v})\mathbf{O}^*(\vec{v}+\vec{\Delta v})\,\overline{\mathcal{H}_S(\vec{v})\,\mathcal{H}_S^*(\vec{v}+\vec{\Delta v})} = \mathbf{C_O}(\vec{v},\vec{\Delta v})\overline{\mathbf{C}}_\mathcal{H}(\vec{v},\vec{v}+\vec{\Delta v}), \tag{9-35}$$

where $\mathbf{C_O}$ is the cross-spectrum of the object and

$$\overline{\mathbf{C}}_\mathcal{H}(\vec{v}, \vec{v}+\vec{\Delta v}) = \overline{\mathcal{H}_S(\vec{v})\,\mathcal{H}_S^*(\vec{v}+\vec{\Delta v})}. \tag{9-36}$$

It is therefore natural to refer to the quantity $\overline{\mathbf{C}}_\mathcal{H}(\vec{v},\vec{v}+\vec{\Delta v})$ as the *cross-spectrum transfer function*.

The cross-spectrum transfer function bears a close resemblance to the speckle transfer function that was seen to apply to speckle interferometry, and indeed reduces to that transfer function when $\vec{\Delta v} = 0$. But the cross-spectral transfer function is a more complicated function, for it depends on two vector variables rather than just one. Like the speckle transfer function, the cross-spectral transfer function is real-valued, although this is not at all obvious at this point. In addition, because of the offset $\vec{\Delta v}$ in its definition, the cross-spectral transfer function is not unity at $\vec{v} = 0$.

Figure 9.9 shows the cross-spectrum transfer function obtained by simulation using the parameters described in footnote 4 of this chapter, and an offset $\vec{\Delta v} =$

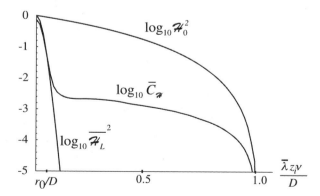

Figure 9.9 Cross-spectrum transfer function $\mathcal{C}_\mathcal{H}$ obtained by simulation, with $r_0 = 0.07D$ and $\Delta v = 0.03D/(\lambda z_i)$. Also shown are the square of the diffraction-limited OTF and the square of the long-exposure OTF.

$(0.03\bar{\lambda}z_i/D)\hat{v}_X$, where $\hat{v}_X$ is a unit vector in the v_X direction. Like the speckle transfer function, the cross-spectral transfer function is seen to initially fall at low frequencies, then level off at mid frequencies, and then fall again at high frequencies. For the particular Δv chosen in this simulation, the level of the plateau at mid-frequencies is similar to the level of the corresponding plateau in the speckle transfer function.

One subtlety that should be mentioned concerns the tilt component of wavefront distortion, which shifts the image from exposure to exposure. This effect has no consequence in astronomical speckle interferometry, but it is readily shown that it will lower the cross-spectral transfer function by an amount $\overline{\exp[j2\pi(\vec{\delta}\cdot\vec{v})]}$ where $\vec{\delta}$ is the random image shift. In practice, the images obtained in the short exposures are shifted to assure that their centroids coincide, thereby largely removing this effect. Such alignment of the images has not been performed in the simulation. However, because r_0 is so small compared with D, the tilt component of wavefront distortion is negligible in this simulation, and therefore the results should not differ appreciably from the aligned-image case.

An exact expression for the cross-spectral transfer function has been obtained (see [137], page 149), but, due to its complexity, will not be pursued here. That exact expression confirms that the cross-spectral transfer function is real-valued.

By showing that a plateau exists in the cross-spectral transfer function, we have demonstrated that there is a possibility of extracting object information at frequencies beyond $r_0/(\lambda z_i)$. However, what information to extract to obtain full reconstruction of the object intensity distribution is as yet unanswered, and it is to this question that we address the following subsection.

9.6.2 Recovering Full Object Information from the Cross-Spectrum

A constraint is found when we consider the effects of the spatial frequency shift $\vec{\Delta v}$. With reference to Fig. 9.4, it can be seen that the correlation width of the fluctuations of the short-exposure MTF (and therefore the OTF) is of the order of $r_0/(\bar{\lambda}z_i)$. Beyond this frequency separation, the fluctuations of the MTF (and the OTF) are approximately uncorrelated. As a consequence, for $\vec{\Delta v} > r_0/(\bar{\lambda}z_i)$, the following will be true:

$$\overline{\mathcal{H}_S(\vec{v})\mathcal{H}_S^*(\vec{v}+\vec{\Delta v})} = \overline{\mathcal{H}_S(\vec{v})}\,\overline{\mathcal{H}_S^*(\vec{v}+\vec{\Delta v})} = 0, \qquad (9\text{-}37)$$

where statistical independence is justified by the lack of correlation between two complex Gaussian variables. We conclude that $|\vec{\Delta v}|$ must be smaller than $r_0/(\bar{\lambda}z_i)$ if the technique is to yield any information.

With the above constraint satisfied, we assume that the average cross-spectrum $\bar{\mathcal{C}}_\mathcal{H}$ of $\mathcal{H}_S$ can be obtained from measurements with a reference star, under atmospheric conditions statistically similar to those present for the more general measurement.

A second requirement is imposed on $\vec{\Delta v}$ by the object itself. Consider the cross-spectrum of the object,

$$\mathbf{C_O}(\vec{v},\vec{v}+\vec{\Delta v}) = \mathbf{O}(\vec{v})\mathbf{O}^*(\vec{v}+\vec{\Delta v}) = |\mathbf{O}(\vec{v})|\,|\mathbf{O}^*(\vec{v}+\vec{\Delta v})|$$
$$\times \exp\{j[\phi_o(\vec{v}) - \phi_o(\vec{v}+\vec{\Delta v})]\}, \tag{9-38}$$

where ϕ_o represents the phase distribution over frequency of the object spectrum. Information about $|\mathbf{O}|$ can be ascertained from Labeyrie-type speckle interferometry, which is obtained when $\vec{\Delta v} = 0$. Therefore, attention is focused on the phase difference

$$\Delta\phi(\vec{v};\vec{\Delta v}) = \phi_o(\vec{v}) - \phi_o(\vec{v}+\vec{\Delta v}). \tag{9-39}$$

Consider a single direction in the frequency plane. If the object has maximum angular width ω in the corresponding direction in angular space, then the magnitude and phase of the object spectrum can be expected to change significantly in a frequency interval equal to the Nyquist sampling interval δv for that object, where $\delta v = \omega/\bar{\lambda}$. If we restrict the frequency increment $|\vec{\Delta v}|$ to be less than or equal to $\omega/\bar{\lambda}$, then we will be assured that the phase will not change too much between frequencies $\vec{v}$ and $\vec{v} + \vec{\Delta v}$. This constitutes the second condition that Δv must satisfy. Thus $|\vec{\Delta v}|$ should satisfy

$$|\vec{\Delta v}| \leq \min\begin{cases} \frac{r_0}{\bar{\lambda} z_i} \\ \frac{\omega}{\bar{\lambda}} \end{cases} \tag{9-40}$$

Note that the interval should be chosen no smaller than absolutely necessary, for if it is chosen too small, the small changes in $\Delta\phi$ will be masked by noise.

To make further progress, consider restricting the vector $\vec{\Delta v}$ to lying either along the v_X-axis or along the v_Y-axis. Thus $\vec{\Delta v}$ is either Δv_X or Δv_Y, and in the two cases we have

$$\Delta\phi_X(v_X, v_Y) = \phi_o(v_X, v_Y) - \phi_o(v_X + \Delta v_X, v_Y) \approx -\frac{\partial \phi_o(\vec{v})}{\partial v_X}\Delta v_X$$
$$\Delta\phi_Y(v_X, v_Y) = \phi_o(v_X, v_Y) - \phi_o(v_X, v_Y + \Delta v_Y) \approx -\frac{\partial \phi_o(\vec{v})}{\partial v_Y}\Delta v_Y, \tag{9-41}$$

where the approximation is valid if Δv_X and Δv_Y are sufficiently small. The two partial derivatives are the two orthogonal components of the gradient of the object phase spectrum, $\nabla\phi_o(\vec{v})$. The problem remaining to be solved is the recovery of the phase function from measurements of the phase differences that approximate the gradient of the phase. This is precisely the same problem encountered in Section 8.3.3 dealing with determining phase maps from speckle interferograms. The reader is referred to that section for a discussion of this problem. It suffices to repeat here that it is possible for different paths in the frequency domain to yield different values of the recovered phase. The critical issue determining whether ambiguities will occur is whether there are phase vortices in the frequency domain, or equivalently whether there are points where the frequency spectrum goes to zero. References to the many approaches for overcoming these problems are given in the cited section. An approach unique to astronomy is that of Nisenson [117], who deals exclusively with averages of complex phasors rather than phases. Such a method manages to avoid the 2π phase ambiguity problem.

Thus, provided the phase recovery is treated with a suitable technique, the crossspectrum approach is capable of reconstructing the entire object spectrum, subject only to adequate signal-to-noise ratio in the measurements. Noise issues are beyond the scope of our treatment here, but some discussion can be found in [137] (page 147). Noise again is the limiting factor in determining the quality of the recovered images.

9.7 The Bispectrum Technique

Another technique that is related to Labeyrie speckle interferometry and to the crossspectrum technique is known as the *bispectrum* technique. For alternative discussions of the bispectrum, see [137], [8] and [102].

The bispectrum $\mathbf{B_I}(\vec{v}_1, \vec{v}_2)$ of a single, short-exposure image $i(\vec{x})$ is defined by

$$\mathbf{B_I}(\vec{v}_1, \vec{v}_2) = \mathbf{I}(\vec{v}_1)\mathbf{I}(\vec{v}_2)\mathbf{I}^*(\vec{v}_1 + \vec{v}_2), \qquad (9\text{-}42)$$

where again $\mathbf{I}(\vec{v})$ is the Fourier transform of $i(\vec{x})$. The four-dimensional inverse Fourier transform of the bispectrum can be shown to be the so-called triple correlation of the image intensity distribution[6]

$$\int d\vec{v}_1 \int d\vec{v}_2 \mathbf{B_I}(\vec{v}_1, \vec{v}_2) e^{j2\pi(\vec{v}_1 \cdot \vec{x}_1 + \vec{v}_2 \cdot \vec{x}_2)} = \int d\vec{x}\, i(\vec{x})i(\vec{x} + \vec{x}_1)i(\vec{x} + \vec{x}_2), \quad (9\text{-}43)$$

where in all cases the integrals are over the infinite plane.

The following symmetry properties of the bispectrum follow from its definition and from the Hermitian properties of the image spectrum:

$$\mathbf{B_I}(\vec{v}_1, \vec{v}_2) = \mathbf{B_I}(\vec{v}_2, \vec{v}_1)$$
$$\mathbf{B_I}(\vec{v}_1, \vec{v}_2) = \mathbf{B_I}(\vec{v}_1 - \vec{v}_2, \vec{v}_2) \qquad (9\text{-}44)$$
$$\mathbf{B_I}(\vec{v}_1, \vec{v}_2) = \mathbf{B_I}^*(-\vec{v}_1, -\vec{v}_2).$$

In addition, like the autocorrelation, the triple correlation and therefore the bispectrum are easily shown to be independent of shifts of the image.

9.7.1 The Bispectrum Transfer Function

The bispectrum of a single image is related to the bispectrum of the object and the bispectrum of a single realization of the short-exposure optical transfer function through

---

6. The proof is easiest if one starts with the right-hand side and performs a four-dimensional Fourier transform, yielding the left-hand side.

$$\mathbf{B_I}(\vec{v}_1, \vec{v}_2) = \mathbf{B}_{\mathcal{H}}(\vec{v}_1, \vec{v}_2)\mathbf{B_O}(\vec{v}_1, \vec{v}_2), \tag{9-45}$$

where

$$\begin{aligned}\mathbf{B_O}(\vec{v}_1, \vec{v}_2) &= \mathbf{O}(\vec{v}_1)\mathbf{O}(\vec{v}_2)\mathbf{O}^*(\vec{v}_1 + \vec{v}_2) \\ \mathbf{B}_{\mathcal{H}}(\vec{v}_1, \vec{v}_2) &= \mathcal{H}_S(\vec{v}_1)\mathcal{H}_S(\vec{v}_2)\mathcal{H}_S^*(\vec{v}_1 + \vec{v}_2).\end{aligned} \tag{9-46}$$

Now consider a sequence of a large number of short-exposure images, each taken with a different realization of the atmospheric turbulence, but with no changes of the object between exposures. The image bispectrum averaged over the ensemble of pictures is, to a good approximation,[7]

$$\overline{\mathbf{B}_I}(\vec{v}_1, \vec{v}_2) = \mathbf{B_O}(\vec{v}_1, \vec{v}_2)\overline{\mathbf{B}_{\mathcal{H}}}(\vec{v}_1, \vec{v}_2), \tag{9-47}$$

where $\overline{\mathbf{B}_{\mathcal{H}}}(\vec{v}_1, \vec{v}_2)$ can reasonably be called the *bispectrum transfer function*.

The first question to be answered is whether the bispectrum transfer function preserves information about the object bispectrum beyond frequency $r_0/(\bar{\lambda}z_i)$. Let $\vec{v}_1 = \vec{v} \gg r_0/(\bar{\lambda}z_i)$ and $\vec{v}_2 = \vec{\Delta v}$ with $|\vec{\Delta v}|$ satisfying Eq. (9-40). Then

$$\overline{\mathbf{B}_{\mathcal{H}}}(\vec{v}, \vec{\Delta v}) = \overline{\mathcal{H}_S(\vec{v})\mathcal{H}_S(\vec{\Delta v})\mathcal{H}_S^*(\vec{v}+\vec{\Delta v})} \approx \overline{\mathcal{H}_S(\vec{\Delta v})}\;\overline{\mathcal{H}_S(\vec{v})\mathcal{H}_S^*(\vec{v}+\vec{\Delta v})} \tag{9-48}$$

where the fact that $\vec{v}$ and $\vec{v} + \vec{\Delta v}$ are much greater than $\vec{\Delta v}$ assures lack of correlation of the first term with the following two terms. The circular complex Gaussian statistics of $\mathcal{H}_S$ assure that lack of correlation also means statistical independence. The quantity $\overline{\mathcal{H}_S(\vec{\Delta v})}$ has significant value for $|\vec{\Delta v}| < r_0/(\bar{\lambda}z_i)$, and the quantity $\overline{\mathcal{H}_S(\vec{v})\mathcal{H}_S^*(\vec{v}+\vec{\Delta v})}$ is simply the cross-spectrum, which we know retains finite value beyond $|\vec{v}| = r_0/(\bar{\lambda}z_i)$. Note that both factors on the right-hand side of the above equation are real-valued, and therefore it is safe to assume that the bispectrum transfer function is real-valued. Thus the bispectral transfer function does preserve information beyond frequency $r_0/(\bar{\lambda}z_i)$. It is assumed that the bispectral transfer function can be found from images of a star under atmospheric conditions statistically similar to those present when the full image is formed.

Derivation of more complete expressions for the bispectral transfer function can be found in [102] and [8], but will not be pursued here.

9.7.2 Recovering Full Object Information from the Bispectrum

Given that the bispectrum transfer function retains nonzero value for mid and high frequencies of the diffraction-limited OTF, the remaining question is how to extract the magnitude and phase of the object spectrum from the object bispectrum. The magnitude $|\mathbf{O}|$ of the object spectrum can be obtained by processing the image data according to the methods appropriate for Labeyrie speckle interferometry. The re-

---

7. The only approximation is that we have represented a finite average over a large number of photographs by a statistical ensemble average.

maining question is how to extract object phase information from the object bispectrum. Note that

$$\mathbf{B_O}(\vec{v}_1, \vec{v}_2) = \mathbf{O}(\vec{v}_1)\mathbf{O}(\vec{v}_2)\mathbf{O}^*(\vec{v}_1 + \vec{v}_2)$$
$$= |\mathbf{O}(\vec{v}_1)||\mathbf{O}(\vec{v}_2)||\mathbf{O}^*(\vec{v}_1 + \vec{v}_2)| \exp[j(\phi_o(\vec{v}_1) + \phi_o(\vec{v}_2) - \phi_o(\vec{v}_1 + \vec{v}_2))]. \quad (9\text{-}49)$$

Thus the phase of the object bispectrum, represented here by the symbol $\Delta\phi(\vec{v}_1;\vec{v}_2)$, which is a measurable quantity, is given by

$$\Delta\phi(\vec{v}_1;\vec{v}_2) = \phi_o(\vec{v}_1) + \phi_o(\vec{v}_2) - \phi_o(\vec{v}_1 + \vec{v}_2). \quad (9\text{-}50)$$

Replacing $\vec{v}_1$ by $\vec{v}$ and $\vec{v}_2$ by $\vec{\Delta v}$, and rearranging, we obtain

$$\phi_o(\vec{v} + \vec{\Delta v}) - \phi_o(\vec{v}) = \Delta\phi(\vec{v};\vec{\Delta v}) - \phi_o(\vec{\Delta v}), \quad (9\text{-}51)$$

providing a recursive equation from which the phase distribution can in principle be obtained, as will now be discussed.

The object spectrum is to be found on a grid of frequency coordinates. The value of ϕ_o at the origin in the frequency domain is known to be zero, since the object itself is real-valued and positive. In addition, the assumption is made that $\phi_o(\pm\Delta v_X, 0)$ and $\phi_o(0, \pm\Delta v_Y)$ are all zero. This assumption can be shown to be equivalent to giving up information regarding the centroid or translation of the object. These five zero values provide a sufficient starting point for finding the phase on the entire array of sample points. For example, to find $\phi_o(2\Delta v_X, 0)$,

$$\phi_o(2\Delta v_X, 0) - \phi_o(\Delta v_X, 0) = \Delta\phi(2\Delta v_X, 0;\Delta v_X, 0) - \phi_o(\Delta v_X, 0), \quad (9\text{-}52)$$

or

$$\phi_o(2\Delta v_X, 0) = \Delta\phi(2\Delta v_X, 0;\Delta v_X, 0). \quad (9\text{-}53)$$

This method of recursively finding the phases of the spectrum was recognized by Roddier [136] as related to the *phase closure* principle known previously in radio astronomy. Earlier application of phase closure principles in imaging through unknown aberrations can be found in [131].

It is possible to take different paths through the (v_X, v_Y) plane to find the spectrum at a particular frequency pair, but again different paths often produce results that differ by integer multiples of 2π, making noise averaging impossible. The issues raised in Section 8.3.3 are again important. Other methods for recovering phase information have also been studied (see, for example, [109]). We have not covered the noise limitations of the technique since it is beyond our scope here. Noise is the most important limiting factor in determining the quality of the recovered images. See [137] and [8] for such considerations.

9.8 Speckle Correlography

Until now, we have considered only the problem of imaging *incoherent* objects in the presence of atmospheric turbulence. We turn now to the subject of laser

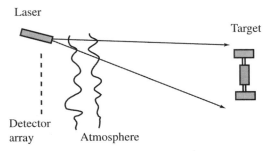

Figure 9.10 Speckle correlography geometry.

correlography, first introduced by Elbaum [38], which requires *coherent* illumination of the object of interest and has immunity to atmospherically induced phase errors of the received wavefront. For additional references on speckle correlography, see [77], [48], [46] and [47].

Consider the detection geometry shown in Fig. 9.10. A distant optically rough object is illuminated by coherent light, and the resulting speckle pattern is directly detected by a detector array. The receiving geometry shown has no conventional optics, but in some cases it may be helpful to introduce optics that image a fixed plane just before the optics onto a detector array. Also shown is atmospheric turbulence before the detector array. If the turbulence is close to the detector, it introduces primarily phase changes in the arriving wavefront, but since it is the intensity of that wavefront that is being detected, the phase has no effect on the detected signal. The atmosphere can therefore be neglected in the discussion that follows.

The speckle pattern intensity produced by the distant object is detected by the detector array, and a Fourier transform operation is performed digitally on that intensity pattern. The squared magnitude of this Fourier transform is taken, yielding the power spectrum of the speckle pattern. Recall the result of Eq. (4-59), namely that the *ensemble average* power spectral density of the speckle pattern is, aside from the spike at the origin, the autocorrelation function of the object brightness distribution. Unfortunately, the power spectrum derived from a single sample function is itself filled with speckle. Figure 9.11 shows an object brightness distribution, one realization of a speckle pattern at the detector, the power spectral density of that sample function, and the ideal ensemble-average power spectral density of the speckle. The average power spectral density of the speckle is the autocorrelation function of the object brightness distribution. In both power spectral densities the spectral spike at the origin has been suppressed.

To obtain a good approximation to the ensemble average power spectral density, rather than the sample power spectral density, some method for averaging speckle must be found. Several such methods have been discussed in Chapter 5: polarization averaging, angle-of-illumination averaging and wavelength averaging. However, in practice, to obtain an average of a large enough number of speckle patterns, one

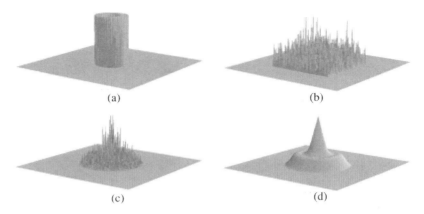

Figure 9.11 (a) Original object, with a donut intensity distribution; (b) speckle pattern produced by the object; (c) power spectral density of the one sample function of speckle; and (d) ensemble-average power spectral density of the speckle.

must probably depend on motion of the object itself. Small translations or rotations of the object can change the speckle pattern sufficiently to decorrelate the speckle, allowing a multitude of independent realizations of the power spectral density to be averaged. If the number of independent realizations is sufficiently large, a reasonable approximation to the ensemble average result will be obtained.

Finally, knowing a reasonably accurate estimate of the object autocorrelation function, is it possible to find the object that gave rise to it? The general answer is "no," but often the practical answer in two or more dimensions is "yes." To do so requires a method for recovering the phase information of the object spectrum from knowledge of the magnitude of that spectrum, plus possible other a priori information. The magnitude information is obtainable from the power spectral density. Fienup [45] has pioneered the subject of phase retrieval for two-dimensional objects, and has demonstrated that, with the right kind of a priori information, such as a spatial bound on the object (which can be determined from the autocorrelation function) and a nonnegativity constraint on the object intensity, phase can be retrieved from magnitude by means of iterative algorithms. To cover such algorithms here would take us beyond the subject matter of primary interest, and therefore the reader is referred to the literature regarding this subject. See, for example, [44] and [45].

Noise limitations of the speckle correlography method have not been considered here, since they are outside of our scope, but the interested reader can find such considerations in [46].

A Linear Transformations of Speckle Fields

In this appendix we explore whether a linear transformation of a vector with components consisting of circular complex Gaussian random variables yields a new vector with components that are likewise circular complex Gaussian random variables. Let the original vector be

$$\underline{\mathbf{A}} = \begin{bmatrix} \mathbf{A}_1 \\ \mathbf{A}_2 \\ \vdots \\ \mathbf{A}_N \end{bmatrix}, \tag{A-1}$$

where $\mathbf{A}_1, \mathbf{A}_2, \ldots, \mathbf{A}_N$ are known to be circular complex Gaussian random variables. Consider a new vector $\underline{\mathbf{A}}'$ defined by

$$\underline{\mathbf{A}}' = \underline{\mathcal{L}}\underline{\mathbf{A}}, \tag{A-2}$$

with

$$\underline{\mathbf{A}}' = \begin{bmatrix} \mathbf{A}'_1 \\ \mathbf{A}'_2 \\ \vdots \\ \mathbf{A}'_N \end{bmatrix} \tag{A-3}$$

and

$$\underline{\mathcal{L}} = \begin{bmatrix} \mathbf{L}_{11} & \mathbf{L}_{12} & \cdots & \mathbf{L}_{1N} \\ \mathbf{L}_{21} & \mathbf{L}_{22} & \cdots & \mathbf{L}_{2N} \\ \vdots & \vdots & & \vdots \\ \mathbf{L}_{N1} & \mathbf{L}_{N2} & \cdots & \mathbf{L}_{NN} \end{bmatrix}. \tag{A-4}$$

APPENDIX A Linear Transformations of Speckle Fields

The Gaussianity of the elements of the transformed matrix $\underline{\mathbf{A}}'$ is guaranteed by the fact that any *linear* transformation of Gaussian random variables yields Gaussian random variables ([70], p. 39). However, there is no obvious guarantee that a linear transformation of circular complex Gaussian random variables yields new random variables that are *circular*.

Before proceeding further, we recall some fundamental properties of circular Gaussian random variables. Let $\mathcal{R}_p$, $\mathcal{I}_p$ represent, respectively, the real and imaginary parts of random variable $\mathbf{A}_p$. Then if $\mathbf{A}_p$ is a circular complex Gaussian random variable, the following relations hold true (see [70], p. 42; indeed these relations define circularity):

$$\overline{\mathcal{R}_p} = \overline{\mathcal{I}_p} = 0, \quad \overline{\mathcal{R}_p^2} = \overline{\mathcal{I}_p^2}, \quad \overline{\mathcal{R}_p \mathcal{I}_p} = 0. \tag{A-5}$$

With the above as background, we are now ready to explore the conditions under which an arbitrary element $\mathbf{A}'_k$ of the transformed vector $\underline{\mathbf{A}}'$ is circular complex Gaussian. The real and imaginary parts of $\underline{\mathbf{A}}'$ can be expressed as

$$\begin{aligned} \mathcal{R}'_k &= \frac{1}{2}[\mathbf{A}'_k + \mathbf{A}'^*_k] \\ \mathcal{I}'_k &= \frac{1}{2j}[\mathbf{A}'_k - \mathbf{A}'^*_k], \end{aligned} \tag{A-6}$$

where * signifies complex conjugate. In addition,

$$\mathbf{A}'_k = \sum_{p=1}^{N} \mathbf{L}_{kp} \mathbf{A}_p, \tag{A-7}$$

and therefore

$$\overline{\mathcal{R}'_k} = \frac{1}{2}\left[\sum_{p=1}^{N} \mathbf{L}_{kp}\overline{\mathbf{A}_p} + \mathbf{L}^*_{kp}\overline{\mathbf{A}^*_p}\right]$$

$$\overline{\mathcal{I}'_k} = \frac{1}{2j}\left[\sum_{p=1}^{N} \mathbf{L}_{kp}\overline{\mathbf{A}_p} - \mathbf{L}^*_{kp}\overline{\mathbf{A}^*_p}\right] \tag{A-8}$$

$$\overline{\mathcal{R}'^2_k} = \frac{1}{4}\left[\sum_{p=1}^{N}\sum_{q=1}^{N} \mathbf{L}_{kp}\mathbf{L}_{kq}\overline{\mathbf{A}_p\mathbf{A}_q} + \mathbf{L}^*_{kp}\mathbf{L}^*_{kq}\overline{\mathbf{A}^*_p\mathbf{A}^*_q} + \mathbf{L}_{kp}\mathbf{L}^*_{kq}\overline{\mathbf{A}_p\mathbf{A}^*_q} + \mathbf{L}^*_{kp}\mathbf{L}_{kq}\overline{\mathbf{A}^*_p\mathbf{A}_q}\right]$$

$$\overline{\mathcal{I}'^2_k} = \frac{1}{4}\left[\sum_{p=1}^{N}\sum_{q=1}^{N} \mathbf{L}_{kp}\mathbf{L}^*_{kq}\overline{\mathbf{A}_p\mathbf{A}^*_q} + \mathbf{L}^*_{kp}\mathbf{L}_{kq}\overline{\mathbf{A}^*_p\mathbf{A}_q} - \mathbf{L}^*_{kp}\mathbf{L}^*_{kq}\overline{\mathbf{A}^*_p\mathbf{A}^*_q} - \mathbf{L}_{kp}\mathbf{L}_{kq}\overline{\mathbf{A}_p\mathbf{A}_q}\right] \tag{A-9}$$

$$\overline{\mathcal{R}'_k \mathcal{I}'_k} = \frac{1}{4}\left[\sum_{p=1}^{N}\sum_{q=1}^{N} \mathbf{L}_{kp}\mathbf{L}_{kq}\overline{\mathbf{A}_p\mathbf{A}_q} + \mathbf{L}^*_{kp}\mathbf{L}_{kq}\overline{\mathbf{A}^*_p\mathbf{A}_q} - \mathbf{L}^*_{kp}\mathbf{L}^*_{kq}\overline{\mathbf{A}^*_p\mathbf{A}^*_q} - \mathbf{L}_{kp}\mathbf{L}^*_{kq}\overline{\mathbf{A}_p\mathbf{A}^*_q}\right]. \tag{A-10}$$

The zero means of the real and imaginary parts, shown in Eqs. (A-5), imply:
$$\overline{\mathbf{A}_p} = \overline{\mathbf{A}_p^*} = 0. \tag{A-11}$$
A second relation that is straightforward to prove is:
$$\overline{\mathbf{A}_p \mathbf{A}_q^*} = (\overline{\mathbf{A}_p^* \mathbf{A}_q})^*. \tag{A-12}$$
A third relation, requiring proof, is
$$\overline{\mathbf{A}_p \mathbf{A}_q} = \overline{\mathbf{A}_p^* \mathbf{A}_q^*} = 0. \tag{A-13}$$
To prove this relationship, we argue that for circular random variables, the statistics are the same in all directions in the complex plane. Therefore the value of $\overline{\mathbf{A}_p \mathbf{A}_q}$ should be unchanged if we rotate the real and imaginary axes 90 degrees in the counterclockwise direction. Under such a rotation we induce the transformation
$$\begin{aligned} R_p &\to I_p \\ I_p &\to -R_p \end{aligned} \tag{A-14}$$
and similarly for R_q and I_q. Equating the two expressions for $\overline{\mathbf{A}_p \mathbf{A}_q}$ that result, we find that this quantity must be zero. A similar argument shows that $\overline{\mathbf{A}_p^* \mathbf{A}_q^*} = 0$.

Finally, we can also make use of the result
$$\overline{|\mathbf{A}_k'|^2} = \sum_{p=0}^{N} \sum_{q=0}^{N} \mathbf{L}_{kp} \mathbf{L}_{kq}^* \overline{\mathbf{A}_p \mathbf{A}_q^*}, \tag{A-15}$$
which we note must be a real and nonnegative quantity.

With these results substituted into the appropriate expressions, we find
$$\overline{\mathcal{R}_k'} = \overline{\mathcal{I}_k'} = 0$$
$$\overline{\mathcal{R}_k'^2} = \overline{\mathcal{I}_k'^2} = \frac{1}{2} \operatorname{Re}\left(\sum_{p=0}^{N} \sum_{q=0}^{N} \mathbf{L}_{kp} \mathbf{L}_{kq}^* \overline{\mathbf{A}_p \mathbf{A}_q^*} \right) = \frac{1}{2} \operatorname{Re}(\overline{|\mathbf{A}_k'|^2}) = \frac{1}{2} \overline{|\mathbf{A}_k'|^2} \tag{A-16}$$
$$\overline{\mathcal{R}_k' \mathcal{I}_k'} = \frac{-j}{2} \operatorname{Im}\left(\sum_{p=0}^{N} \sum_{q=0}^{N} \mathbf{L}_{kp} \mathbf{L}_{kq}^* \overline{\mathbf{A}_p \mathbf{A}_q^*} \right) = \frac{-j}{2} \operatorname{Im}(\overline{|\mathbf{A}_k'|^2}) = 0.$$

We see that the means and the variances of the real and imaginary parts of the $\mathbf{A}_k'$ satisfy the requirements for a circular complex Gaussian variable, and in addition we see that the real and imaginary parts of that random variable are also uncorrelated. The conclusion follows that *any linear transformation of a set of circular complex Gaussian random variables yields a new set of circular complex Gaussian random variables.*

B Contrast of Partially Developed Speckle

In this appendix, we present some of the details of the derivation that leads to the expressions for the moments and the contrast of partially developed speckle, especially Eqs. (3-92) and (3-94).

Specifically, we wish to first calculate the second moment of the intensity,

$$\overline{I^2} = \frac{1}{N^2} \sum_{n=1}^{N} \sum_{m=1}^{N} \sum_{p=1}^{N} \sum_{q=1}^{N} \overline{a_n a_m a_p a_q} \overline{e^{j(\phi_n - \phi_m + \phi_p - \phi_q)}} \qquad \text{(B-1)}$$

where we have assumed, as usual, that the amplitudes and phases of the component phasors are independent.

For this sum, there are 15 different cases that must be considered, as indicated below:

1. $n = m = p = q$, giving N identical terms. The sum of these terms yield a contribution

$$\frac{1}{N^2} N \overline{a^4} = \frac{1}{N} \overline{a^4}, \qquad \text{(B-2)}$$

where the a_n have been assumed to be identically distributed for all n.

2. $n = m$, $p = q$, $n \neq p$, giving $N(N-1)$ identical terms. The sum of these terms yields a contribution

$$\frac{1}{N^2} N(N-1)(\overline{a^2})^2 = \left(1 - \frac{1}{N}\right)(\overline{a^2})^2. \qquad \text{(B-3)}$$

3. $n = m$, $p \neq q \neq n$, giving $N(N-1)(N-2)$ identical terms. The sum of these terms yields

$$\frac{1}{N^2} N(N-1)(N-2)\overline{a^2}(\bar{a})^2 \mathbf{M}_\phi(1)\mathbf{M}_\phi(-1)$$
$$= \left(1 - \frac{1}{N}\right)(N-2)\overline{a^2}(\bar{a})^2 \mathbf{M}_\phi(1)\mathbf{M}_\phi(-1), \quad \text{(B-4)}$$

where $\mathbf{M}_\phi(\omega)$ is the characteristic function[1] of the random phase ϕ, evaluated at argument ω.

4. $n = p$, $m = q$, $n \neq m$, giving $N(N-1)$ identical terms. The sum of these terms is

$$\frac{N(N-1)}{N^2}(\overline{a^2})^2 \mathbf{M}_\phi(2)\mathbf{M}_\phi(-2) = \frac{1}{N}(N-1)(\overline{a^2})^2 \mathbf{M}_\phi(2)\mathbf{M}_\phi(-2). \quad \text{(B-5)}$$

5. $n = p$, $m \neq q \neq n$, giving $N(N-1)(N-2)$ identical terms. The sum of these terms is

$$\frac{N(N-1)(N-2)}{N^2}\overline{a^2}(\bar{a})^2 \mathbf{M}_\phi(2)\mathbf{M}_\phi(-1)\mathbf{M}_\phi(-1)$$
$$= \left(1 - \frac{1}{N}\right)(N-2)\overline{a^2}(\bar{a})^2 \mathbf{M}_\phi(-1)\mathbf{M}_\phi(-1)\mathbf{M}_\phi(2). \quad \text{(B-6)}$$

6. $n = q$, $m = p$, $n \neq m$, giving $N(N-1)$ identical terms. The sum of these terms is

$$\frac{N(N-1)}{N^2}\overline{a^2}(\bar{a})^2 = \left(1 - \frac{1}{N}\right)(\overline{a^2})^2. \quad \text{(B-7)}$$

7. $n = q$, $m \neq p \neq n$, giving $N(N-1)(N-2)$ identical terms. The sum of these terms is

$$\frac{N(N-1)(N-2)}{N^2}\overline{a^2}(\bar{a})^2 \mathbf{M}_\phi(1)\mathbf{M}_\phi(-1)$$
$$= \left(1 - \frac{1}{N}\right)(N-2)\overline{a^2}(\bar{a})^2 \mathbf{M}_\phi(1)\mathbf{M}_\phi(-1). \quad \text{(B-8)}$$

8. $n = m = p$, $n \neq q$, giving $N(N-1)$ identical terms. The sum of these terms is

$$\frac{N(N-1)}{N^2}\overline{a^3}\bar{a}\mathbf{M}_\phi(1)\mathbf{M}_\phi(-1) = \left(1 - \frac{1}{N}\right)\overline{a^3}\bar{a}\mathbf{M}_\phi(1)\mathbf{M}_\phi(-1). \quad \text{(B-9)}$$

9. $n = m = q$, $n \neq p$, giving $N(N-1)$ identical terms. The sum of these terms is

$$\frac{N(N-1)}{N^2}\overline{a^3}\bar{a}\mathbf{M}_\phi(1)\mathbf{M}_\phi(-1) = \left(1 - \frac{1}{N}\right)\overline{a^3}\bar{a}\mathbf{M}_\phi(1)\mathbf{M}_\phi(-1). \quad \text{(B-10)}$$

---

1. Note that, due to the fact that the probability density function of ϕ is real-valued, $\mathbf{M}_\phi(-\omega) = \mathbf{M}_\phi^*(\omega)$.

10. $n = p = q$, $n \neq m$, giving $N(N - 1)$ identical terms. The sum of these terms is

$$\frac{N(N - 1)}{N^2}\overline{a^3}\bar{a}\mathbf{M}_\phi(1)\mathbf{M}_\phi(-1) = \left(1 - \frac{1}{N}\right)\overline{a^3}\bar{a}\mathbf{M}_\phi(1)\mathbf{M}_\phi(-1). \quad \text{(B-11)}$$

11. $p = q = m$, $n \neq q$, giving $N(N - 1)$ identical terms. The sum of these terms is

$$\frac{N(N - 1)}{N^2}\overline{a^3}\bar{a}\mathbf{M}_\phi(1)\mathbf{M}_\phi(-1) = \left(1 - \frac{1}{N}\right)\overline{a^3}\bar{a}\mathbf{M}_\phi(1)\mathbf{M}_\phi(-1). \quad \text{(B-12)}$$

12. $n \neq m \neq p \neq q$, giving $N(N - 1)(N - 2)(N - 3)$ identical terms. The sum of these terms is

$$\frac{N(N - 1)(N - 2)(N - 3)}{N^2}(\bar{a})^4 \mathbf{M}_\phi^2(1)\mathbf{M}_\phi^2(-1)$$

$$= \left(1 - \frac{1}{N}\right)(N - 2)(N - 3)(\bar{a})^4 \mathbf{M}_\phi^2(1)\mathbf{M}_\phi^2(-1). \quad \text{(B-13)}$$

13. $p = q$, $n \neq m \neq p$, giving $N(N - 1)(N - 2)$ identical terms. The sum of these terms is

$$\frac{N(N - 1)(N - 2)}{N^2}\overline{a^2}(\bar{a})^2 \mathbf{M}_\phi(1)\mathbf{M}_\phi(-1)$$

$$= \left(1 - \frac{1}{N}\right)(N - 2)\overline{a^2}(\bar{a})^2 \mathbf{M}_\phi(1)\mathbf{M}_\phi(-1). \quad \text{(B-14)}$$

14. $m = q$, $n \neq m \neq p$, giving $N(N - 1)(N - 2)$ identical terms. The sum of these terms is

$$\frac{N(N - 1)(N - 2)}{N^2}\overline{a^2}(\bar{a})^2 \mathbf{M}_\phi(1)\mathbf{M}_\phi(1)\mathbf{M}_\phi(-2)$$

$$= \left(1 - \frac{1}{N}\right)(N - 2)\overline{a^2}(\bar{a})^2 \mathbf{M}_\phi(1)\mathbf{M}_\phi(1)\mathbf{M}_\phi(-2). \quad \text{(B-15)}$$

15. $m = p$, $n \neq m \neq q$, giving $N(N - 1)(N - 2)$ identical terms. The sum of these terms is

$$\frac{N(N - 1)(N - 2)}{N^2}\overline{a^2}(\bar{a})^2 \mathbf{M}_\phi(1)\mathbf{M}_\phi(-1)$$

$$= \left(1 - \frac{1}{N}\right)(N - 2)\overline{a^2}(\bar{a})^2 \mathbf{M}_\phi(1)\mathbf{M}_\phi(-1). \quad \text{(B-16)}$$

The total number of terms represented by the above 15 cases is N^4.

The next task is to sum all 15 terms denumerated above. A symbolic manipulation program, such as *Mathematica*, can be a great help in this task. The result is found to be

$$\overline{I^2} = \frac{1}{N}\overline{a^4} + 2\left(1 - \frac{1}{N}\right)(\overline{a^2})^2$$
$$+ 4\left(1 - \frac{1}{N}\right)(N-2)\overline{a^2}(\bar{a})^2 \mathbf{M}_\phi(1)\mathbf{M}_\phi(-1) + 4\left(1 - \frac{1}{N}\right)\overline{a^3}\bar{a}\mathbf{M}_\phi(1)\mathbf{M}_\phi(-1)$$
$$+ \left(1 - \frac{1}{N}\right)(N-2)\overline{a^2}(\bar{a})^2 \mathbf{M}_\phi(-1)\mathbf{M}_\phi(-1)\mathbf{M}_\phi(2)$$
$$+ \left(1 - \frac{1}{N}\right)(N-2)\overline{a^2}(\bar{a})^2 \mathbf{M}_\phi(1)\mathbf{M}_\phi(1)\mathbf{M}_\phi(-2)$$
$$+ \left(1 - \frac{1}{N}\right)(N-2)(N-3)(\bar{a})^4 \mathbf{M}_\phi^2(1)\mathbf{M}_\phi^2(-1)$$
$$+ \left(1 - \frac{1}{N}\right)(\overline{a^2})^2 \mathbf{M}_\phi(2)\mathbf{M}_\phi(-2). \tag{B-17}$$

To find the variance of the intensity, we must subtract the square of the mean intensity (i.e. the square of Eq. (3-91)),

$$\bar{I}^2 = [\overline{a^2} + (N-1)(\bar{a})^2 \mathbf{M}_\phi(1)\mathbf{M}_\phi(-1)]^2$$
$$= (\overline{a^2})^2 + 2(N-1)\overline{a^2}(\bar{a})^2 \mathbf{M}_\phi(1)\mathbf{M}_\phi(-1) + (N-1)^2(\bar{a})^4 \mathbf{M}_\phi^2(1)\mathbf{M}_\phi^2(-1). \tag{B-18}$$

The resulting expression is sufficiently complex that we prefer to make some simplifying assumptions at this point. Let the lengths of all the elementary phasors be unity ($a_k = 1$, all k), in which case $\overline{a^4} = \overline{a^3} = \overline{a^2} = \bar{a} = 1$. In addition, we assume that the probability density function for all the independent phases ϕ_k is Gaussian with zero mean,

$$p_\phi(\phi) = \frac{1}{\sqrt{2\pi}\sigma_\phi} \exp\left(-\frac{\phi^2}{2\sigma_\phi^2}\right), \tag{B-19}$$

in which case the characteristic function is given by

$$\mathbf{M}_\phi(\omega) = \exp\left(-\frac{1}{2}\omega^2\sigma_\phi^2\right). \tag{B-20}$$

Use of these assumptions allows us (after considerable simplification) to express the variance of the intensity as

$$\sigma_I^2 = 8\left(1 - \frac{1}{N}\right)e^{-2\sigma_\phi^2}[N - 1 + \cosh(\sigma_\phi^2)]\sinh^2\left(\frac{\sigma_\phi^2}{2}\right). \tag{B-21}$$

The contrast of the speckle follows directly as

$$C = \frac{\sigma_I}{\bar{I}} = \sqrt{\frac{8(N-1)[N - 1 + \cosh(\sigma_\phi^2)]\sinh^2\left(\frac{\sigma_\phi^2}{2}\right)}{N(N - 1 + e^{\sigma_\phi^2})^2}}, \tag{B-22}$$

as claimed in Eq. (3-94).

C Calculations Leading to the Statistics of the Derivatives of Intensity and Phase

C.1 The Correlation Matrix

In this section, we aim to find the detailed form of the correlation matrix of Eq. (4-174),

$$\mathcal{C} = \begin{bmatrix} \Gamma_{\mathcal{RR}} & \Gamma_{\mathcal{RI}} & \Gamma_{\mathcal{RR}_x} & \Gamma_{\mathcal{RI}_x} & \Gamma_{\mathcal{RR}_y} & \Gamma_{\mathcal{RI}_y} \\ \Gamma_{\mathcal{IR}} & \Gamma_{\mathcal{II}} & \Gamma_{\mathcal{IR}_x} & \Gamma_{\mathcal{II}_x} & \Gamma_{\mathcal{IR}_y} & \Gamma_{\mathcal{II}_y} \\ \Gamma_{\mathcal{R}_x\mathcal{R}} & \Gamma_{\mathcal{R}_x\mathcal{I}} & \Gamma_{\mathcal{R}_x\mathcal{R}_x} & \Gamma_{\mathcal{R}_x\mathcal{I}_x} & \Gamma_{\mathcal{R}_x\mathcal{R}_y} & \Gamma_{\mathcal{R}_x\mathcal{I}_y} \\ \Gamma_{\mathcal{I}_x\mathcal{R}} & \Gamma_{\mathcal{I}_x\mathcal{I}} & \Gamma_{\mathcal{I}_x\mathcal{R}_x} & \Gamma_{\mathcal{I}_x\mathcal{I}_x} & \Gamma_{\mathcal{I}_x\mathcal{R}_y} & \Gamma_{\mathcal{I}_x\mathcal{I}_y} \\ \Gamma_{\mathcal{R}_y\mathcal{R}} & \Gamma_{\mathcal{R}_y\mathcal{I}} & \Gamma_{\mathcal{R}_y\mathcal{R}_x} & \Gamma_{\mathcal{R}_y\mathcal{I}_x} & \Gamma_{\mathcal{R}_y\mathcal{R}_y} & \Gamma_{\mathcal{R}_y\mathcal{I}_y} \\ \Gamma_{\mathcal{I}_y\mathcal{R}} & \Gamma_{\mathcal{I}_y\mathcal{I}} & \Gamma_{\mathcal{I}_y\mathcal{R}_x} & \Gamma_{\mathcal{I}_y\mathcal{I}_x} & \Gamma_{\mathcal{I}_y\mathcal{R}_y} & \Gamma_{\mathcal{I}_y\mathcal{I}_y} \end{bmatrix}. \quad \text{(C-1)}$$

Similar analyses have been carried out by Blackman (see [18], p. 60) in one dimension and by Ochoa and Goodman [119] in two dimensions.

To evaluate the terms in the matrix, we first make use of the known facts that, for circular complex Gaussian variables,

$$\Gamma_{\mathcal{RR}} = \Gamma_{\mathcal{II}} = \sigma^2 \qquad \Gamma_{\mathcal{RI}} = \Gamma_{\mathcal{IR}} = 0$$
$$\Gamma_{\mathcal{RR}}(\Delta x, \Delta y) = \Gamma_{\mathcal{II}}(\Delta x, \Delta y) \qquad \Gamma_{\mathcal{RI}}(\Delta x, \Delta y) = -\Gamma_{\mathcal{IR}}(\Delta x, \Delta y) \quad \text{(C-2)}$$

where $\Gamma_{\mathcal{R},\mathcal{I}}(\Delta x, \Delta y) = \overline{\mathcal{R}(x,y)\mathcal{I}(x-\Delta x, y-\Delta y)}$ and the other auto- and cross-correlation functions are defined analogously. Following from these relations, we have

$$\Gamma_\mathbf{A} = 2\Gamma_{\mathcal{RR}} - 2j\Gamma_{\mathcal{RI}}. \quad \text{(C-3)}$$

Further progress requires developing some results regarding auto- and cross-correlations involving derivatives of random processes. We begin by recalling the van Cittert–Zernike theorem (cf. Eq. (4-55)), namely that

APPENDIX C Statistics of the Derivatives of Intensity and Phase

$$\mathbf{\Gamma}_A(\Delta x, \Delta y) = \frac{\kappa}{\lambda^2 z^2} \iint_{-\infty}^{\infty} I(\alpha, \beta) e^{-j\frac{2\pi}{\lambda z}[\alpha \Delta x + \beta \Delta y]} \, d\alpha \, d\beta.$$

From Eq. (C-3), we must have

$$\mathbf{\Gamma}_{\mathcal{R}\mathcal{R}}(\Delta x, \Delta y) = \frac{\kappa}{2\lambda^2 z^2} \iint_{-\infty}^{\infty} I(\alpha, \beta) \cos\left[\frac{2\pi}{\lambda z}(\alpha \Delta x + \beta \Delta y)\right] d\alpha \, d\beta$$

$$\mathbf{\Gamma}_{\mathcal{R}\mathcal{I}}(\Delta x, \Delta y) = \frac{\kappa}{2\lambda^2 z^2} \iint_{-\infty}^{\infty} I(\alpha, \beta) \sin\left[\frac{2\pi}{\lambda z}(\alpha \Delta x + \beta \Delta y)\right] d\alpha \, d\beta.$$
(C-4)

In addition, we need the general result (cf. [124], p. 317, and [119]) that

$$\mathbf{\Gamma}_{\mathcal{P}_{x(n)} \mathcal{Q}_{y(m)}} = \left[(-1)^{n+m} \frac{\partial^n}{\partial \Delta x^n} \frac{\partial^m}{\partial \Delta y^m} \mathbf{\Gamma}_{\mathcal{P}\mathcal{Q}}(\Delta x, \Delta y)\right]_{\substack{\Delta x=0 \\ \Delta y=0}}, \quad \text{(C-5)}$$

where the symbol $\mathbf{\Gamma}_{\mathcal{P}_{x(n)} \mathcal{Q}_{y(m)}}$ signifies the cross-correlation[1] of the nth derivative of the random process $\mathcal{P}$ and the mth derivative of the random process $\mathcal{Q}$. Applying Eq. (C-5) to the equations above, and interchanging orders of differentiation and integration, yields the following relations:

$$\mathbf{\Gamma}_{\mathcal{R}\mathcal{R}} = \mathbf{\Gamma}_{\mathcal{I}\mathcal{I}} = \sigma^2$$

$$\mathbf{\Gamma}_{\mathcal{R}\mathcal{I}} = \mathbf{\Gamma}_{\mathcal{I}\mathcal{R}} = 0$$

$$\mathbf{\Gamma}_{\mathcal{R}_x \mathcal{R}} = \mathbf{\Gamma}_{\mathcal{R}\mathcal{R}_x} = -\left[\frac{\partial}{\partial \Delta x} \mathbf{\Gamma}_{\mathcal{R}\mathcal{R}}(\Delta x, \Delta y)\right]_{\substack{\Delta x=0 \\ \Delta y=0}} = 0$$

$$\mathbf{\Gamma}_{\mathcal{R}\mathcal{R}_y} = \mathbf{\Gamma}_{\mathcal{R}_y \mathcal{R}} = -\left[\frac{\partial}{\partial \Delta y} \mathbf{\Gamma}_{\mathcal{R}\mathcal{R}}(\Delta x, \Delta y)\right]_{\substack{\Delta x=0 \\ \Delta y=0}} = 0$$

$$\mathbf{\Gamma}_{\mathcal{I}\mathcal{I}_y} = \mathbf{\Gamma}_{\mathcal{I}_y \mathcal{I}} = -\left[\frac{\partial}{\partial \Delta y} \mathbf{\Gamma}_{\mathcal{I}\mathcal{I}}(\Delta x, \Delta y)\right]_{\substack{\Delta x=0 \\ \Delta y=0}} = 0$$

$$\mathbf{\Gamma}_{\mathcal{I}_x \mathcal{I}} = \mathbf{\Gamma}_{\mathcal{I}\mathcal{I}_x} = -\left[\frac{\partial}{\partial \Delta x} \mathbf{\Gamma}_{\mathcal{I}\mathcal{I}}(\Delta x, \Delta y)\right]_{\substack{\Delta x=0 \\ \Delta y=0}} = 0$$

$$\mathbf{\Gamma}_{\mathcal{R}_x \mathcal{I}_y} = -\mathbf{\Gamma}_{\mathcal{I}_y \mathcal{R}_x} = \left[\frac{\partial^2}{\partial \Delta x \partial \Delta y} \mathbf{\Gamma}_{\mathcal{R}\mathcal{I}}(\Delta x, \Delta y)\right]_{\substack{\Delta x=0 \\ \Delta y=0}} = 0$$

---

1. When a symbol Γ with any subscripts appears without the arguments $(\Delta x, \Delta y)$, it implies that it is the value of the correlation function with arguments Δx and Δy set equal to zero.

C.1 The Correlation Matrix

$$\Gamma_{\mathcal{R}_y\mathcal{I}_x} = -\Gamma_{\mathcal{I}_x\mathcal{R}_y} = \left[\frac{\partial^2}{\partial\Delta x\Delta y}\Gamma_{\mathcal{R}\mathcal{I}}(\Delta x,\Delta y)\right]_{\substack{\Delta x=0\\\Delta y=0}} = 0$$

$$\Gamma_{\mathcal{R}_x\mathcal{I}_x} = -\Gamma_{\mathcal{I}_x\mathcal{R}_x} = \left[\frac{\partial^2}{\partial\Delta x^2}\Gamma_{\mathcal{R}\mathcal{I}}(\Delta x,\Delta y)\right]_{\substack{\Delta x=0\\\Delta y=0}} = 0$$

$$\Gamma_{\mathcal{R}_y\mathcal{I}_y} = -\Gamma_{\mathcal{I}_y\mathcal{R}_y} = \left[\frac{\partial^2}{\partial\Delta y^2}\Gamma_{\mathcal{R}\mathcal{I}}(\Delta x,\Delta y)\right]_{\substack{\Delta x=0\\\Delta y=0}} = 0$$

$$\Gamma_{\mathcal{R}_x\mathcal{I}} = -\Gamma_{\mathcal{I}\mathcal{R}_x} = -\left[\frac{\partial}{\partial\Delta x}\Gamma_{\mathcal{R}\mathcal{I}}(\Delta x,\Delta y)\right]_{\substack{\Delta x=0\\\Delta y=0}} = \frac{\pi\kappa}{\lambda^3 z^3}\iint\limits_{-\infty}^{\infty}\alpha I(\alpha,\beta)\,d\alpha\,d\beta$$

$$\Gamma_{\mathcal{R}_y\mathcal{I}} = -\Gamma_{\mathcal{I}\mathcal{R}_y} = \left[\frac{\partial}{\partial\Delta y}\Gamma_{\mathcal{R}\mathcal{I}}(\Delta x,\Delta y)\right]_{\substack{\Delta x=0\\\Delta y=0}} = \frac{\pi\kappa}{\lambda^3 z^3}\iint\limits_{-\infty}^{\infty}\beta I(\alpha,\beta)\,d\alpha\,d\beta$$

$$\Gamma_{\mathcal{R}\mathcal{I}_x} = -\Gamma_{\mathcal{I}_x\mathcal{R}} = -\left[\frac{\partial}{\partial\Delta x}\Gamma_{\mathcal{R}\mathcal{I}}(\Delta x,\Delta y)\right]_{\substack{\Delta x=0\\\Delta y=0}} = \frac{\pi\kappa}{\lambda^3 z^3}\iint\limits_{-\infty}^{\infty}\alpha I(\alpha,\beta)\,d\alpha\,d\beta$$

$$\Gamma_{\mathcal{R}\mathcal{I}_y} = -\Gamma_{\mathcal{I}_y\mathcal{R}} = -\left[\frac{\partial}{\partial\Delta y}\Gamma_{\mathcal{R}\mathcal{I}}(\Delta x,\Delta y)\right]_{\substack{\Delta x=0\\\Delta y=0}} = \frac{\pi\kappa}{\lambda^3 z^3}\iint\limits_{-\infty}^{\infty}\beta I(\alpha,\beta)\,d\alpha\,d\beta$$

$$\Gamma_{\mathcal{R}_x\mathcal{R}_x} = \Gamma_{\mathcal{I}_x\mathcal{I}_x} = \left[\frac{\partial^2}{\partial\Delta x^2}\Gamma_{\mathcal{R}\mathcal{R}}(\Delta x,\Delta y)\right]_{\substack{\Delta x=0\\\Delta y=0}} = \frac{2\kappa\pi^2}{\lambda^4 z^4}\iint\limits_{-\infty}^{\infty}\alpha^2 I(\alpha,\beta)\,d\alpha\,d\beta$$

$$\Gamma_{\mathcal{R}_y\mathcal{R}_y} = \Gamma_{\mathcal{I}_y\mathcal{I}_y} = \left[\frac{\partial^2}{\partial\Delta y^2}\Gamma_{\mathcal{R}\mathcal{R}}(\Delta x,\Delta y)\right]_{\substack{\Delta x=0\\\Delta y=0}} = \frac{2\kappa\pi^2}{\lambda^4 z^4}\iint\limits_{-\infty}^{\infty}\beta^2 I(\alpha,\beta)\,d\alpha\,d\beta$$

$$\Gamma_{\mathcal{R}_x\mathcal{R}_y} = \Gamma_{\mathcal{R}_y\mathcal{R}_x} = \left[\frac{\partial^2}{\partial\Delta x\Delta y}\Gamma_{\mathcal{R}\mathcal{R}}(\Delta x,\Delta y)\right]_{\substack{\Delta x=\\\Delta y=0}} = \frac{2\kappa\pi^2}{\lambda^4 z^4}\iint\limits_{-\infty}^{\infty}\alpha\beta I(\alpha,\beta)\,d\alpha\,d\beta$$

$$\Gamma_{\mathcal{I}_x\mathcal{I}_y} = \Gamma_{\mathcal{I}_y\mathcal{I}_x} = \left[\frac{\partial^2}{\partial\Delta x\Delta y}\Gamma_{\mathcal{I}\mathcal{I}}(\Delta x,\Delta y)\right]_{\substack{\Delta x=\\\Delta y=0}} = \frac{2\kappa\pi^2}{\lambda^4 z^4}\iint\limits_{-\infty}^{\infty}\alpha\beta I(\alpha,\beta)\,d\alpha\,d\beta.$$

To simplify further, note that the (α,β) coordinates can be chosen to have their origin at the centroid of the scattering spot, in which case the equations with integrands containing $\alpha I(\alpha,\beta)$ and $\beta I(\alpha,\beta)$ all evaluate to zero. For simplicity, the remaining nonzero correlations are represented by

$$\sigma^2 = \frac{\kappa}{2\lambda^2 z^2} \iint_{-\infty}^{\infty} I(\alpha,\beta)\, d\alpha\, d\beta \tag{C-6}$$

$$b_x = \frac{2\kappa\pi^2}{\lambda^4 z^4} \iint_{-\infty}^{\infty} \alpha^2 I(\alpha,\beta)\, d\alpha\, d\beta \tag{C-7}$$

$$b_y = \frac{2\kappa\pi^2}{\lambda^4 z^4} \iint_{-\infty}^{\infty} \beta^2 I(\alpha,\beta)\, d\alpha\, d\beta \tag{C-8}$$

$$d = \frac{2\kappa\pi^2}{\lambda^4 z^4} \iint_{-\infty}^{\infty} \alpha\beta I(\alpha,\beta)\, d\alpha\, d\beta. \tag{C-9}$$

In addition, if $I(\alpha,\beta)$ is symmetric and the α and β axes are aligned with orthogonal axes of symmetry of the scattering spot, the parameter d is zero.

The correlation matrix can then be written in the comparatively simple form

$$C = \begin{bmatrix} \sigma^2 & 0 & 0 & 0 & 0 & 0 \\ 0 & \sigma^2 & 0 & 0 & 0 & 0 \\ 0 & 0 & b_x & 0 & d & 0 \\ 0 & 0 & 0 & b_x & 0 & d \\ 0 & 0 & d & 0 & b_y & 0 \\ 0 & 0 & 0 & d & 0 & b_y \end{bmatrix} = \begin{bmatrix} \sigma^2 & 0 & 0 & 0 & 0 & 0 \\ 0 & \sigma^2 & 0 & 0 & 0 & 0 \\ 0 & 0 & b_x & 0 & 0 & 0 \\ 0 & 0 & 0 & b_x & 0 & 0 \\ 0 & 0 & 0 & 0 & b_y & 0 \\ 0 & 0 & 0 & 0 & 0 & b_y \end{bmatrix}. \tag{C-10}$$

C.2 Joint Density Function of the Derivatives of Phase

Our goal in this section is to perform the integrals indicated in

$$p(\theta_x, \theta_y) = \int_0^\infty dI \int_{-\infty}^\infty dI_x \int_{-\infty}^\infty dI_y \int_{-\pi}^\pi d\theta$$
$$\times \frac{\exp\left[-\frac{4b_x b_y I^2 + \sigma^2(b_y I_x^2 + b_x I_y^2) + 4\sigma^2 I^2(b_y \theta_x^2 + b_x \theta_y^2)}{8I\sigma^2 b_x b_y}\right]}{64\pi^3 \sigma^2 b_x b_y}. \tag{C-11}$$

Since the integrand does not depend on θ, the first integration simply yields a factor of 2π, reducing the required integrations to

$$p(\theta_x, \theta_y) = \int_0^\infty dI \int_{-\infty}^\infty dI_x \int_{-\infty}^\infty dI_y$$
$$\times \frac{\exp\left[-\frac{4b_x b_y I^2 + \sigma^2(b_y I_x^2 + b_x I_y^2) + 4\sigma^2 I^2(b_y \theta_x^2 + b_x \theta_y^2)}{8I\sigma^2 b_x b_y}\right]}{32\pi^2 \sigma^2 b_x b_y}. \tag{C-12}$$

Next we integrate with respect to I_y, yielding

$$p(I, I_x, \theta_x, \theta_y) = \frac{\sqrt{I}\exp\left[-\frac{1}{2}I\left(\frac{\theta_y^2}{b_y}+\frac{1}{\sigma^2}\right)-\left(\frac{I_x^2+4I^2\theta_x^2}{8Ib_x}\right)\right]}{8\sqrt{2}\pi^{3/2}\sigma^2 b_x\sqrt{b_y}}. \quad \text{(C-13)}$$

Integration with respect to I_x is next, yielding

$$p(I, \theta_x, \theta_y) = \frac{I\exp\left[-\frac{1}{2}I\left(\frac{1}{\sigma^2}+\frac{\theta_x^2}{b_x}+\frac{\theta_y^2}{b_y}\right)\right]}{4\pi\sigma^2\sqrt{b_xb_y}}. \quad \text{(C-14)}$$

The final integration is with respect to I, after which we have our final result,

$$p(\theta_x, \theta_y) = \frac{\sigma^2/\pi\sqrt{b_xb_y}}{\left(1+\frac{\sigma^2}{b_x}\theta_x^2+\frac{\sigma^2}{b_y}\theta_y^2\right)^2}. \quad \text{(C-15)}$$

C.3 Joint Density Function of the Derivatives of Intensity

Our goal in this section is to perform the integrations required to evaluate the joint density function of the derivatives of intensity, as indicated in

$$p(I_x, I_y) = \int_0^\infty dI \int_{-\infty}^\infty d\theta_y \int_{-\infty}^\infty d\theta_x \int_{-\pi}^\pi d\theta$$

$$\times \frac{\exp\left[-\frac{4b_xb_yI^2+\sigma^2(b_yI_x^2+b_xI_y^2)+4\sigma^2I^2(b_y\theta_x^2+b_x\theta_y^2)}{8I\sigma^2 b_xb_y}\right]}{64\pi^3\sigma^2 b_xb_y}. \quad \text{(C-16)}$$

The integration over θ again yields a factor of 2π, and the required integrations become

$$p(I_x, I_y) = \int_0^\infty dI \int_{-\infty}^\infty d\theta_y \int_{-\infty}^\infty d\theta_x$$

$$\times \frac{\exp\left[-\frac{4b_xb_yI^2+\sigma^2(b_yI_x^2+b_xI_y^2)+4\sigma^2I^2(b_y\theta_x^2+b_x\theta_y^2)}{8I\sigma^2 b_xb_y}\right]}{32\pi^2\sigma^2 b_xb_y}. \quad \text{(C-17)}$$

Integrating next with respect to θ_x, we obtain

$$p(I_x, I_y) = \int_0^\infty dI \int_{-\infty}^\infty d\theta_y \frac{\exp\left[-\frac{\frac{4I^2}{\sigma^2}+\frac{I_x^2}{b_x}+\frac{I_y^2}{b_y}+\frac{4I^2\theta_y^2}{b_y}}{8I}\right]}{16\sqrt{2}\pi^{3/2}\sigma^2\sqrt{Ib_x}b_y}. \quad \text{(C-18)}$$

The integration with respect to θ_y can now be performed, yielding

$$p(I_x, I_y) = \int_0^\infty dI \frac{\exp\left[-\frac{\frac{4I^2}{\sigma^2}+\frac{I_x^2}{b_x}+\frac{I_y^2}{b_y}}{8I}\right]}{16\pi I\sigma^2\sqrt{b_xb_y}}. \quad \text{(C-19)}$$

The integral over I can be performed with the help of integral tables ([73], 3.471, 9), with the result

$$p(I_x, I_y) = \frac{K_0\left(\frac{1}{2\sigma}\sqrt{\frac{I_x^2}{b_x} + \frac{I_y^2}{b_y}}\right)}{8\pi\sigma^2\sqrt{b_x b_y}}, \qquad \text{(C-20)}$$

where $K_0(x)$ is a modified Bessel function of the second kind, order zero.

The marginal density of I_x can be found by returning to Eq. (C-19) and integrating with respect to I_y first, and then with respect to I. We obtain

$$p(I_x) = \int_0^\infty dI \, \frac{\exp\left(-\frac{I_x^2}{8Ib_x} - \frac{I}{2\sigma^2}\right)}{4\sqrt{2\pi}\sigma^2\sqrt{Ib_x}}$$

$$= \frac{1}{4\sigma\sqrt{b_x}} \exp\left[-\frac{|I_x|}{2\sigma\sqrt{b_x}}\right], \qquad \text{(C-21)}$$

where again [73] (3.471, 9) has been used to evaluate the final integral. This result agrees with that of Ebeling [36], who was the first to derive it, in a purely one-dimensional analysis. Likewise the marginal density function of I_y is of the form

$$p(I_y) = \frac{1}{4\sigma\sqrt{b_y}} \exp\left[-\frac{|I_y|}{2\sigma\sqrt{b_y}}\right]. \qquad \text{(C-22)}$$

D Analysis of Wavelength and Angle Dependence of Speckle

Our goal in this appendix is to find the cross-correlation of an observed complex speckle field $\mathbf{A}(x_1, y_1)$ and a second complex speckle field $\mathbf{A}(x_2, y_2)$, where the differences between $\mathbf{A}_1$ and $\mathbf{A}_2$ arise due to changes of wavelength, angle of illumination, and/or angle of observation. Two different geometries are considered, free space and imaging.

D.1 Free-Space Geometry

The geometry is that illustrated in Fig. 5.8, with free-space propagation from the (α, β) plane to the parallel (x, y) plane. The distance between these planes is taken to be z.

Adopting the notation introduced before Eqs. (5-25) through (5-33), the field in the (α, β) plane can be written

$$\begin{aligned}\mathbf{a}(\alpha, \beta) &= \mathbf{r}\,\mathbf{S}(\alpha, \beta)e^{j\phi(\alpha,\beta)}e^{-j\frac{2\pi}{\lambda}[(-\hat{i}\cdot\hat{\alpha})\alpha + (-\hat{i}\cdot\hat{\beta})\beta]}\\ &= \mathbf{r}\,\mathbf{S}(\alpha, \beta)e^{jq_z h(\alpha,\beta)}e^{-j\frac{2\pi}{\lambda}[(-\hat{i}\cdot\hat{\alpha})\alpha + (-\hat{i}\cdot\hat{\beta})\beta]},\end{aligned} \quad \text{(D-1)}$$

where $\mathbf{r}$ is the average amplitude reflectivity of the surface, and $\mathbf{S}$ represents the shape of the amplitude distribution across the incident spot, excluding phase factors associated with illumination tilt, which are included in the last term. The relationship between the scattered field $\mathbf{a}$ and the observed field $\mathbf{A}$ is taken to be the Fresnel diffraction equation,

$$\mathbf{A}(x, y) = \frac{1}{j\lambda z}\iint\limits_{-\infty}^{\infty}[\mathbf{a}(\alpha, \beta)e^{j\frac{\pi}{\lambda z}(\alpha^2 + \beta^2)}]e^{-j\frac{2\pi}{\lambda z}(x\alpha + y\beta)}\,d\alpha\,d\beta, \quad \text{(D-2)}$$

where we have dropped certain phase factors preceding the integral since ultimately they will not affect the result for the speckle intensity cross-correlation.

The cross-correlation between two observed fields $\mathbf{A}_1(x, y)$ and $\mathbf{A}_2(x_2, y_2)$ can now be written (initially in a rather daunting form, but later simplified),

$$\boldsymbol{\Gamma}_\mathbf{A}(x_1, y_1; x_2, y_2) = \frac{1}{\lambda_1 \lambda_2 z^2} \iint_{-\infty}^{\infty} \iint_{-\infty}^{\infty} \boldsymbol{\Gamma}_\mathbf{a}(\alpha_1, \beta_1; \alpha_2, \beta_2)$$

$$\times e^{j\frac{\pi}{\lambda_1 z}(\alpha_1^2 + \beta_1^2) - j\frac{\pi}{\lambda_2 z}(\alpha_2^2 + \beta_2^2)}$$

$$\times e^{-j\frac{2\pi}{\lambda_1 z}(x_1 \alpha_1 + y_1 \beta_1) + j\frac{2\pi}{\lambda_2 z}(x_2 \alpha_2 + y_2 \beta_2)} \, d\alpha_1 \, d\beta_1 \, d\alpha_2 \, d\beta_2 \quad \text{(D-3)}$$

where

$$\boldsymbol{\Gamma}_\mathbf{a}(\alpha_1, \beta_1; \alpha_2, \beta_2) = \overline{\mathbf{a}_1(\alpha_1, \beta_1) \mathbf{a}_2^*(\alpha_2, \beta_2)}$$

$$= |\mathbf{r}|^2 \mathbf{S}(\alpha_1, \beta_1) \mathbf{S}^*(\alpha_2, \beta_2) \overline{e^{j(q_{z1} h(\alpha_1, \beta_1) - q_{z2} h(\alpha_2, \beta_2))}}$$

$$\times e^{-j2\pi \left[(-\hat{i}_1 \cdot \hat{\alpha}) \frac{\alpha_1}{\lambda_1} - (-\hat{i}_2 \cdot \hat{\alpha}) \frac{\alpha_2}{\lambda_2} + (-\hat{i}_1 \cdot \hat{\beta}) \frac{\beta_1}{\lambda_1} - (-\hat{i}_2 \cdot \hat{\beta}) \frac{\beta_2}{\lambda_2} \right]}$$

$$= |\mathbf{r}|^2 \mathbf{S}(\alpha_1, \beta_1) \mathbf{S}^*(\alpha_2, \beta_2) \mathbf{M}_h(q_{z1}, -q_{z2})$$

$$\times e^{-j2\pi \left[(-\hat{i}_1 \cdot \hat{\alpha}) \frac{\alpha_1}{\lambda_1} - (-\hat{i}_2 \cdot \hat{\alpha}) \frac{\alpha_2}{\lambda_2} + (-\hat{i}_1 \cdot \hat{\beta}) \frac{\beta_1}{\lambda_1} - (-\hat{i}_2 \cdot \hat{\beta}) \frac{\beta_2}{\lambda_2} \right]}. \quad \text{(D-4)}$$

Here $\mathbf{M}_h(\omega_1, \omega_2)$ is the second-order characteristic function of the surface-height function $h(\alpha, \beta)$, which is assumed to be statistically stationary.

Now, in the same spirit as Eq. (4-53), we assume that the correlation width of the scattered wave is sufficiently small that we can approximate $\boldsymbol{\Gamma}_\mathbf{a}$ as follows:

$$\boldsymbol{\Gamma}_\mathbf{a}(\alpha_1, \beta_1; \alpha_2, \beta_2) \approx \kappa |\mathbf{r}|^2 |\mathbf{S}(\alpha_1, \beta_1)|^2 \overline{e^{j(q_{z1} - q_{z2}) h(\alpha_1, \beta_1)}}$$

$$\times e^{-j2\pi \left[(-\hat{i}_1 \cdot \hat{\alpha}) \frac{\alpha_1}{\lambda_1} - (-\hat{i}_2 \cdot \hat{\alpha}) \frac{\alpha_2}{\lambda_2} + (-\hat{i}_1 \cdot \hat{\beta}) \frac{\beta_1}{\lambda_1} - (-\hat{i}_2 \cdot \hat{\beta}) \frac{\beta_2}{\lambda_2} \right]}$$

$$\times \delta(\alpha_1 - \alpha_2, \beta_1 - \beta_2)$$

$$= \kappa |\mathbf{r}|^2 |\mathbf{S}(\alpha_1, \beta_1)|^2 \mathbf{M}_h(\Delta q_z)$$

$$\times e^{-j2\pi \left[\left(\frac{(-\hat{i}_1 \cdot \hat{\alpha})}{\lambda_1} - \frac{(-\hat{i}_2 \cdot \hat{\alpha})}{\lambda_2} \right) \alpha_1 + \left(\frac{(-\hat{i}_1 \cdot \hat{\beta})}{\lambda_1} - \frac{(-\hat{i}_2 \cdot \hat{\beta})}{\lambda_2} \right) \beta_1 \right]}$$

$$\times \delta(\alpha_1 - \alpha_2, \beta_1 - \beta_2), \quad \text{(D-5)}$$

where $\Delta q_z = q_{z1} - q_{z2}$, $\mathbf{M}_h(\omega)$ is now a first-order characteristic function of h, and κ is a constant with dimensions (length)^2. With this substitution, the general equation for $\boldsymbol{\Gamma}_\mathbf{A}$ can be simplified to

$$\Gamma_A(x_1, y_1; x_2, y_2) = \frac{\kappa |\mathbf{r}|^2}{\lambda_1 \lambda_2 z^2} \mathbf{M}_h(\Delta q_z) \iint\limits_{-\infty}^{\infty} |\mathbf{S}(\alpha, \beta)|^2 e^{j\frac{\pi}{z}\left(\frac{1}{\lambda_1} - \frac{1}{\lambda_2}\right)(\alpha^2 + \beta^2)}$$

$$\times e^{-j2\pi \left[\left(\frac{(-\hat{i}_1 \cdot \hat{x})}{\lambda_1} - \frac{(-\hat{i}_2 \cdot \hat{x})}{\lambda_2}\right)\alpha + \left(\frac{(-\hat{i}_1 \cdot \hat{\beta})}{\lambda_1} - \frac{(-\hat{i}_2 \cdot \hat{\beta})}{\lambda_2}\right)\beta\right]}$$

$$\times e^{-j\frac{2\pi}{\lambda_1 z}(x_1\alpha + y_1\beta) + j\frac{2\pi}{\lambda_2 z}(x_2\alpha + y_2\beta)} \, d\alpha \, d\beta, \tag{D-6}$$

where there is no need for the subscripts on (α, β) any longer.

For further simplification, let $\lambda_2 = \lambda_1 + \Delta\lambda$, where $\Delta\lambda$ is the change of wavelength, and note

$$\frac{1}{\lambda_1} - \frac{1}{\lambda_2} = \frac{1}{\lambda_1} - \frac{1}{\lambda_1 + \Delta\lambda} = \frac{1}{\lambda_1}\left(1 - \frac{1}{1 + \Delta\lambda/\lambda_1}\right) \approx \frac{\Delta\lambda}{\lambda_1^2} \tag{D-7}$$

where the approximation is valid for $|\Delta\lambda| \ll \lambda_1$. If we apply this approximation to the quadratic-phase term in the integrand, we find

$$e^{j\frac{\pi}{z}\left(\frac{1}{\lambda_1} - \frac{1}{\lambda_2}\right)(\alpha^2 + \beta^2)} \approx e^{j\frac{\pi}{\lambda_1 z}\left(\frac{\Delta\lambda}{\lambda_1}\right)(\alpha^2 + \beta^2)}. \tag{D-8}$$

If the linear dimension of the scattering spot is D, then this exponential can be replaced by unity provided

$$\frac{\pi D^2}{\lambda_1 z} \frac{|\Delta\lambda|}{\lambda_1} \ll 1. \tag{D-9}$$

Recall that the far-field or Fraunhofer approximation (which we have not made here) is valid for $\frac{\pi D^2}{\lambda_1 z} < 1$. The condition assumed here is much more easily met due to the presence of the term $\Delta\lambda/\lambda_1$. Therefore we drop the quadratic-phase term in what follows. In addition, under the same condition, in the multiplier of the double integral, the term $1/(\lambda_1 \lambda_2)$ can be replaced with $1/\lambda_1^2$, without loss of significant accuracy. The result becomes

$$\Gamma_A(x_1, y_1; x_2, y_2) = \frac{\kappa |\mathbf{r}|^2}{\lambda_1^2 z^2} \mathbf{M}_h(\Delta q_z) \iint\limits_{-\infty}^{\infty} |\mathbf{S}(\alpha, \beta)|^2$$

$$\times e^{-j2\pi \left[\left(\frac{(-\hat{i}_1 \cdot \hat{x})}{\lambda_1} - \frac{(-\hat{i}_2 \cdot \hat{x})}{\lambda_2}\right)\alpha + \left(\frac{(-\hat{i}_1 \cdot \hat{\beta})}{\lambda_1} - \frac{(-\hat{i}_2 \cdot \hat{\beta})}{\lambda_2}\right)\beta\right]}$$

$$\times e^{-j\frac{2\pi}{\lambda_1 z}(x_1\alpha + y_1\beta) + j\frac{2\pi}{\lambda_2 z}(x_2\alpha + y_2\beta)} \, d\alpha \, d\beta. \tag{D-10}$$

Substituting Eq. (5-33) in the last line of this equation we obtain

$$\Gamma_A(x_1, y_1; x_2, y_2) = \frac{\kappa |\mathbf{r}|^2}{\lambda_1^2 z^2} M_h(\Delta q_z) \iint\limits_{-\infty}^{\infty} |S(\alpha, \beta)|^2$$

$$\times e^{-j2\pi \left[\left(\frac{(-\hat{i}_1 \cdot \hat{\alpha})}{\lambda_1} - \frac{(-\hat{i}_2 \cdot \hat{\alpha})}{\lambda_2} \right) \alpha + \left(\frac{(-\hat{i}_1 \cdot \hat{\beta})}{\lambda_1} - \frac{(-\hat{i}_2 \cdot \hat{\beta})}{\lambda_2} \right) \beta \right]}$$

$$\times e^{-j2\pi \left[\left(\frac{(\hat{o}_1 \cdot \hat{\alpha})}{\lambda_1} - \frac{(\hat{o}_2 \cdot \hat{\alpha})}{\lambda_2} \right) \alpha + \left(\frac{(\hat{o}_1 \cdot \hat{\beta})}{\lambda_1} - \frac{(\hat{o}_2 \cdot \hat{\beta})}{\lambda_2} \right) \beta \right]} \, d\alpha \, d\beta. \quad \text{(D-11)}$$

This result can now be greatly simplified, at least notationally, with the help of Eqs. (5-28) and (5-25) and by recognizing that

$$q_{\alpha 1} = 2\pi \frac{(\hat{o}_1 \cdot \hat{\alpha})}{\lambda_1} - 2\pi \frac{(-\hat{i}_1 \cdot \hat{\alpha})}{\lambda_1}$$

$$q_{\alpha 2} = 2\pi \frac{(\hat{o}_2 \cdot \hat{\alpha})}{\lambda_2} - 2\pi \frac{(-\hat{i}_2 \cdot \hat{\alpha})}{\lambda_2}$$

$$q_{\beta 1} = 2\pi \frac{(\hat{o}_1 \cdot \hat{\beta})}{\lambda_1} - 2\pi \frac{(-\hat{i}_1 \cdot \hat{\beta})}{\lambda_1} \quad \text{(D-12)}$$

$$q_{\beta 2} = 2\pi \frac{(\hat{o}_2 \cdot \hat{\beta})}{\lambda_2} - 2\pi \frac{(-\hat{i}_2 \cdot \hat{\beta})}{\lambda_2}$$

$$\vec{q}_{t1} = q_{\alpha 1} \hat{\alpha} + q_{\beta 1} \hat{\beta}$$

$$\vec{q}_{t2} = q_{\alpha 2} \hat{\alpha} + q_{\beta 2} \hat{\beta}.$$

Substituting in Eq. (D-11), we find the following result,

$$\Gamma_A(x_1, y_1; x_2, y_2) = \frac{\kappa |\mathbf{r}|^2}{\lambda_1^2 z^2} M_h(\Delta q_z) \Psi(\Delta \vec{q}_t) \iint\limits_{-\infty}^{\infty} |S(\alpha, \beta)|^2 \, d\alpha \, d\beta, \quad \text{(D-13)}$$

where $\Delta \vec{q}_t = \vec{q}_{t1} - \vec{q}_{t2}$ and

$$\Psi(\Delta \vec{q}_t) = \frac{\iint\limits_{-\infty}^{\infty} |S(\alpha, \beta)|^2 e^{-j2\pi \Delta \vec{q}_t \cdot \vec{\alpha}_t} \, d\alpha \, d\beta}{\iint\limits_{-\infty}^{\infty} |S(\alpha, \beta)|^2 \, d\alpha \, d\beta} \quad \text{(D-14)}$$

is the normalized Fourier transform of the intensity distribution across the scattering spot.

The normalized cross-correlation function, μ_A is found by dividing Γ_A by its value when $\vec{q}_1 = \vec{q}_2$, for which $\Delta q_z = 0$ and $\Delta \vec{q}_t = 0$. The result is[1]

---

1. The characteristic function **M** is, by definition, unity at the origin.

$$\boldsymbol{\mu}_\mathbf{A}(x_1, y_1; x_2, y_2) = \boldsymbol{\mu}_\mathbf{A}(\vec{q}_1, \vec{q}_2) = \mathbf{M}_h(\Delta q_z)\boldsymbol{\Psi}(\Delta \vec{q}_t). \tag{D-15}$$

This equation represents the final result of our analysis of the free-space geometry.

D.2 Imaging Geometry

In this section, we calculate the cross-correlation function

$$\boldsymbol{\Gamma}_\mathbf{A}(x_1, y_1; x_2, y_2) = \overline{\mathbf{A}_1(x_1, y_1)\mathbf{A}_2^*(x_2, y_2)}, \tag{D-16}$$

for the imaging geometry shown in Fig. 5.16, with

$$\mathbf{A}(x, y) = \iint_{-\infty}^{\infty} \mathbf{k}(x, y; \alpha, \beta)\mathbf{a}(\alpha, \beta)\, d\alpha\, d\beta \tag{D-17}$$

and, for $i = 1, 2$,

$$\mathbf{k}_i(x_i, y_i; \alpha, \beta) = \frac{1}{\lambda_i^2 z^2} e^{j\frac{\pi}{\lambda_i z}(\alpha^2 + \beta^2)} \iint_{-\infty}^{\infty} \mathbf{P}(\xi, \eta) e^{-j\frac{2\pi}{\lambda_i z}[\xi(\alpha + x_i) + \eta(\beta + y_i)]} d\xi\, d\eta. \tag{D-18}$$

Substituting the equation for $\mathbf{A}$ into the expression for $\boldsymbol{\Gamma}_\mathbf{A}$, we obtain

$$\boldsymbol{\Gamma}_\mathbf{A}(x_1, y_1; x_2, y_2) = \iint_{-\infty}^{\infty} \iint_{-\infty}^{\infty} \mathbf{k}_1(x_1, y_1; \alpha_1, \beta_1)\mathbf{k}_2^*(x_2, y_2; \alpha_2, \beta_2)$$
$$\times \overline{\mathbf{a}_1(\alpha_1, \beta_1)\mathbf{a}_2^*(\alpha_2, \beta_2)}\, d\alpha_1\, d\beta_1\, d\alpha_2\, d\beta_2. \tag{D-19}$$

The averaged quantity under the integral signs is the cross-correlation of the fields in the (α, β) plane; we will apply the approximation of Eq. (D-5) to this average, as we did in the previous analysis,

$$\boldsymbol{\Gamma}_\mathbf{a}(\alpha_1, \beta_1; \alpha_2, \beta_2) \approx \kappa|\mathbf{r}|^2|\mathbf{S}(\alpha_1, \beta_1)|^2 \overline{e^{j(q_{z1} - q_{z2})h(\alpha_1, \beta_1)}}$$
$$\times e^{-j2\pi\left[(-\hat{i}_1\cdot\hat{\alpha})\frac{\alpha_1}{\lambda_1} - (-\hat{i}_2\cdot\hat{\alpha})\frac{\alpha_2}{\lambda_2} + (-\hat{i}_1\cdot\hat{\beta})\frac{\beta_1}{\lambda_1} - (-\hat{i}_2\cdot\hat{\beta})\frac{\beta_2}{\lambda_2}\right]}$$
$$\times \delta(\alpha_1 - \alpha_2, \beta_1 - \beta_2)$$
$$= \kappa|\mathbf{r}|^2|\mathbf{S}(\alpha_1, \beta_1)|^2 \mathbf{M}_h(\Delta q_z)$$
$$\times e^{-j2\pi\left[\left(\frac{(-\hat{i}_1\cdot\hat{\alpha})}{\lambda_1} - \frac{(-\hat{i}_2\cdot\hat{\alpha})}{\lambda_2}\right)\alpha_1 + \left(\frac{(-\hat{i}_1\cdot\hat{\beta})}{\lambda_1} - \frac{(-\hat{i}_2\cdot\hat{\beta})}{\lambda_2}\right)\beta_1\right]}$$
$$\times \delta(\alpha_1 - \alpha_2, \beta_1 - \beta_2), \tag{D-20}$$

where $\hat{i}_1, \hat{i}_2$ are unit vectors in the two directions of illumination, $\mathbf{r}$ is the average reflectivity of the surface, $\mathbf{S}$ is the amplitude distribution incident on the scattering spot,

κ is a constant with dimensions length^2, $\mathbf{M}_h$ is the characteristic function of the surface height fluctuations, and Δq_z is the z-component of the change in the vector $\vec{q}$. Substituting this result in Eq. (D-19), we obtain

$$\Gamma_A(x_1, y_1; x_2, y_2) = \kappa |\mathbf{r}|^2 \mathbf{M}_h(\Delta q_z)$$

$$\times \iint_{-\infty}^{\infty} |\mathbf{S}(\alpha,\beta)|^2 \mathbf{k}_1(x_1, y_1; \alpha, \beta) \mathbf{k}_2^*(x_2, y_2; \alpha, \beta)$$

$$\times e^{-j2\pi\left[\left(\frac{(-\hat{i}_1 \cdot \hat{\alpha})}{\lambda_1} - \frac{(-\hat{i}_2 \cdot \hat{\alpha})}{\lambda_2}\right)\alpha + \left(\frac{(-\hat{i}_1 \cdot \hat{\beta})}{\lambda_1} - \frac{(-\hat{i}_2 \cdot \hat{\beta})}{\lambda_2}\right)\beta\right]} d\alpha \, d\beta. \quad \text{(D-21)}$$

The widths of the point-spread functions $\mathbf{k}_1$ and $\mathbf{k}_2$ are determined by the scaled width of the Fourier transform of the pupil function $\mathbf{P}$, and as such are very compact or narrow functions in the (α, β) plane. This allows an approximation, namely for a comparatively uniform intensity of illumination of the surface, the point-spread function product samples the term $|\mathbf{S}(\alpha,\beta)|^2$ at coordinates[2] $\alpha = -(x_1 + x_2)/2$, $\beta = -(y_1 + y_2)/2$, i.e. at the midpoint between the centers of $\mathbf{k}_1$ and $\mathbf{k}_2$, allowing $\left|\mathbf{S}\left(-\frac{x_1+x_2}{2}, -\frac{y_1+y_2}{2}\right)\right|^2$ to be pulled outside the (α, β) integration. Thus

$$\Gamma_A(x_1, y_1; x_2, y_2) = \kappa |\mathbf{r}|^2 \mathbf{M}_h(\Delta q_z) \left|\mathbf{S}\left(-\frac{x_1+x_2}{2}, -\frac{y_1+y_2}{2}\right)\right|^2$$

$$\times \iint_{-\infty}^{\infty} \mathbf{k}_1(x_1, y_1; \alpha, \beta) \mathbf{k}_2^*(x_2, y_2; \alpha, \beta)$$

$$\times e^{-j2\pi\left[\left(\frac{(-\hat{i}_1 \cdot \hat{\alpha})}{\lambda_1} - \frac{(-\hat{i}_2 \cdot \hat{\alpha})}{\lambda_2}\right)\alpha + \left(\frac{(-\hat{i}_1 \cdot \hat{\beta})}{\lambda_1} - \frac{(-\hat{i}_2 \cdot \hat{\beta})}{\lambda_2}\right)\beta\right]} d\alpha \, d\beta. \quad \text{(D-22)}$$

A similar approximation is possible for the exponential term under the integral sign.[3] That is, we can make the same substitutions for (α, β), and pull the exponential outside the integral signs. In addition, since we are ultimately interested in the magnitude of the cross-correlation, we can drop the complex exponential completely, since it has unit magnitude. Thus we obtain

$$\Gamma_A(x_1, y_1; x_2, y_2) = \kappa |\mathbf{r}|^2 \mathbf{M}_h(\Delta q_z) \left|\mathbf{S}\left(-\frac{x_1+x_2}{2}, -\frac{y_1+y_2}{2}\right)\right|^2$$

$$\times \iint_{-\infty}^{\infty} \mathbf{k}_1(x_1, y_1; \alpha, \beta) \mathbf{k}_2^*(x_2, y_2; \alpha, \beta) \, d\alpha \, d\beta. \quad \text{(D-23)}$$

---

2. The minus signs are a result of image inversion.
3. This approximation effectively assumes that the angle between $\hat{i}_1$ and $\hat{i}_2$ is small.

D.2 Imaging Geometry

Considering now the product $\mathbf{k}_1\mathbf{k}_2^*$, this product contains a term $e^{j\frac{\pi}{z}\left(\frac{1}{\lambda_1}-\frac{1}{\lambda_2}\right)(\alpha^2+\beta^2)}$. From the same argument used in Eq. (D-9), we can drop this term under the assumption that $\Delta\lambda \ll \lambda_1$. With the first of these approximations, the point-spread functions become a function of $x + \alpha$ and $y + \beta$, rather than x, y, α and β independently. Thus

$$\Gamma_A(x_1, y_1; x_2, y_2) = \kappa |\mathbf{r}|^2 M_h(\Delta q_z) \left| S\left(-\frac{x_1+x_2}{2}, -\frac{y_1+y_2}{2}\right) \right|^2$$

$$\times \iint_{-\infty}^{\infty} \mathbf{k}_1(x_1+\alpha, y_1+\beta) \mathbf{k}_2^*(x_2+\alpha, y_2+\beta)\, d\alpha\, d\beta. \quad \text{(D-24)}$$

With a further approximation that $\frac{1}{\lambda_1^2 \lambda_2^2} \approx \frac{1}{\lambda_1^4}$ in front of the expression for the product for $\mathbf{k}_1\mathbf{k}_2^*$ (valid again for $\Delta\lambda \ll \lambda_1$), we have

$$\mathbf{k}_1(x_1+\alpha, y_1+\beta)\mathbf{k}_2^*(x_2+\alpha, y_2+\beta)$$

$$\approx \frac{1}{\lambda_1^4 z^4} \iint_{-\infty}^{\infty} \iint_{-\infty}^{\infty} \mathbf{P}(\xi_1, \eta_1) \mathbf{P}^*(\xi_2, \eta_2)$$

$$\times e^{-j\frac{2\pi}{\lambda_1 z}[\xi_1(\alpha+x_1)+\eta_1(\beta+y_1)]} e^{j\frac{2\pi}{\lambda_2 z}[\xi_2(\alpha+x_2)+\eta_2(\beta+y_2)]}\, d\xi_1\, d\eta_1\, d\xi_2\, d\eta_2. \quad \text{(D-25)}$$

The next step is to substitute Eq. (D-25) into the expression for Γ_A and integrate first with respect to (α, β). When this is done, the resulting integral over (α, β) takes the form

$$\iint_{-\infty}^{\infty} e^{-j\frac{2\pi}{z}\left[\left(\frac{\xi_1}{\lambda_1}-\frac{\xi_2}{\lambda_2}\right)\alpha + \left(\frac{\eta_1}{\lambda_1}-\frac{\eta_2}{\lambda_2}\right)\beta\right]}\, d\alpha\, d\beta = \delta\left(\frac{\xi_1}{\lambda_1 z}-\frac{\xi_2}{\lambda_2 z}, \frac{\eta_1}{\lambda_1 z}-\frac{\eta_2}{\lambda_2 z}\right)$$

$$= \delta\left[\frac{1}{\lambda_1 z}\left(\xi_1 - \frac{\lambda_1}{\lambda_2}\xi_2\right), \frac{1}{\lambda_1 z}\left(\eta_1 - \frac{\lambda_1}{\lambda_2}\eta_2\right)\right]$$

$$= (\lambda_1 z)^2 \delta\left(\xi_1 - \frac{\lambda_1}{\lambda_2}\xi_2, \eta_1 - \frac{\lambda_1}{\lambda_2}\eta_2\right). \quad \text{(D-26)}$$

When this result is substituted into Eq. (D-25), the sifting property of the delta function can be employed to reduce that result to

$$\Gamma_A(x_1, y_1; x_2, y_2) = \frac{\kappa}{\lambda_1^2 z^2} |\mathbf{r}|^2 M_h(\Delta q_z) \left| S\left(-\frac{x_1+x_2}{2}, -\frac{y_1+y_2}{2}\right) \right|^2$$

$$\times \iint_{-\infty}^{\infty} \mathbf{P}\left(\frac{\lambda_1}{\lambda_2}\xi, \frac{\lambda_2}{\lambda_1}\eta\right) \mathbf{P}^*(\xi, \eta) e^{-j\frac{2\pi}{\lambda_2 z}(\xi \Delta x + \eta \Delta y)}\, d\xi\, d\eta, \quad \text{(D-27)}$$

where $\Delta x = x_1 - x_2$, $\Delta y = y_1 - y_2$, and we have dropped the subscripts on (ξ, η) since they are no longer needed. The normalized cross-correlation function μ_A is obtained by dividing Γ_A by its value when $\Delta x = 0$ and $\Delta y = 0$, with the result

$$\mu_A(x_1, y_1; \Delta x, \Delta y) = \frac{\Gamma_A(x_1, y_1; x_2, y_2)}{\Gamma_A(x_1, y_1; x_1, y_1)}$$

$$= \mathbf{M}_h(\Delta q_z) \frac{\left|\mathbf{S}\left(-\frac{x_1+x_2}{2}, -\frac{y_1+y_2}{2}\right)\right|^2}{|\mathbf{S}(-x_1, -y_1)|^2}$$

$$\times \frac{\iint\limits_{-\infty}^{\infty} \mathbf{P}\left(\frac{\lambda_1}{\lambda_2}\xi, \frac{\lambda_2}{\lambda_1}\eta\right) \mathbf{P}^*(\xi, \eta) e^{-j\frac{2\pi}{\lambda_2 z}(\xi \Delta x + \eta \Delta y)} d\xi\, d\eta}{\iint\limits_{-\infty}^{\infty} \mathbf{P}\left(\frac{\lambda_1}{\lambda_2}\xi, \frac{\lambda_2}{\lambda_1}\eta\right) \mathbf{P}^*(\xi, \eta)\, d\xi\, d\eta}. \quad \text{(D-28)}$$

A final set of approximations leads to a comparatively simple result. First we note that the values of $(\Delta x, \Delta y)$ for which μ_A has significant value are quite small, compared with the size of the scattering spot, thus allowing us to write

$$\left|\mathbf{S}\left(-\frac{x_1 + x_2}{2}, -\frac{y_1 + y_2}{2}\right)\right|^2 \approx |\mathbf{S}(-x_1, -y_1)|^2 \quad \text{(D-29)}$$

and to drop the term involving $|\mathbf{S}|^2$, the intensity distribution across the scattering spot. Second, for small changes of wavelength,

$$\mathbf{P}\left(\frac{\lambda_1}{\lambda_2}\xi, \frac{\lambda_2}{\lambda_1}\eta\right) \approx \mathbf{P}(\xi, \eta), \quad \text{(D-30)}$$

allowing us to write a final result with good approximation,

$$\mu_A(\Delta x, \Delta y) = \mathbf{M}_h(\Delta q_z) \frac{\iint\limits_{-\infty}^{\infty} |\mathbf{P}(\xi, \eta)|^2 e^{-j\frac{2\pi}{\lambda_2 z}(\xi \Delta x + \eta \Delta y)} d\xi\, d\eta}{\iint\limits_{-\infty}^{\infty} |\mathbf{P}(\xi, \eta)|^2 d\xi\, d\eta}$$

$$= \mathbf{M}_h(\Delta q_z) \mathbf{\Psi}(\Delta x, \Delta y), \quad \text{(D-31)}$$

where $\mathbf{\Psi}$ is now the normalized Fourier transform of the intensity distribution across the pupil of the imaging system,

$$\mathbf{\Psi}(\Delta x, \Delta y) = \frac{\iint\limits_{-\infty}^{\infty} |\mathbf{P}(\xi, \eta)|^2 e^{-j\frac{2\pi}{\lambda_2 z}(\xi \Delta x + \eta \Delta y)} d\xi\, d\eta}{\iint\limits_{-\infty}^{\infty} |\mathbf{P}(\xi, \eta)|^2 d\xi\, d\eta}. \quad \text{(D-32)}$$

As an aside, note that when $\lambda_2 = \lambda_1$, the term $\mathbf{M}_h$ is unity, and our result is a proof of the assertion in Section 4.4.2 that, in an imaging geometry, the pupil of the optical system can be imagined to be a scattering surface when calculating the autocorrelation of the speckle in the image plane using the van Cittert–Zernike theorem.

E Speckle Contrast when a Dynamic Diffuser Is Projected onto a Random Screen

Our goal in this appendix is to find expressions for speckle contrast when an object to be projected is first imaged onto a changing diffuser, and then projected onto a screen for viewing. We consider two types of random diffuser, one for which the diffuser uniformly fills the projection lens but does not overfill it, and a second that overfills the projection lens. In the former case, for a uniformly transmitting object, a pure phase diffuser image is projected onto the screen, while for the second case, each resolution cell of the projector lens produces a fully developed speckle amplitude. A third case, a deterministic diffuser with certain orthogonality properties, is considered in the main text but not in this appendix. In both cases considered here, the optically rough viewing screen introduces speckle within each resolution element of the eye. The symbols and geometry assumed here are discussed in Sect. 6.4.

E.1 Random Phase Diffusers

A single eye resolution element integrates, over time, M different statistically independent intensity patterns, one for each realization of the random phase diffuser. Thus the total intensity seen at one point on the observer's retina is

$$I = \sum_{m=1}^{M} I_m. \tag{E-1}$$

One such pattern is the squared magnitude of a sum of K different fields produced by the K projector lens resolution elements lying within one eye resolution element,

$$I_m = \left| \sum_{k=1}^{K} \mathbf{A}_k \mathbf{B}_k^{(m)} \right|^2 = \sum_{k=1}^{K} \sum_{l=1}^{K} \mathbf{A}_k \mathbf{A}_l^* \mathbf{B}_k^{(m)} \mathbf{B}_l^{(m)*}, \tag{E-2}$$

where $\mathbf{B}_k^{(m)}$ represents the field projected onto the screen by one projector-lens resolution element during the mth diffuser realization, and $\mathbf{A}_k$ is the random speckle field that would be projected onto the retina by the screen in that one projector lens resolution element if the field $\mathbf{B}_k^{(m)}$ were unity and nonrandom. The total intensity integrated over time is therefore

$$I = \sum_{m=1}^{M} \sum_{k=1}^{K} \sum_{l=1}^{K} \mathbf{A}_k \mathbf{A}_l^* \mathbf{B}_k^{(m)} \mathbf{B}_l^{(m)*}. \tag{E-3}$$

We first calculate the mean intensity $\bar{I}$. To do so requires averaging in two steps: first we average over the statistics of the imaged diffuser, then we average over the statistics of the screen. The average intensity, conditioned by knowledge of the $\mathbf{A}_k$s, which we represent by $E[I \mid \mathbf{A}]$, is

$$E[I \mid \mathbf{A}] = \sum_{m=1}^{M} \sum_{k=1}^{K} \sum_{l=1}^{K} \mathbf{A}_k \mathbf{A}_l^* \overline{\mathbf{B}_k^{(m)} \mathbf{B}_l^{(m)*}}. \tag{E-4}$$

Under the assumption that $\overline{\mathbf{B}_k^{(m)}} = 0$ (which is true for the two cases of interest here), if $k \neq l$, then $\overline{\mathbf{B}_k^{(m)} \mathbf{B}_l^{(m)*}} = \overline{\mathbf{B}_k^{(m)}} \, \overline{\mathbf{B}_l^{(m)*}} = 0$, and thus only the K terms with $k = l$ survive the average, yielding

$$E[I \mid \mathbf{A}] = \sum_{m=1}^{M} \sum_{k=1}^{K} |\mathbf{A}_k|^2 \overline{|\mathbf{B}|^2} = M J_B \sum_{k=1}^{K} |\mathbf{A}_k|^2, \tag{E-5}$$

where $J_B = \overline{|\mathbf{B}|^2}$. The superscript (m) and the subscript k are no longer needed on $\mathbf{B}$ since it is assumed that all $\mathbf{B}_k^{(m)}$ have the same average intensity (i.e. the average illumination intensity does not change with time and is uniform across the resolution element of the eye on the screen). Now we remove the conditioning on $\mathbf{A}$ by averaging over the statistics of the $\mathbf{A}_k$s. All $|\mathbf{A}_k|^2$ have the same average value, which we represent by J_A, yielding

$$\bar{I} = M K J_A J_B. \tag{E-6}$$

In order to find the contrast of the observed speckle, we will have to find the standard deviation of the observed intensity, which in turn requires finding the second moment of I. Again averaging first with respect to the $\mathbf{B}$s,

$$E[I^2 \mid \mathbf{A}] = \sum_{m=1}^{M} \sum_{n=1}^{M} \sum_{k=1}^{K} \sum_{l=1}^{K} \sum_{p=1}^{K} \sum_{q=1}^{K} \mathbf{A}_k \mathbf{A}_l^* \mathbf{A}_p \mathbf{A}_q^* \overline{\mathbf{B}_k^{(m)} \mathbf{B}_l^{(m)*} \mathbf{B}_p^{(n)} \mathbf{B}_q^{(n)*}}. \tag{E-7}$$

Now we assume that $\mathbf{B}$s with superscripts (m) and (n) are independent when $m \neq n$ (because the diffuser has changed), in which case we collect M terms with $m = n$ and $M^2 - M$ terms with $m \neq n$,

$$E[I^2 \mid \mathbf{A}] = M \sum_{k=1}^{K} \sum_{l=1}^{K} \sum_{p=1}^{K} \sum_{q=1}^{K} \mathbf{A}_k \mathbf{A}_l^* \mathbf{A}_p \mathbf{A}_q^* \overline{\mathbf{B}_k^{(m)} \mathbf{B}_l^{(m)*} \mathbf{B}_p^{(m)} \mathbf{B}_q^{(m)*}}$$
$$+ (M^2 - M) \sum_{k=1}^{K} \sum_{l=1}^{K} \sum_{p=1}^{K} \sum_{q=1}^{K} \mathbf{A}_k \mathbf{A}_l^* \mathbf{A}_p \mathbf{A}_q^* \overline{\mathbf{B}_k^{(m)} \mathbf{B}_l^{(m)*}} \, \overline{\mathbf{B}_p^{(n)} \mathbf{B}_q^{(n)*}}. \quad \text{(E-8)}$$

We refer to the two lines of this equation as "line 1" and "line 2". Focusing first on line 1, there are three conditions under which the average yields nonzero results: $k = l = p = q$ (K terms), $k = l$, $p = q$, $k \neq p$ ($K^2 - K$ terms), and $k = q$, $l = p$, $k \neq l$ ($K^2 - K$ terms). Using this fact, line 1 reduces to

$$\text{line 1} = M \sum_{k=1}^{K} |\mathbf{A}_k|^4 \overline{|\mathbf{B}|^4} + 2M \sum_{k=1}^{K} \sum_{p=1, p \neq k}^{K} |\mathbf{A}_k|^2 |\mathbf{A}_p|^2 (\overline{|\mathbf{B}|^2})^2. \quad \text{(E-9)}$$

For line 2, the only nonzero terms come from $k = l$ and $p = q$, regardless of the relation between k and p, so we have

$$\text{line 2} = (M^2 - M) \sum_{k=1}^{K} \sum_{p=1}^{K} |\mathbf{A}_k|^2 |\mathbf{A}_p|^2 (\overline{|\mathbf{B}|^2})^2. \quad \text{(E-10)}$$

Now we average with respect to the statistics of the $\mathbf{A}$s. We have

$$\text{line 1} = MK \overline{|\mathbf{A}|^4} \, \overline{|\mathbf{B}|^4} + 2M(K^2 - K)(\overline{|\mathbf{A}|^2})^2 (\overline{|\mathbf{B}|^2})^2$$
$$= MK \overline{|\mathbf{A}|^4} \, \overline{|\mathbf{B}|^4} + 2M(K^2 - K) J_A^2 J_B^2, \quad \text{(E-11)}$$

and

$$\text{line 2} = (M^2 - M) K \overline{|\mathbf{A}|^4} (\overline{|\mathbf{B}|^2})^2 + (M^2 - M)(K^2 - K)(\overline{|\mathbf{A}|^2})^2 (\overline{|\mathbf{B}|^2})^2$$
$$= (M^2 - M) K \overline{|\mathbf{A}|^4} J_B^2 + (M^2 - M)(K^2 - K) J_A^2 J_B^2. \quad \text{(E-12)}$$

Thus we conclude that

$$\overline{I^2} = MK \overline{|\mathbf{A}|^4} \, \overline{|\mathbf{B}|^4} + 2M(K^2 - K) J_A^2 J_B^2$$
$$+ (M^2 - M) K \overline{|\mathbf{A}|^4} J_B^2 + (M^2 - M)(K^2 - K) J_A^2 J_B^2. \quad \text{(E-13)}$$

The variance σ_I^2 is found by subtracting the square of the mean from the second moment, with the result

$$\sigma_I^2 = MK \overline{|\mathbf{A}|^4} \, \overline{|\mathbf{B}|^4} + 2M(K^2 - K) J_A^2 J_B^2$$
$$+ (M^2 - M) K \overline{|\mathbf{A}|^4} J_B^2 + (M^2 - M)(K^2 - K) J_A^2 J_B^2 - M^2 K^2 J_A^2 J_B^2. \quad \text{(E-14)}$$

Finally, we note that the quantity $\mathbf{A}$ is the contribution of the screen ensemble to the observed speckle, and for a rough screen, will have the statistics of fully developed speckle. In this case, we have that the second moment of the intensity of **A** is twice the square of the first moment (a property of negative-exponential statistics), with the result that $\overline{|\mathbf{A}|^4} = 2J_A^2$, and

$$\sigma_I^2 = 2MKJ_A^2\overline{|\mathbf{B}|^4} + 2M(K^2 - K)J_A^2 J_B^2$$
$$+ 2(M^2 - M)KJ_A^2 J_B^2 + (M^2 - M)(K^2 - K)J_A^2 J_B^2 - M^2 K^2 J_A^2 J_B^2. \quad (E\text{-}15)$$

We are now ready to consider the two cases of interest, a diffuser that just fills the projection optics, and a diffuser that overfills the projection optics.

E.2 Diffuser that Just Fills the Projection Optics

In this case, the phase diffuser is imaged onto the projection optics as a pure phase function, with the result that

$$\mathbf{B}_k^{(m)} = e^{j\phi_k^{(m)}}. \quad (E\text{-}16)$$

For such a **B** we have $\overline{|\mathbf{B}|^4} = 1$, $(\overline{|\mathbf{B}|^2})^2 = J_B^2 = 1$, and $J_B = 1$. Thus

$$(\bar{I})^2 = M^2 K^2 J_A^2 \quad (E\text{-}17)$$

and

$$\sigma_I^2 = 2MKJ_A^2 + 2M(K^2 - K)J_A^2 + 2(M^2 - M)KJ_A^2$$
$$+ (M^2 - M)(K^2 - K)J_A^2 - M^2 K^2 J_A^2 = KM(M + K - 1)J_A^2. \quad (E\text{-}18)$$

The speckle contrast is given by[1]

$$C = \frac{\sigma_I}{\bar{I}} = \sqrt{\frac{M + K - 1}{KM}}, \quad (E\text{-}19)$$

as asserted in Eq. (6-41).

E.3 Diffuser that Overfills the Projection Optics

In this case, the phase diffuser produces a fully developed speckle field at each resolution cell of the projector optics on the screen. For such statistics we have $\overline{|\mathbf{B}|^4} = 2J_B^2$ (from the negative-exponential statistics of $|\mathbf{B}|^2$). The squared mean and the variance of I are, in this case,

$$(\bar{I})^2 = M^2 K^2 J_A^2 J_B^2 \quad (E\text{-}20)$$

---

1. A similar result has been found for a related problem by Ivakin et al. in Ref. [81].

E.3 Diffuser that Overfills the Projection Optics

and

$$\begin{aligned}\sigma_I^2 &= 4MKJ_A^2J_B^2 + 2M(K^2-K)J_A^2J_B^2 \\ &+ 2(M^2-M)KJ_A^2J_B^2 + (M^2-M)(K^2-K)J_A^2J_B^2 \\ &- M^2K^2J_A^2J_B^2 = KM(M+K+1)J_A^2J_B^2.\end{aligned} \qquad \text{(E-21)}$$

The speckle contrast is given in this case by

$$C = \sqrt{\frac{M+K+1}{KM}}, \qquad \text{(E-22)}$$

as asserted in Eq. (6-45).

F Statistics of Constrained Speckle

Through the vast majority of this book, we have dealt with the statistics of what we will call *classical* speckle, by which we mean speckle that is the result of an unconstrained random walk in the complex plane. Most speckle encountered in practice is of this kind. However, there are certain applications in which the classical statistics do not apply, and it is to these cases that we devote this appendix. The results will be useful in dealing with speckle in multimode fibers.

In this appendix we assume that the speckle arises from a superposition of a finite but large number M_T of CW modes, and that the total power W_T contained in the totality of those modes is constant. The modes are assumed to add together on a complex-amplitude basis to produce speckle at all coordinates (x, y) on a finite area $\mathcal{A}_T$. By a conservation of degrees-of-freedom argument, the total number of speckle intensity correlation cells across the area $\mathcal{A}_T$ is also M_T. We assume that the number of modes M_T is large, and that the speckle is spatially statistically stationary over the area $\mathcal{A}_T$.

Let the powers associated with $M \leq M_T$ of the correlation cells of the speckle be summed to produce a power W that is $\leq W_T$. The statistical problem posed is to find the probability density function $p(W \mid W_T)$, namely, the conditional density function of W given that the total power summed over all correlation cells is W_T. To find this probability density function, we invoke Bayes' rule (see [151], p. 59) to write

$$p(W \mid W_T) = \frac{p(W_T \mid W)p(W)}{p(W_T)}, \qquad \text{(F-1)}$$

where $p(W)$ is the probability density function of W without a constraint, $p(W_T)$ is the probability density function of W_T without a constraint, and $p(W_T \mid W)$ is the probability density function of W_T subject to the constraint of a known value of W.

The total number of correlation cells across the beam is M_T. M represents the number of speckle correlation cells over which the power is summed. To a good approximation,

$$M/M_T = \mathcal{A}/\mathcal{A}_T = \kappa, \qquad \text{(F-2)}$$

where $\mathcal{A}$ is the area of the beam that over which the summation takes place, $\mathcal{A}_T$ is the total area of the beam, and κ will be called the area ratio. Thus

$$M = \kappa M_T. \qquad \text{(F-3)}$$

In addition, we define the average power per degree of freedom to be $s^2 = \overline{W}/M = \overline{W}_T/M_T$.

With these definitions, the unconstrained probability density functions of W and W_T are given by gamma density functions:[1]

$$p(W_T) = \frac{\left(\frac{M_T}{\overline{W}_T}\right)^{M_T} W_T^{M_T-1} \exp\left[-\frac{M_T W_T}{\overline{W}_T}\right]}{\Gamma(M_T)} = \frac{\left(\frac{1}{s^2}\right)^{M_T} W_T^{M_T-1} \exp\left[-\frac{W_T}{s^2}\right]}{\Gamma(M_T)}$$

$$p(W) = \frac{\left(\frac{M}{\overline{W}}\right)^{M} W^{M-1} \exp\left[-\frac{MW}{\overline{W}}\right]}{\Gamma(M)} = \frac{\left(\frac{1}{s^2}\right)^{\kappa M_T} W^{\kappa M_T-1} \exp\left[-\frac{W}{s^2}\right]}{\Gamma(\kappa M_T)},$$

where these equations hold for arguments ≥ 0 and are zero otherwise and in the last line we have used $M = \kappa M_T$.

The probability density function of W_T conditioned by knowledge of W is also a gamma density with $M_T - M = M_T(1-\kappa)$ degrees of freedom. Thus

$$p(W_T \mid W) = \frac{\left(\frac{1}{s^2}\right)^{(1-\kappa)M_T} (W_T - W)^{(1-\kappa)M_T - 1} \exp\left[-\frac{(W_T - W)}{s^2}\right]}{\Gamma((1-\kappa)M_T)}. \qquad \text{(F-4)}$$

This expression is valid only for $W_T - W \geq 0$ and is zero otherwise.

Combining all of the above results, the expression for the conditional probability density of the summed power for constrained speckle is given by the following density function:

$$p(W \mid W_T) = \frac{1}{W_T}\left(\frac{W}{W_T}\right)^{\kappa M_T - 1}\left(1 - \frac{W}{W_T}\right)^{(1-\kappa)M_T - 1} \frac{\Gamma(M_T)}{\Gamma(\kappa M_T)\Gamma((1-\kappa)M_T)}, \qquad \text{(F-5)}$$

---

1. It may help the reader's understanding to think about the problem in the following way. We first construct an ensemble of unconstrained, fully developed, speckle sample functions. Each sample function has a random integrated intensity W_T, and over the ensemble, W_T is gamma distributed with M_T degrees of freedom. Now select a sub-ensemble of speckle sample functions from the initial ensemble, choosing only those sample functions that have a total integrated intensity that lies within a very small increment $\pm \varepsilon$ of a particular value W_T, which in this case is a constant representing the total integrated intensity coupled into the fiber. The statistics of the integrated intensity W are the statistics of the integral over a finite sub-area the new sub-ensemble just created. This is in fact the meaning of a conditional probability density function $p(W \mid W_T)$.

valid for $0 \leq W \leq W_T$ and zero otherwise. In the statistics literature, this density function is known as the *beta* density function. The nth normalized moment of such a density is given by

$$\frac{\overline{W^n}}{W_T^n} = \frac{\Gamma(M_T)\Gamma(\kappa M_T + n)}{\Gamma(\kappa M_T)\Gamma(M_T + n)}. \qquad \text{(F-6)}$$

More properties of the speckle described by this density function are explored in Section 7.1.2. Here it suffices to present expressions for the contrast C and the rms signal-to-noise ratio, $(S/N)_{\text{rms}}$, of the summed speckle. From the expression for the moments, we find

$$C = \frac{\sigma_W}{\overline{W}} = \sqrt{\frac{1-\kappa}{\kappa}} \, \frac{1}{\sqrt{M_T}} \qquad \text{(F-7)}$$

and

$$\left(\frac{S}{N}\right)_{\text{rms}} = \frac{\overline{W}}{\sigma_W} = \sqrt{\frac{\kappa}{1-\kappa}} \sqrt{M_T}. \qquad \text{(F-8)}$$

Note that the number of speckles influencing the contrast and rms signal-to-noise ratio can never be less than one (a single degree of freedom). Therefore, κ can never be smaller than $1/M_T$ in these expressions, in which case, for large M_T, $C \to 1$ and $(S/N)_{\text{rms}} \to 1$. At the other extreme, when κ approaches unity, $C \to 0$ and $(S/N)_{\text{rms}} \to \infty$.

G | Sample *Mathematica* Programs for Simulating Speckle

On the chance that they may be useful for the reader, we supply two simple, fully commented *Mathematica* programs for simulating the effects of speckle.

G.1 Speckle Simulation With Free Space Propagation

This program assumes that the scattering spot is rectangular and that observation is in the far field (or equivalently in the focal plane of a positive lens). It allows the user to control the size of the array used in the computation, and the size of the scattering spot. The program is presented in Fig. G.1.

G.2 Speckle Simulation With an Imaging Geometry

The program simulates the formation of speckle in an imaging system with unity magnification. It creates an object consisting of three vertical bars with normalized intensities 1.0, 0.5, and 0.1. The pupil of the imaging lens is assumed to be circular. The program is presented in Fig. G.2. A typical result from this program is shown in Fig. G.3.

APPENDIX G Sample *Mathematica* Programs for Simulating Speckle

```
(* Program for Simulating Speckle Formation
by Free Space Propagation *)

n=1024;
(* Linear dimension of the nxn array to be used.
The user can change this number. *)

k=8;
(* Number of samples (in one dimension) per speckle.
The user can change this number.*)

start = Table[Exp[I*2*Pi*Random[]],{n/k},{n/k}];
(* Generate an n/k x n/k array of random phasors,*)

scatterarray = PadRight[start,{n,n}];
(* Pad the phasor array with zeros.
The scattering spot is square of size n/k x n/k. *)

specklefield=Fourier[scatterarray];
(* Find the FFT of the padded array.
This is the speckle field in the
observation plane.*)

speckleintensity=Abs[specklefield]^2;
(* Find the intensity of the observed
speckle field *)

ListDensityPlot[speckleintensity,Mesh?False]
```

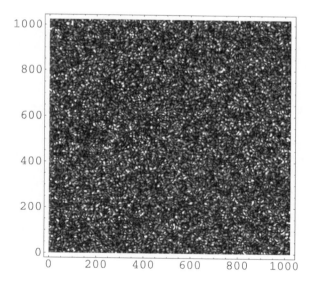

– DensityGraphics –

Figure G.1 *Mathematica* program simulating speckle formation with free space propagation and far field observation.

G.2 Speckle Simulation With an Imaging Geometry

```
(* Program Simulating Speckle Formation in Imaging *)

circ[r_]:=If[r<=1,1,0]
(* Define the circle function. *)

rect[x_]:=If[-1/2<=x<1/2,1,0]
(* Define the rectangle function *)

n=1024;
(* Linear dimension of the nxn array to be used.
This number can be changed by the user.*)

k=16;
(* Number of samples per speckle.
This number can be changed by the user.*)

r0 = n/k;
(* Radius of the lens pupil function in pixels.*)

objectintensity = Table[(1.*rect[(x-n/6)/(n/3)]
+0.5*rect[(x-n/2)/(n/3)]+0.1*rect[(x-5*n/6)/(n/3)])
*rect[y/(2*n)],{x,n},{y,n}];
(* This function defines the object intensity distribution
in the absence of speckle.
Here it consists of three rectangular regions of
different brightnesses.
You can change it if you wish.*)

randomfield = Table[Exp[I*2*Pi*Random[]],{n},{n}];
(* Generate an nxn array of random phasors. *)

scatterfield=Sqrt[objectintensity]*randomfield;
(* The object amplitude distribution is
multiplied by the array of random phasors *)

p1=ListDensityPlot[Abs[scatterfield]^2,Mesh->False,
DisplayFunction->Identity];
(* Plot the intensity distribution across the scattering surface,
but save the plot for later *)

bandpass=Table[circ[Sqrt[(p-n/2)^2+(q-n/2)^2]/r0]
,{p,1,n},{q,1,n}];
(* Define circular pupil function of the lens. *)

pupilfield=bandpass*Fourier[scatterfield];
(* Calculate the field transmitted by the lens pupil.*)

imagefield=InverseFourier[pupilfield];
(* Calculate the image field. *)

imageintensity=Abs[imagefield]^2;
(* Calculate the image intensity *)

p2=ListDensityPlot[imageintensity,Mesh->False,
DisplayFunction->Identity];
(* Plot the intensity of the speckled image,
but save the plot for the next step *)

Show[GraphicsArray[{{p1},{p2}}]]
(* Show the original object intensity
distribution and the speckled image intensity
distribution *)
```

Figure G.2 *Mathematica* program for simulating speckle in an imaging system.

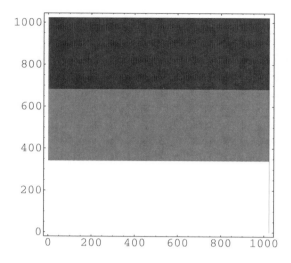

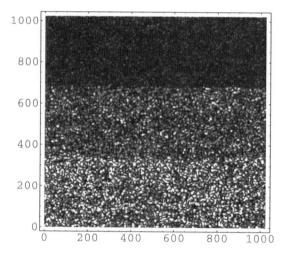

Figure G.3 Typical output from the speckle imaging program of the previous figure. The top density plot shows the original object intensity distribution, and the bottom density plot shows the speckled image of that object.

Bibliography

[1] M. Abramowitz and I.A. Stegun. *Handbook of Mathematical Functions.* Dover Publications, Inc., New York, NY, 1965.

[2] M.A. Al-Habash, L.C. Andrews, and R.L. Phillips. Mathematical model for the irradiance probability density function of a laser beam propagating through turblent media. *Opt. Eng.*, 40:1554–1562, 2001.

[3] I.M. Andrews and J.A. Leendertz. Speckle pattern interferometry of vibration modes. *IBM J. Res. Develop.*, 20:285–289, 1976.

[4] E. Archibold, J.M. Burch, and A.E. Ennos. Recording of in-plane displacement by double-exposure speckle photography. *Optica Acta*, 17:883–898, 1970.

[5] E. Archibold, A.E. Ennos, and P.A. Taylor. A laser speckle interferometer for the detection of surface movements and vibration. In J.H. Dickson, editor, *Optical Instruments and Techniques*, page 265. Oriel, Newcastle upon Tyne, England, 1969.

[6] J.M. Artigas, A. Felipe, and M.J. Buades. Contrast sensitivity of the visual system in speckle imagery. *J. Opt. Soc. Am. A*, 11:2345–2349, 1994.

[7] J.R. Artigas and A. Felipe. Effect of luminance on photopic visual acuity in the presence of laser speckle. *J. Opt. Soc. Am. A*, 5:1767–1771, 1988.

[8] G.R. Ayers, M.J. Northcott, and J.C. Dainty. Knox-Thompson and triple correlation imaging through atmospheric turbulence. *J. Opt. Soc. Am.*, 5:963–985, 1988.

[9] R.D. Bahuguna, K.K. Gupta, and K. Singh. Expected number of intensity level crossings in a normal speckle pattern. *J. Opt. Soc. Am.*, 70:874–876, 1980.

[10] R.D. Bahuguna and D. Malacara. Stationarity of speckle in laser refraction. *J. Opt. Soc. Am.*, 73:1213–1215, 1983.

[11] R.D. Bahuguna and D. Malacara. Speckle motion: The apparent source position for a plane diffuser. *J. Opt. Soc. Am. A*, 1:420–422, 1984.

[12] R. Barakat. First-order probability densities of laser speckle patterns observed through finite-size scanning apertures. *Optica Acta*, 20:729–740, 1973.

[13] R. Barakat. The level-crossing rate and above level duration time of the intensity of a gaussian random process. *Inf. Sci. (New York)*, 20:83–87, 1980.

[14] N.B. Baranova, B. Ya. Zel'dovich, A.V. Mamaev, N. Pilipetskii, and V.V. Shukov. Dislocations of the wavefront of a speckle-inhomogeneous field (theory and experiment). *JETP Letters*, 33:195–199, 1981.

[15] J. Barton and S. Stromski. Flow measurement without phase information in optical coherence tomography images. *Optics Express*, 13:5234–5239, 2005.

[16] P. Beckmann and A. Spizzichino. *The Scattering of Electromagnetic Waves from Rough Surfaces*. Artech House, Norwood, MA, 1987.

[17] M.V. Berry. Disruption of wavefronts: Statistics of dislocations in incoherent gaussian random waves. *J. Phys. A*, 11:27–37, 1978.

[18] N.M. Blackman. *Noise and Its Effect on Communication*. McGraw-Hill Book Co., New York, NY, 1963.

[19] B.E. Bouma and G.J. Tearney, editors. *Handbook of Optical Coherence Tomography*. Marcel Dekker, Inc., New York, NY, 2002.

[20] R.N. Bracewell. *The Fourier Transform and Its Applications*. McGraw-Hill Book Company, New York, NY, 3rd edition, 1999.

[21] D. Brogioli, A. Vailati, and M. Giglio. Heterodyne near-field scattering. *Applied Phys. Lett.*, 81:4109, 2002.

[22] J.M. Burch and J.M.J. Tokarski. Production of multiple beam fringes from photographic scatterers. *Optica Acta*, 15:101–111, 1968.

[23] J.N. Butters. Speckle pattern interferometry using video techniques. *J. Soc. PhotoOpt. Instrum. Eng.*, 10:5–9, 1971.

[24] W.N. Charman. Speckle movement in laser refraction. I. Theory. *Am. J. Optom. Physiol. Opt.*, 56:219–227, 1979.

[25] W.N. Charman and H. Whitefoot. Speckle motion in laser refraction. II. Experimental. *Am. J. Optom. Physiol. Opt.*, 56:295–304, 1979.

[26] Y.-C. Chiang, Y.-P. Chiou, and H.-C. Chang. Improved full-vectorial finite-difference mode solver for optical waveguides with step-index profiles. *J. Lightwave Tech.*, 20:1609–1618, 2002.

[27] M.A. Condie. *An Experimental Investigation of the Statistics of Diffusely Reflected Coherent Light*. Engineer degree thesis, Stanford University, 1966.

[28] K. Creath. Phase-shifting speckle interferometry. *Appl. Opt.*, 24:3053–3058, 1985.

[29] J.C. Dainty. Detection of images immersed in speckle noise. *Optica Acta*, 18:327–339, 1971.

[30] J.C. Dainty, editor. *Laser Speckle and Related Phenomena*. Springer Verlag, Berlin, Heidelberg, New York, Tokyo, 2nd edition, 1984.

[31] J.C. Dainty and W.T. Welford. Reduction of speckle in image plane hologram reconstruction by moving pupils. *Optics Commun.*, 3:289–294, 1971.

[32] W.B. Davenport and W.L. Root. *Random Signals and Noise*. McGraw-Hill Book Co., New York, NY, 1958.

[33] M.R. Dennis. *Topological Singularities in Wave Fields*. PhD thesis, University of Bristol, 2001.

[34] S. Donati and G. Martini. Speckle pattern intensity and phase: Second-order conditional statistics. *J. Opt. Soc. Am.*, 69:1690–1694, 1979.

[35] D. Dudley, W.M. Duncan, and J. Slaughter. Emerging digital micromirror device (DMD) applications. In H. Urey, editor, *Proc. SPIE, MOEMS Display and Imaging Systems*, volume 4985, pages 14–26, 2003.

[36] K.J. Ebeling. Statistical properties of spatial derivatives of amplitude and intensity of monochromatic speckle patterns. *Optica Acta*, 26:1505–1521, 1979.

[37] M. Elbaum, M. Greenebaum, and M. King. A wavelength diversity technique for reduction of speckle size. *Optics Commun.*, 5:171–174, 1972.

[38] M. Elbaum, M. King, and M. Greenbaum. Laser correlography: Transmission of high-resolution object signatures through the turbulent atmosphere. Technical Report Report T-1/306-3-11, Riverside Research Institute, New York, NY, 1974.

[39] A.E. Ennos. Speckle interferometry. In J.C. Dainty, editor, *Laser Speckle and Related Phenomena*, pages 203–253. Springer-Verlag, 2nd edition, 1984.

[40] R.E. Epworth. The phenomenon of modal noise in analog and digital optical fibre systems. In *Proceedings of the Fourth European Conference on Optical Communications*, pages 492–501, Genoa, Italy, September 1978.

[41] R.K. Erf, editor. *Speckle Metrology*. Academic Press, New York, NY, 1978.

[42] Leonard Eyges. *The Classical Electromagnetic Field*. Dover Publications, Inc., New York, NY, 1980.

[43] A.F. Fercher, W. Drexler, C.K. Hitzenberger, and T. Lasser. Optical coherence tomography—principles and applications. *Rep. Prog. Phys.*, 66:239–303, 2003.

[44] J.R. Fienup. Reconstruction of an object from the modulus of its Fourier transform. *Opt. Lett.*, 3:27–29, 1978.

[45] J.R. Fienup. Phase retrieval algorithms: A comparison. *Appl. Opt.*, 21:2758–2769, 1982.

[46] J.R. Fienup. Unconventional imaging systems: Image formation from non-imaged laser speckle patterns. In S.R. Robinson, editor, *The Infrared & Electro-Optical Systems Handbook*, volume 8. Environmental Research Institute of Michigan, 1993.

[47] J.R. Fienup and P.S. Idell. Imaging correlography with sparse arrays of detectors. *Opt. Engin.*, 27:778–784, 1988.

[48] J.R. Fienup, R.G. Paxman, M.F. Reiley, and B.J. Thelen. 3-D imaging correlography and coherent image reconstruction. In *Digital Image Recovery and Synthesis IV*, volume 3815 of *Proc. SPIE*, Bellingham, WA, 1999. SPIE.

[49] M. Françon. *Laser Speckle and Applications in Optics*. Academic Press, New York, San Francisco, London, 1979. English translation by H.H. Arsenault.

[50] I. Freund. Optical vortices in gaussian random wave fields: Statistical probability densities. *J. Opt. Soc. Am. A*, 11:1644–1652, 1994.

[51] I. Freund and N. Shvartsman. Wave-field phase singularities: The sign principle. *Phys. Rev. A*, 50:5164–5172, 1994.

[52] T. Fricke-Begemann and J. Burke. Speckle interferometry: Three-dimensional deformation field measurement with a single interferogram. *Appl. Opt.*, 40:5011–5022, 2001.

[53] D.L. Fried. Optical resolution through a randomly inhomogeneous medium for very long and very short exposures. *J. Opt. Soc. Am.*, 56:1372–1379, 1966.

[54] H. Fuji and T. Asakura. Effects of surface roughness on the statistical distribution of image speckle intensity. *Optics Commun.*, 11:35–38, 1974.

[55] N. George, C.R. Christensen, J.S. Benneth, and B.D. Guenther. Speckle noise in displays. *J. Opt. Soc. Am.*, 66:1282–1289, 1976.

[56] N. George and A. Jain. Speckle reduction using multiple tones of illumination. *Appl. Opt.*, 12:1202–1212, 1973.

[57] N. George and A. Jain. Space and wavelength dependence of speckle intensity. *Appl. Phys.*, 4:201–212, 1974.

[58] D.Y. Gezari, A. Labeyrie, and R.V. Stachnik. Speckle interferometry: diffraction-limited measurements of nine stars with the 200-inch telescope. *Astrophys. J.*, 173:L1–L5, 1972.

[59] A. Ghatak and K. Thyagarajan. Graded index optical waveguides: A review. In E. Wolf, editor, *Progress in Optics*, pages 1–109. North Holland, 1980.

[60] D.C. Ghiglia and M.D. Pritt. *Two-Dimensional Phase Unwrapping*. John Wiley & Sons, New York, NY, 1998.

[61] M. Giglio, M. Carpineti, and A. Vailati. Space intensity correlations in the near field of scattered light: A direct measurement of the density correlation function $g(r)$. *Phys. Rev. Lett.*, 85:1416–1419, 2000.

[62] M. Giglio, M. Carpineti, A. Vailati, and D. Brogioli. Near-field intensity correlations of scattered light. *Appl. Opt.*, 40:4036–4040, 2001.

[63] D. Gloge and E.A. Marcatili. Multimode theory of graded-core fibers. *Bell Syst. Tech. J.*, 52:1563–1578, 1973.

[64] L.I. Goldfischer. Autocorrelation function and power spectral density of laser-produced speckle patterns. *J. Opt. Soc. Am.*, 55:247, 1965.

[65] J.W. Goodman. Statistical properties of laser sparkle patterns. Technical Report T.R. 2303-1, Stanford Electronics Laboratories, Stanford, CA, 1963.

[66] J.W. Goodman. Some effects of target-induced scintillation on optical radar performance. *Proc. IEEE*, 53:1688–1700, 1965.

[67] J.W. Goodman. Comparative performance of optical radar detection techniques. *IEEE Trans. Aerospace and Electronic Systems*, AES-2:526–535, 1966.

[68] J.W. Goodman. Dependence of image speckle contrast on surface roughness. *Optics Commun.*, 14:324–327, 1975.

[69] J.W. Goodman. Role of coherence concepts in the study of speckle. *Proc. SPIE*, 194:86–94, 1979.

[70] J.W. Goodman. *Statistical Optics*. John Wiley & Sons, New York, NY, 1985.

[71] J.W. Goodman. *Introduction to Fourier Optics*. Roberts & Company, Englewood, CO, 3rd edition, 2005.

[72] J.W. Goodman and E.G. Rawson. Statistics of modal noise in fibers: A case of constrained speckle. *Opt. Lett.*, 6:324–326, 1981.

[73] I.I. Gradshteyn and I.M. Ryzhik. *Table of Integrals Series and Products*. Academic Press, New York, NY, 1965.

[74] D.A. Gregory. Basic physical principles of defocused speckle photography: A tilt topology inspection technique. *Opt. Laser Technol.*, 8:201–213, 1976.

[75] G. Groh. Engineering uses of laser produced speckle patterns. In *Proc. Symposium on the Engineering Uses of Holography*, pages 483–497, London, 1970. University of Strathclyde, Cambridge University Press.

[76] M.R. Hee. Optical coherence tomography: Theory. In B.E. Bouma and G.J. Tearney, editors, *Handbook of Optical Coherence Tomography*, chapter 2, pages 41–66. Marcell Dekker, Inc., New York, NY, 2002.

[77] P.S. Idell, J.R. Fienup, and R.S. Goodman. Image synthesis from nonimaged laser-speckle patterns. *Opt. Lett.*, 12:858–860, 1987.

[78] K. Iizuka. *Elements of Photonics in Free Space Media and Special Media*, volume 1. Wiley-Interscience, New York, NY, 2002.

[79] E. Ingelstam and S. Ragnarsson. Eye refraction examined by aid of speckle pattern produced by coherent light. *Vision Res.*, 12:411–420, 1972.

[80] A. Ishimaru. *Wave Propagation and Scattering in Random Media*. Academic Press, New York, NY, 1978.

[81] E.V. Ivakin, A.I. Kitsak, N.V. Karelin, A.M. Lazaruk, and A.S. Rubanov. Approaches to coherence destruction of short laser pulses. In K.N. Drabovich, N.S. Kazak, V.A. Makarov, and A.P. Voitovich, editors, *Proceedings of the SPIE: Nonlinear Optical Phenomena and Nonlinear Dynamics of Optical Systems*, volume 4751, pages 34–41, 2002.

[82] E. Jakeman. On the statistics of K-distributed noise. *J. Phys. A: Math. Gen.*, 13:31–48, 1980.

[83] E. Jakeman. K-distributed noise. *J. Opt. A., Pure Appl. Opt.*, 1:784–789, 1999.

[84] E. Jakeman and P.N. Pusey. A model for non-Rayleigh sea echo. *IEEE Trans. Antennas and Propag.*, 24:806–814, 1976.

[85] E. Jakeman and P.N. Pusey. Significance of K-distributions in scattering experiments. *Phys. Rev. Lett.*, 40:546–548, 1978.

[86] R. Jones and C. Wykes. *Holographic and Speckle Interferometry*. Cambridge University Press, Cambridge, UK, 1989.

[87] R.C. Jones. A new calculus for the treatment of optical systems. *J. Opt. Soc. Am.*, 31:488–503, 1941.

[88] T.R. Judge and P.J. Bryanston-Cross. A review of phase unwrapping techniques in fringe analysis. *Opt. Lasers Eng.*, 21:199–239, 1994.

[89] D.A. Kessler and I. Freund. Level-crossing densities in random wave fields. *J. Opt. Soc. Am. A*, 15:1608–1618, 1998.

[90] K. Khare and N. George. Sampling theory approach to prolate spheroidal wavefunctions. *J. Phys. A*, 36:10011–10021, 2003.

[91] K.T. Knox and B.J. Thompson. Recovery of images from atmospherically degraded short exposure images. *Astrophys. J.*, 193:L45–L48, 1974.

[92] A.N. Kolmogorov. The local structure of turbulence in incompressible viscous fluids for very large Reynolds numbers. In S.K. Friedlander and L. Topper, editors, *Turbulence, Classic Papers on Statistical Theory*. Wiley Interscience, New York, NY, 1961.

[93] D. Korff. Analysis of a method for obtaining near-diffraction-limited information in the presence of atmospheric turbulence. *J. Opt. Soc. Am.*, 63:971–980, 1973.

[94] M. Kowalczyk. Density of 2D gradient of intensity in fully developed speckle patterns. *Optics Commun.*, 48:233–236, 1883.

[95] V.N. Kowar and J.R. Pierce. Detection of gratings and small features in speckle imagery. *Appl. Opt.*, 20:312–319, 1981.

[96] A. Kozma and C.R. Christensen. Effects of speckle on resolution. *J. Opt. Soc. Am.*, 66:1257–1260, 1976.

[97] A. Labeyrie. Attainment of diffraction limited resolution in large telescopes by Fourier analyzing speckle patterns in star images. *Astron. Astrophys.*, 6:85, 1970.

[98] J.A. Leendertz. Interferometric displacement measurement. *J. Phys. E: Scientific Instruments*, 3:214–218, 1970.

[99] D. Léger, E. Mathieu, and J.C. Perrin. Optical surface roughness determination using speckle correlation technique. *Appl. Opt.*, 14:872–877, 1975.

[100] L. Leushacke and M. Kirchner. Three-dimensional correlation of speckle intensity for rectangular and circular apertures. *J. Opt. Soc. Am. A*, 7:827–832, 1990.

[101] H.J. Levinson. *Principles of Lithography*. SPIE Press, Bellingham, WA, 2nd edition, 2005.

[102] A.W. Lohmann, G. Weigelt, and B. Wirnitzer. Speckle masking in astonomy: Triple correlation theory and applications. *Appl. Opt.*, 22:4028–4037, 1983.

[103] S. Lowenthal and D. Joyeux. Speckle removal by a slowly moving diffuser associated with a motionless diffuser. *J. Opt. Soc. Am.*, 61:847, 1971.

[104] A. Macovski, S.D. Ramsey, and L.F. Schaefer. Time-lapse interferometry and contouring using television systems. *Appl. Opt.*, 10:2722–2727, 1971.

[105] L. Mandel. Fluctuations of photon beams: The distribution of photo-electrons. *Proc. Phys. Soc.*, 74:233–243, 1959.

[106] M. Mansuripur. *Classical Optics and Its Applications*. Cambridge University Press, Cambridge, UK, 2002.

[107] J.I. Marcum. A statistical theory of target detection by pulsed radar. *IRE Trans. Information Theory*, IT-6:59, 1960.

[108] W. Martienssen and S. Spiller. Holographic reconstruction without granulation. *Phys. Lett.*, 24A:126–128, 1967.

[109] C.L. Matson. Weighted-least-squares phase reconstruction from the bispectrum. *J. Opt. Soc. Am. A*, 8:1905–1913, 1991.

[110] T.S. McKechnie. Speckle reduction. In J.C. Dainty, editor, *Laser Speckle and Related Phenomena*. Springer-Verlag, 2nd edition, 1984.

[111] David Middleton. *An Introduction to Statistical Communication Theory*. IEEE Press, Piscataway, NJ, 1996. IEEE Press Classic Re-issue.

[112] B.D. Millikan and J.S. Wellman. Digital projection systems based on LCOS. In C.M. Lampert, C.-G. Granqvist, and K.L. Lewis, editors, *Proc. SPIE, Solar and Switching Materials*, volume 4458, pages 226–229, 2001.

[113] N. Mohon and A. Rodemann. Laser speckle for determining ametropia and accomodation response of the eye. *Appl. Opt.*, 12:783–787, 1973.

[114] M.N. Morris, J. Millerd, N. Brock, J. Hayes, and B. Saif. Dynamic phase-shifting electronic speckle pattern interferometer. In H.P. Stahl, editor, *Optical Manufacturing and Testing IV*, volume 5869, pages 1B–1–9. SPIE, Bellingham, WA, 2005.

[115] B. Moslehi, J.W. Goodman, and E.G. Rawson. Bandwidth estimation for multimode optical fibers using the frequency correlation function of speckle patterns. *Appl. Opt.*, 22:995–999, 1983.

[116] P.K. Murphy, J.P. Allebach, and N.C. Gallagher. Effect of optical aberrations on laser speckle. *J. Opt. Soc. Am. A*, 3:215–222, 1986.

[117] P. Nisenson. Speckle imaging with the PAPA detector and the Knox-Thompson algorithm. In D.M. Alloin and J.M. Mariotti, editors, *Diffraction-Limited Imaging*, pages 157–169. Kluwer Academic Publshers, Boston, MA, 1989.

[118] J.F. Nye and M.V. Berry. Dislocations in wave trains. *Proc. Roy. Soc. Lond.*, A 366:165–190, 1974.

[119] E. Ochoa and J.W. Goodman. Statistical properties of ray directions in a monochromatic speckle pattern. *J. Opt. Soc. Am.*, 73:943–949, 1983.

[120] J. Ohtsubo and T. Asakura. Statistical properties of speckle intensity variations in the diffraction field under illumination of coherent light. *Optics Commun.*, 14:30, 1975.

[121] B.M. Oliver. Sparkling spots and random diffraction. *Proc. IEEE*, 51:220, 1963.

[122] G.R. Osche. *Optical Detection Theory for Laser Applications*. John Wiley & Sons, Inc., New York, NY, 2002.

[123] G.C. Papen and G.M. Murphy. Modal noise in multimide fibers under restricted launch conditions. *J. Lightwave Tech.*, 17:817–822, 1999.

[124] A. Papoulis. *Probability, Random Variables and Stochastic Processes*. McGraw-Hill Book Co., New York, NY, 1965.

[125] G. Parry. Speckle patterns in partially coherent light. In J.C. Dainty, editor, *Laser Speckle and Related Phenomena*, volume 9 of *Topics in Applied Physics*, pages 77–122. Springer-Verlag, second edition, 1984.

[126] H.M. Pedersen. On the contrast of polychromatic speckle patterns and its dependence on surface roughness. *Optica Acta*, 22:15–24, 1975.

[127] Y. Piederrièrre, J. Cariou, Y. Guern, B. LeJeune, G. LeBrun, and J. Lotrian. Scattering through fluids: Speckle size measurement and Monte Carlo simulations close to and into multiple scattering. *Optics Express*, 12:176–188, 2004.

[128] P.K. Rastogi, editor. *Digital Speckle Pattern Interferometry and Related Techniques*. John Wiley & Sons, Ltd., New York, NY, 2001.

[129] E.G. Rawson, J.W. Goodman, and R.E. Norton. Frequency dependence of modal noise in multimode optical fibers. *J. Opt. Soc. Am.*, 70:968–976, 1980.

[130] Lord Rayleigh. On the problem of random vibrations and of random flights in one, two and three dimensions. *Phil. Mag.*, 37:321–397, 1919.

[131] W.T. Rhodes and J.W. Goodman. Interferometric technique for recording and restoring images degraded by unknown aberrations. *J. Opt. Soc. Am.*, 63:647–657, 1973.

[132] S.O. Rice. Mathematical analysis of random noise. In N. Wax, editor, *Selected Papers on Noise and Stochastic Processes*, pages 133–294. Dover Press, New York, NY, 1954.

[133] J.D. Rigden and E.I. Gordon. The granularity of scattered optical maser light. *Proc. IRE*, 50:2367, 1962.

[134] D.W. Robinson. Phase unwrapping methods. In D.W. Robinson and G.T. Reid, editors, *Interferogram Analysis*, pages 194–229. Inst. of Phys. Pub., 1993.

[135] F. Roddier. The effects of atmospheric turbulence in optical astronomy. In E. Wolf, editor, *Progress in Optics*, volume XIX, chapter V, pages 281–376. Elsevier North-Holland, Inc., New York, NY, 1981.

[136] F. Roddier. Triple correlation as a phase closure technique. *Optics Commun.*, 60:145–148, 1986.

[137] M.C. Roggemann and B. Welsh. *Imaging Through Turbulence*. CRC Press, Boca Raton, FL, 1996.

[138] C. Rydberg, J. Bengtsson, and T. Sandstrom. Dynamic laser speckle as a detrimental phenomenon in optical projection lithography. *J. Microlith. Microfab. Microsyst.*, 5:033004-1—8, 2006.

[139] B.E.A. Saleh and M.C. Teich. *Fundamentals of Photonics*. John Wiley & Sons, Inc., New York, NY, 1991.

[140] T. Sandstrom, C. Rydberg, and J. Bengtsson. Dynamic laser speckle in optical projection lithography: Causes, effects on CDU and LER, and possible remedies. In B.W. Smith, editor, *Optical Microlithography XVIII*, volume 5754 of *S.P.I.E. Proceedings*, pages 274–284. SPIE, Bellingham, WA, 2005.

[141] J.M. Schmitt. Array detection for speckle reduction in optical coherence microscopy. *Phys. Med. Biol.*, 42:1427–1439, 1997.

[142] J.M. Schmitt. Optical coherence tomography (OCT): A review. *IEEE J. Selected Topics in Quantum Electronics*, 5:1205–1215, 1999.

[143] A.A. Scribot. First-order probability density functions of speckle measured with a finite aperture. *Optics Commun.*, 11:238–295, 1974.

[144] J.M. Senior. *Optical Fiber Communications*. Pearson Education, Upper Saddle River, NJ, 2nd edition, 1992.

[145] N. Shvartsman and I. Freund. Wave-field phase singularities: Near-neighbor correlations and anticorrelations. *J. Opt. Soc. Am. A*, 11:2710–2718, 1994.

[146] R.S. Sirohi, editor. *Speckle Metrology*. Marcel Dekker, New York, NY, 1993.

[147] M. Slack. The probability distribution of sinusoidal oscillations combined in random phase. *J.I.E.E.*, 93, Part III:76–86, 1946.

[148] D. Slepian. Prolate spheroidal wave functions, Fourier analysis and uncertainty—I. *Bell Syst. Tech. J.*, 40:43–63, 1961.

[149] D. Slepian. Prolate spheroidal wave functions, Fourier analysis and uncertainty—II. *B. Syst. Tech. J.*, 40:65–84, 1961.

[150] R.A. Sprague. Surface roughness measurement using white light speckle. *Appl. Opt.*, 11:2811–2816, 1972.

[151] Henry Stark and John W. Woods. *Probability, Random Processes and Estimation Theory for Engineers*. Prentice-Hall, Englewood Cliffs, NJ, 1986.

[152] W. Steinchen and L. Yang. *Digital Shearography: Theory and Applications of Digital Speckle Pattern Shearing Interferometry*. SPIE Press, Bellingham, WA, 2003.

[153] G. Strang. *Linear Algebra and its Applications*. Academic Press, New York, NY, 1976. Page 224.

[154] J.W. Strohbehn and S. Clifford. Polarization and angle-of-arrival fluctuations for a plane wave propagated through a turbulent medium. *IEEE Trans. Ant. Prop.*, AP-15:416–421, 1967.

[155] Y. Surrel. Customized phase shift algorithms. In P. Rastogi and D. Inaudi, editors, *Trends in Optical Non-Destructive Testing and Inspection*, chapter 5, pages 71–83. Elsevier, Amsterdam, 2000.

[156] H. Takajo and T. Takahashi. Least-squares phase estimation from the phase difference. *J. Opt. Soc. Am. A*, 5:416–425, 1988.

[157] M. Takeda. Recent progress in phase unwrapping techniques. In C. Gorecki, editor, *Optical Inspection and Measurements*, volume 2782, pages 334–343, Bellingham, WA, 1996. S.P.I.E.

[158] M. Takeda, H. Ina, and S. Kobayashi. Fourier-transform method for fringe-pattern analysis for computer-based topography and interferometry. *J. Opt. Soc. Am.*, 72:156–160, 1982.

[159] V.I. Tatarski. *Wave Propagation in a Turbulent Medium*. McGraw-Hill Book Co., New York, NY, 1961.

[160] C.A. Thompson, K.J. Webb, and A.M. Weiner. Diffusive media characterization with laser speckle. *Appl. Opt.*, 36:3726–3734, 1997.

[161] C.A. Thompson, K.J. Webb, and A.M. Weiner. Imaging in scattering media by use of laser speckle. *J. Opt. Soc. Am. A*, 14:2269–2277, 1997.

[162] H.J. Tiziani. Application of speckling for in-plane vibration analysis. *Opt. Acta*, 18:891–902, 1971.

[163] H.J. Tiziani. A study of the use of laser speckle to measure small tilts of optically rough surfaces accurately. *Optics Commun.*, 5:271–274, 1972.

[164] Y. Tremblay, B.S. Kawasaki, and K.O. Hill. Modal noise in optical fibers: Open and closed speckle pattern regimes. *Appl. Opt.*, 20:1652–1655, 1981.

[165] J.I. Trisnadi. Speckle contrast reduction in laser projection displays. In M.H. Wu, editor, *SPIE Proceedings—Projection Displays VIII*, volume 4657, pages 131–137, 2002.

[166] J.I. Trisnadi. Hadamard speckle contrast reduction. *Optics Lett.*, 29:11–13, 2004.

[167] R.F. van Ligten. Speckle reduction by simulation of partially coherent object illumination in holography. *Appl. Optics*, 12:255–265, 1973.

[168] B.A. Wandell. *Foundations of Vision*. Sinauer Associates Inc., 1995.

[169] L. Wang, T. Tschudi, T. Halldórsson, and P.R. Pétursson. Speckle reduction in laser projection systems by diffractive optical elements. *Appl. Opt.*, 37:1770–1775, 1998.

[170] T.R. Watts, K.I. Hopcroft, and T.R. Faulkner. Single measurements on probability density functions and their use in non-gaussian light scattering. *J. Phys. A:Math. Gen.*, 29:7501–7517, 1996.

[171] M.A. Webster, K.J. Webb, and A.M. Weiner. Temporal response of a random medium from third-order laser speckle frequency correlations. *Phys. Rev. Lett.*, 88:033901/1–033901/4, 2002.

[172] M.A. Webster, K.J. Webb, A.M. Weiner, J. Xu, and H. Cao. Temporal response of a random medium from speckle intensity frequency correlations. *J. Opt. Soc. Am.*, 20:2057–2070, 2003.

[173] A.J. Weierholt, E.G. Rawson, and J.W. Goodman. Frequency-correlation properties of optical waveguide intensity patterns. *J. Opt. Soc. Am. A*, 1:201–205, 1984.
[174] G. Westheimer. The eye as an optical instrument. In J. Thomas et al., editor, *Handbook of Perception*, chapter 4, pages 4.1–4.20. John Wiley & Sons, 1986.
[175] D.R. Williams, D.H. Brainard, M.J. McMahon, and R. Navarro. Double-pass and interferometric measures of the optical quality of the eye. *J. Opt. Soc. Am. A*, 11:3123–3135, 1994.
[176] E. Wolf. A macroscopic theory of interference and diffraction of light from finite sources, I: Fields with a narrow spectral range. *Proc. Roy. Soc.*, A225:96–111, 1954.
[177] E. Wolf. Optics in terms of observable quantities. *Nuovo Cimento*, 12:884–888, 1954.
[178] E. Wolf. Coherence properties of partially polarized electromagnetic radiation. *Nuovo Cimento*, 13:1165–1181, 1959.
[179] I. Yamaguchi. Speckle displacement and decorrelation—theory and applications. In P. Rastogi and D. Inaudi, editors, *Trends in Optical Non-Destructive Testing and Inspection*, chapter 11, pages 151–170. Elsevier, Amsterdam, 2000.
[180] M. Young, B. Faulkner, and J. Cole. Resolution in optical systems using coherent illumination. *J. Opt. Soc. Am.*, 60:137–139, 1970.
[181] N. Youssef, T. Munakata, and M. Takeda. Rice probability functions for level-crossing intervals of speckle intensity fields. *Optics Commun.*, 123:55–62, 1996.
[182] B.Y. Zel'dovich, A.V. Mamaev, and V.V. Shkunov. *Speckle-Wave Interactions in Application to Holography and Nonlinear Optics*. CRC Press, Inc., Boca Raton, FL, 1995.
[183] F. Zernike. The concept of degree of coherence and its application to optical problems. *Physica*, 5:785–795, 1938.

Index

Adaptive optics 301
Angle Diversity 153–158, 162–166
   in projection displays 211–213
   in holography 193, 194
Angle sensitivity of speckle 162–164, 351–358
Aperture masks for suppressing speckle
   in holography 194
   in optical coherence tomography 202, 203
Archibald, E. 275, 301
Asakura, T. 305
Atmospheric turbulence 311–313
Autocorrelation function, of speckle 73–84
   axial 83–84
   free-space geometry 73–80
   imaging geometry 80–82
Autocovariance matrix 61, 121, 122, 345–348

Barakat, R. 113, 116, 117
Beam ratio, in holography 30
Beckmann, P. 305
Beta probability density function 240, 241
   derivation of 365–367
Bispectrum
   definition 331
   information retrievable from 332, 333
   transfer function 331, 332
Blackman, N. 345
Boiling, of speckle 188, 209
Bose-Einstein
   distribution of photoevents 254
   signal in Poisson noise 257–259
Boxcar approximation 111, 112
Briers, J.D. 305
Bryanston-Cross, P.J. 299
Burch, J.M. 275

Carrier frequency 5
   in holography 192
Central limit theorem 10, 20, 117
Changing diffuser 143–153
   in projection TV 214–222
Characteristic function 6, 18
   of negative exponential distribution 29
   of sums of speckle patterns 39, 42
Circular complex Gaussian statistics 10, 11, 46, 337–339
Coherence 170–172
   area 172, 174
   diameter 315
   of excimer lasers 228, 229
   time 171, 173, 174
Coherence tomography 195–203
   analysis of 196–200
   overview of 195, 196
   speckle in 200–203
Coherency matrix 45
Coherent detection 249
Coherent receiver 261, 270–272
Comparison of detection techniques 270–272
Complex degree of coherence 170
Complex signal representation 6
Compounding of speckle suppression techniques 186
Condie, M. 113, 117
Conditional probability of speckle intensity 71
Constant phasor plus a random phasor sum 30–34
Constrained speckle 239–243, 365–367
Contrast of speckle 28, 37, 40, 41, 98–102
Correlation area of speckle 80
Correlation coefficient 21
Counting functional 130

Index

Covariance
    definition 59
    matrix (see autocovariance matrix)
    of surface roughness 305, 306, 308, 309
Creath, K. 298
Cross-spectrum 327–331
    definition 327
    information retrievable from 329–331
    technique 327–331
    transfer function 327–329

Dainty, J.C. 2, 113, 194
Degree of polarization 48, 49, 142
Degree of coherence (see complex degree of coherence)
Dennis, M. 133
Density of vortices 136–140
Derivatives of speckle intensity 120–124, 127–133
Detection probability
    Incoherent detection 257–260, 270–272
    heterodyne detection 266, 269, 270–272
Detection statistics 256–273
    comparison of detection techniques 270–273
    direct detection 256–260
    heterodyne detection 260–270
Difference of speckle intensities 71, 72
Difference of speckle phases 68
Diffuse target 249, 267–273
Digital light projection device (DLP) 204, 205, 223
Discrete Fourier transform 3, 5
Displacement, in-plane 277, 278, 284, 285
Displacement, out-of-plane 286
Doppler filter 261, 262, 265
Double diffuser 143–149, 151–153
Dynamic diffuser 359–363

Ebeling, K.J. 350
Effective area, in heterodyne detection 268–271
Effective size, of scattering spot 91
Eigenfunctions 114
Eigenvalues 114, 116
    of the coherency matrix 46, 48
Elbaum, M. 334
Electronic speckle pattern interferometry (ESPI) 291–293
Engineered diffusers 225
Ennos, A.E. 275, 301
Epworth, R.E. 237
Error function, definition 15
Excimer laser 228–231
Eye, human
    line spread function 220
    point spread function of 220
    speckle in 187–190

False alarm probability
    definition 256
    in a coherent receiver 265, 266
    in Poisson noise 256, 257

Farsighted eye 188, 190
Finite number of phasors, sum of 34–37
Frequency covariance function 245–248
    relation to transfer function of a fiber 247, 248
Fresnel integral 83, 84, 246
Freund, I. 133
Fried, D.L. 315
Fried parameter 315, 316
Fringe contrast 304
Fringe pattern 278, 280, 303, 304, 325
    extracting phase from 296–301
Fourier-Bessel transform 19
Fourier transform, definition 6
Fourier transform method for phase estimation 297, 298
Françon, M. 2
Full-frame display 204–206

Gamma density function 112, 254
Gaussian phase distribution 16
Gaussian phase statistics 22
Gaussian pulse 230
Gaussian spectrum 230
Gaussian statistics 10
Gaussian surface height correlation function 95
Generalized van Cittert-Zernike theorem 88
Gezari, D.Y. 322
Ghiglia, C.C. 299
Giglio, M. 90
Goodman, J.W. 111, 124, 345
Gradient of speckle intensity 128, 130
Gradient of speckle phase 125–127
Grating light valve (GLV) 207
Groh, G. 275

Hadamard phase mask 223
Hankel transform 19
HDTV video 204
Hermitian matrix 46
Heterodyne radar system 261
Hologram 191, 192
Holographic interferometry 275
Holography 30
    principles of 190–192
    and speckle 190–195
Human observer, effects of speckle on 141
Hypermetropic eye 189, 190

Ina, H. 297
Incoherent detection 249
Integrated speckle 105–120
    approximate density function 111–113
    contrast 108
    exact density function 113–120
Intensity, definition 25–27
Intensity derivative 345–350
Intensity transmission vs. exposure characteristic 289
Interference 3, 30, 190, 287, 292, 294

Jacobian 10, 63, 137
Jakeman, E. 54
Joint density function of
 speckle amplitude 65–68
 speckle intensity 68–72
 speckle phase 67, 68
Joyeux, D. 153
Judge, T.R. 299

K distribution 54, 55, 58
Karhunen-Loéve expansion 113
Known phasor plus a random phasor sum 33
Knox-Thompson technique (see cross spectrum)
Kobayashi, S. 297
Kolmogorov, A.N. 312
Kolmogorov turbulence 312
Korff, D. 325
Kowalczyk, M. 128

Labeyrie, A. 322
Laplace probability density 71
Leendertz, J.A. 287
Lehmann, M. 2
Level crossings of speckle intensity 130–132
Level crossing rate 130–137
Linear transformations of speckle amplitude 337–339
Line edge fluctuations, in microlithography 231–233
Line-scanned display 206, 207
Liquid crystal on silicon (LCOS) 205, 206
Local oscillator 261–264
Løkberg, O.J. 291
Long-exposure OTF 315–317
Long-exposure PSF 313–315
Lowenthal, S. 153

McKechnie, T.S. 194
Mamaev, A. 2
Mandel, L. 111
Marcum Q Function 34, 266
Martienssen, W. 194
Mathematica 11, 15, 125, 258, 299, 343
 programs for simulating speckle 369–372
Meridional rays 244, 245
Michelson interferometer 195, 196
Micro-display 225
Microstructured screen 225, 226
Middleton, D. 130
Modal coupling in fibers 244
Modal noise in fibers 237–248
 causes of 238
 frequency dependence of 243–248
Modes, number of
 step index fiber 236
 graded index fiber 237
Modified Rician density function 31, 32
Moment theorem
 real Gaussian variables 60
 complex circular Gaussian variables 62

Moving diffuser (see also changing diffuser) 143–153
 and coherence reduction 172–175
Moving screen 209, 210
Multidimensional Gaussian distribution 121, 122
Multimode fibers 235–237, 243–245
Multiple specklegram windows 285
Multiplex holography 194
Multivariate Gaussian statistics 59, 60
Mutual coherence function 170
Mutual intensity 171
Myopic eye 190

Nearsighted eye 188–190
Negative binomial distribution 254, 255
Negative binomial signal in Poisson noise 260, 261
Negative exponential density 28
 moments of 28
Negative vortex 135
Nisenson, P. 331
Non-circular statistics 306
Non-laser source in projection display 223–225
NTSC video 204
Numerical aperture 185
 of fiber 236

Ochoa, E. 124, 345
Optical path length probability density function 104, 105
Optical singularity 133
Optical transfer function 184
 with atmospheric turbulence 315–322
Optical vortex 133
Origin of speckle 1
Orthogonal code, in projection TV 222, 223
Orthogonal functions 6
Out-of-plane motion 290, 296

Paraxial wave 26
Partially developed speckle 50–53
 contrast of 51, 341–344
Partially polarized speckle 47–50
 contrast of 49, 50, 120
 integration of 118–120
 intensity probability density function of 49
Path dependence in phase unwrapping 301
Phase aberration function 318
Phase ambiguity 330, 331, 333
Phase closure 333
Phase derivative, statistics 345–350
Phase difference, statistics 68
Phase gradient 125–127
Phase map 297, 298
Phase retrieval 335
Phase shifting speckle interferometry 298, 299
Phase statistics
 nonuniform 15, 16
 uniform 12, 13
Phase unwrapping 299–301
Phase vortex 301

Photocount threshold 257
Photocounting receiver 262
Photographic detection 287
Photoresist threshold 231, 232
Piezoelectric transducer (PZT) 298
Point-spread functions with atmospheric turbulence 313–315
Poisson distribution 253–255
Poisson signal in Poisson noise 256–258
Polarization diversity 142, 143
    in optical coherence tomography 202
    in projection displays 208, 209
Positive vortex 135
Power spectral density
    definition 76
    of speckle intensity, free space 73–80
    of speckle intensity, imaging 80–82
Pritt, M.D. 299
Projection displays 203–228
Projection optics
    overdesign of 213, 214
    overfill of 215, 216
Pupil of the eye 209
Pusey, P.N. 54

Quantum efficiency 253

Random phasor plus known phasor 13
Random phasor sum 7
    sums of 17
    with finite number of elements 17
    with nonuniform phases 19
    with uniformly distributed phases 7–13
Random variables, transformations of 27
Random walk 3, 4, 8, 10
    with a large number of steps 10
    with a finite number of steps 17–20
Raster-scanned display 207, 208
Rastogi, P.K. 294
Ray directions in a speckle pattern 124–127
Raleigh probability density function 27, 28, 65
Rayleigh distributed amplitude 281, 320
Rayleigh statistics 11, 264, 265, 267
Reference wave, in holography 191
Refractive index inhomogeneities 312
Relation between scattered wave and surface heights 92–98
Rice, S.O. 111, 130
Rician density function 14, 15, 71
Rician phasor 265
Robinson, D.W. 299
Roddier, F. 333
Roggemann, M.C. 327
Rotation, measurement of 286
Rough object 151–153

Scattering spot shape 123, 124
Scribot, A.A. 117

Semiclassical theory of photoemission 252–256
    assumptions underlying 253
Shearing interferometer 294
Shkunov, V. 2
Shot noise 265
Short-exposure optical transfer function 315–322
Short-exposure point-spread function 313–315
Sign principle of optical vortices 136
Signal to noise ratio of speckle 28
Skew rays in multimode fibers 244
Slepian, D. 116
Smooth object 149–151
Spatial coherence reduction 178–185
    by time delay 185
    with circular incoherent source 179–181
    with two incoherent point sources 181–185
Spatial correlation, of optical radar returns 250–252
Spatially incoherent light 172
Speckle
    and optical radars 248–273
    at low light levels 252–256
    in microlithography 228–233
    in multimode fibers 235–248
    in optical coherence tomography 195–203
    in the eye 187–190
Speckle contrast
    and surface roughness 305, 306
    with two diffusers 359–363
Speckle correlography 333–335
Speckle decorrelation
    with angle change 307, 308
    with wavelength change 306, 307
Speckle interferometry
    astronomical 322–327
    in metrology 287–296
Speckle photography 275–286
Speckle reduction in optical radars 273
Speckle shearing interferometry 294–296
Speckle simulation 369–372
Speckled speckle 53–58
Speckle suppression 141–186
    by spatial coherence reduction 178–185
    by temporal coherence reduction 175–178
    in holography 192–195
    in projection displays 208
Speckle transfer function (in astronomy) 324–327
Speckled image examples
    coherence tomography 201
    microwave 3
    optical 2
    ultrasound 4
Specklegram 276, 277, 279, 280
    with multiple windows 285
Specular target 249
    and coherent detection 264–266
Spiller, S. 194
Spizzichino, A. 305
Stachnik, R.V. 322

Steinchen, W. 294
Structure constant 312, 315
Structure function 312
Sums of speckle patterns 37–47
    correlated 44–47
    independent 38–44
    on an amplitude basis 38
Surface scattering 85, 176, 177
Surface roughness measurement 305–309
Synthetic aperture radar 2, 3

Takeda, M. 297, 299, 301
Tatarski, V.I. 311
Taylor, P.A. 275, 301
Temporal coherence reduction 175–178
Temporal speckle 224, 225, 229–231
Thermal light 224
Threshold, coherent detection 266
Threshold, incoherent detection 256, 257
Tiziani, H.J. 301, 305
Tokarski, J.M.J. 275
Transverse displacement, measurement of 277–281, 287–290
    limitations in speckle photography 284, 285
Trisnadi, J.I. 217, 222, 223
Tropopause 313
Turbulence, atmospheric 311–313

Ultrasound 2, 4
Unitary linear transformation 46

van Cittert-Zernike theorem 75, 172, 250, 282
    generalized 84–90, 308
van Ligten, R.F. 194
Vibration measurement 301–305
Volume scattering 85, 102–105, 177, 178
Vortex circulation 135
Vortex, optical 133–139

Wavelength dependence of speckle 351–358
Wavefront dislocation 133
Wavelength diversity 153–162, 164–169
    in holography 193, 194
    in optical coherence tomography 202
    in projection displays 211
Wavelength sensitivity of speckle 159, 160, 166, 167, 175–178
Welford, W.T. 194
Westheimer, G. 220
Wolf, E. 170
Wrapped phase 297, 299–301

Yang, L. 294

Zel'dovich, B. 2
Zeros, of speckle intensity 133–139